Communications in Computer and Information Science 934

Commenced Publication in 2007
Founding and Former Series Editors:
Phoebe Chen, Alfredo Cuzzocrea, Xiaoyong Du, Orhun Kara, Ting Liu,
Dominik Ślęzak, and Xiaokang Yang

T0214520

More information about this series at http://www.springer.com/series/7899

Sergei O. Kuznetsov · Gennady S. Osipov
Vadim L. Stefanuk (Eds.)

Artificial Intelligence

16th Russian Conference, RCAI 2018
Moscow, Russia, September 24–27, 2018
Proceedings

 Springer

Editors
Sergei O. Kuznetsov ⓘD
Department of Data Analysis and Artificial
  Intelligence
National Research University Higher School
  of Economics
Moscow, Russia

Vadim L. Stefanuk
Institute for Information Transmission
  Problems
Russian Academy of Sciences
Moscow, Russia

Gennady S. Osipov ⓘD
Federal Research Center Computer Science
  and Control
Institute of Informatics Problems
Moscow, Russia

ISSN 1865-0929   ISSN 1865-0937 (electronic)
Communications in Computer and Information Science
ISBN 978-3-030-00616-7   ISBN 978-3-030-00617-4 (eBook)
https://doi.org/10.1007/978-3-030-00617-4

Library of Congress Control Number: 2018954547

This Springer imprint is published by the registered company Springer Nature Switzerland AG
The registered company address is: Gewerbestrasse 11, 6330 Cham, Switzerland

Preface

You are presented with the proceedings of the 16th Russian Conference on Artificial Intelligence, RCAI 2018. The organizers of the conference were the Russian Association for Artificial Intelligence, the Federal Research Center "Computer Science and Control" of the Russian Academy of Sciences, the Institute for Control Problems of Russian Academy of Sciences, and the National Research University Higher School of Economics. The conference was supported by the Russian Foundation for Basic Research and National Research University Higher School of Economics.

Being a long standing member of the European Association for Artificial Intelligence (EurAI, formely – ECCAI), the Russian Association for Artificial Intelligence has a great deal of experience in running important international AI events. The first such conference was held in the Soviet Union in 1975. It was the fourth highly recognized International Joint Conference on Artificial Intelligence - IJCAI-IV. It was held by the Academy of Sciences in Tbilisi, Georgia. A well-known Firbush Meeting on AI was held in 1976 in St. Petersburg (Leningrad), Russia. In 1994, the first Joint Conference on Knowledge Based Software Engineering (JCKBSE) was held in Pereslavl-Zalessky as well as the 6th conference JCKBSE, which was held in Protvino (near Moscow) in 2004.

All these international conferences provided important venues for exchanging opinions with leading international experts on Artificial Intelligence and demonstrating some important achievements in Artificial Intelligence obtained in Soviet Union Republics and Russia, in particular.

Russian conferences on Artificial Intelligence has a 30-year history and RCAI 2018 celebrated its jubilee. The first such conference was held in Pereslavl-Zalessky in 1988. Since then it was held every other year. The conference gathers the leading specialist from Russia, Ex-Soviet Republics, and foreign countries in the field of Artificial Intelligence.

Among the participants there were mainly the members of academic institutes and universities, and some other research establishments from Moscow, St. Petersburg, Kaliningrad, Apatite, Tver, Smolensk, Nizhniy Novgorod, Belgorod, Taganrog, Rostov-on-Don, Voronezh, Samara, Saratov, Kazan, Ulyanovsk, Kaluga, Ufa, Yekaterinburg, Tomsk, Krasnoyarsk, Novosibirsk, Khabarovsk, and Vladivostok. Several submitted papers came from Belarus, Ukraine, Azerbaijan, Armenia, Germany, Vietnam, Thailand, and Ecuador.

Topics of the conference included data mining and knowledge discovery, text mining, reasoning, decision making, natural language processing, vision, intelligent robotics, multi-agent systems, machine learning, AI in applied systems, ontology engineering, etc.

Each submitted paper was reviewed by three reviewers, either by the members of the Program Committee or by other invited experts in the field of Artificial Intelligence, to whom we would like to express our thanks. The final decision on the acceptance was based following the results of the reviewing process and was made during a special meeting of the RCAI Program Committee. The conference received 130 submissions in total, 25 of them were selected by the International Program Committee and are featured in this volume.

The editors of the volume would like to express their special thanks to Alexander Panov and Konstantin Yakovlev for their active participation in forming the volume and preparing it for publication. We hope that the appearance of the present volume will stimulate the further research in various domains of Artificial Intelligence.

September 2018   Sergei O. Kuznetsov
                                                                 Gennady S. Osipov
                                                                 Vadim L. Stefanuk

Organization

16 Russian Conference on Artificial Intelligence, RCAI 2018

RCAI is the biennial conference organized by the Russian Association for Artificial Intelligence since 1988. This time the conference was co-organized by the Federal Research Center "Computer Science and Control" of the Russian Academy of Sciences and the National Research University Higher School of Economics. RCAI covers a whole range of AI sub-disciplines: Machine learning, reasoning, planning, natural language processing, etc. Before 2018, despite the international status of the conference, the proceedings were published in Russian and were indexed in the Russian Science Citation Index. 2018 was the first year the selected high-quality papers of RCAI were published in English. Recent conference history includes RCAI 2016 in Smolensk, Russia, RCAI 2014 in Kazan, Russia, and goes back to RCAI 1988 held in Pereslavl-Zalessky. The conference was supported by the Russian Foundation for Basic Research (grant no. 18-07-20067) and the National Research University Higher School of Economics.

General Chair

Igor A. Sokolov — Federal Research Center "Computer Science and Control" of the Russian Academy of Sciences, Russia

Co-chairs

Stanislav N. Vasil'ev — Institute of Control Sciences of the Russian Academy of Sciences, Russia

Gennady S. Osipov — Federal Research Center "Computer Science and Control" of the Russian Academy of Sciences, Russia

Organizing Committee

Sergei O. Kuznetsov (Chair) — National Research University Higher School of Economics, Russia

Committee Members

Aleksandr I. Panov — Federal Research Center "Computer Science and Control" of the Russian Academy of Sciences, Russia

| Konstantin Yakovlev | Federal Research Center "Computer Science and Control" of the Russian Academy of Sciences, Russia |
| Oksana Dohoyan | National Research University Higher School of Economics, Russia |

International Program Committee

| Vadim Stefanuk (Chair) | Institute for Information Transmission Problems, Russia |

Committee Members

Toby Walsh	National ICT Australia and University of New South Wales, Australia
Vasil Sgurev	Institute of Information and Communication Technologies, Bulgaria
Georg Gottlob	University of Oxford, UK
Gerhard Brewka	University of Leipzig, Germany
Franz Baader	Dresden University of Technology, Germany
Javdet Suleymanov	Institute of Applied Semiotics, Russia, Tatarstan
Alla Kravets	Volgograd State University, Russia
Yves Demazeau	Laboratoire d'Informatique de Grenoble, France
Sergey Kovalev	Rostov State Railway University, Russia
Shahnaz Shahbazova	Azerbaijan Technical University, Azerbaijan
Boris Stilman	University of Colorado Denver, USA
Ildar Batyrshin	Instituto Politecnico Nacional, Mexico
Leonid Perlovsky	Harvard University, USA
Valeriya Gribova	Institute for Automation and Control Processes, Russia
Sergei O. Kuznetsov	National Research University Higher School of Economics, Russia
Alexey Averkin	Federal Research Center "Computer Science and Control" of the Russian Academy of Sciences, Russia
Vladimir Pavlovsky	Keldysh Institute of Applied Mathematics, Russia
Alexey Petrovsky	Federal Research Center "Computer Science and Control" of the Russian Academy of Sciences, Russia
Valery Tarassov	Bauman Moscow State Technical University, Russia
Vladimir Khoroshevsky	Federal Research Center "Computer Science and Control" of the Russian Academy of Sciences, Russia
Vladimir Golenkov	Belarusian State University of Informatics and Radioelectronics, Belarus
Vadim Vagin	National Research University, "Moscow Power Engineering Institute," Russia
Tatyana Gavrilova	St. Petersburg University, Russia
Alexander Kolesnikov	Kaliningrad branch of FRC CSC RAS, Russia
Yuri Popkov	Federal Research Center "Computer Science and Control" of the Russian Academy of Sciences, Russia

Contents

Spatial Reasoning and Planning in Sign-Based World Model

Gleb Kiselev[1,2], Alexey Kovalev[2], and Aleksandr I. Panov[1,3(✉)]

[1] Federal Research Center "Computer Science and Control" of the Russian Academy
of Sciences, Moscow, Russia
panov.ai@mipt.ru
[2] National Research University Higher School of Economics, Moscow, Russia
[3] Moscow Institute of Physics and Technology, Moscow, Russia

Abstract. The paper discusses the interaction between methods of
modeling reasoning and behavior planning in a sign-based world model
for the task of synthesizing a hierarchical plan of relocation. Such interaction is represented by the formalism of intelligent rule-based dynamic
systems in the form of alternate use of transition functions (planning)
and closure functions (reasoning). Particular attention is paid to the ways
of information representation of the object spatial relationships on the
local map and the methods of organizing pseudo-physical reasoning in a
sign-based world model. The paper presents a number of model experiments on the relocation of a cognitive agent in different environments
and replenishment of the state description by means of the variants of
logical inference.

Keywords: Sign · Sign-based world model · Relocation planning
Reasoning modeling · Pseudo-physical logic

1 Introduction

One of the long-standing problems in artificial intelligence is the problem of
the formation or setting of the goal of actions by an intelligent agent, for the
achievement of which it synthesizes the plan of its behavior. The study of the
goal-setting process [1,2] showed that the formation of a new goal in many important cases is connected to the reasoning in the sign-based world model of the
actor. In other words, reasoning is an integral part of the process of generating
a new goal and hence the planning process. A number of artificial intelligence
studies related to goal-driven autonomy [3] also indicate that an important step
in the planning process is some formal conclusion aimed at eliminating cognitive
dissonance caused by new conditions that require a change or the formation of
a new goal.

This work is devoted to the study of one type of interaction. Consideration
of such interaction is conducted in the context of intelligent rule-based dynamic
systems [4–6]. We consider the problem of spatial planning and reasoning using

© Springer Nature Switzerland AG 2018
S. O. Kuznetsov et al. (Eds.): RCAI 2018, CCIS 934, pp. 1–10, 2018.
https://doi.org/10.1007/978-3-030-00617-4_1

elements of pseudo-physical logic [7]. There is a well-known approach to representing information about an environment as a semantic map [8–10] that mixes different structures such as a metric map, a grid map, a topological graph, etc. The papers [11,12] describe a hierarchical approach to planning the behavior of an intelligent agent, in which abstract geometric reasoning is used to describe the current situation. Also, the algorithm uses a probabilistic representation of the location of objects. The hierarchical refinement of the surrounding space used in the article is justified from the viewpoint of reducing the time spent by the agent for recognizing the surrounding space, but preserving all refined knowledge and generating possible actions leads to unnecessary load on the processor of the agent, which negatively affects its speed. The approach in [13] describes the activity of an agent that uses logic derived from studies of rat brain activity in performing tasks related to spatial representation. The hierarchy of the map views reduces the noise caused by the remoteness from the agent of some parts of the map, to which linear search trajectories were built (paths to the target from the current location). Keeping all possible trajectories to any part of the environment requires additional resources from the agent, significantly reducing the speed of decision making by the agent. If there is a dynamic space in which other agents work and the location of the objects can change, the approach will require too much resources to calculate all possible outcomes of activities. These problems were partially addressed in [14,15], which led to the creation of the RatSLAM system, which allowed the agent to travel long distances in real terrain.

In our case, the representation of spatial knowledge, planning processes and reasoning is formalized in terms of a sign-based world model [1]. As a demonstration of the proposed approach a number of model experiments on the relocation of a cognitive agent in various environments and state replenishment with one of the variants of logical inference are presented.

2   Sign Approach to Spatial Knowledge Representation

The concept of a sign-based world model for describing the knowledge of a cognitive agent about the environment and himself was introduced in [1,2]. The main component of the sign world model is the sign represented at the structural description level (according to [16]) as a quadruple $s = \langle n, p, m, a \rangle$, where $n \in N$, $p \subset P, m \subset M, a \subset A$. N is a set of names, i.e. a set of words of finite length over some alphabet, P is a set of closed atomic formulas of the first-order predicate calculus language, which is called the set of images. M is a set of significances. A is a set of personal meanings.

In the case of the so-called everyday sign-based world model, which we will consider below, the image component of the sign participates in the process of recognition and categorization. Significances represent fixed script knowledge of the intellectual agent about the subject area and the environment, and personal meanings characterize his preferences and current activity context. The name component binds the remaining components of the sign into a single unit (naming).

At the structural level of the sign-based world model description each component of the sign is a set of causal matrices that are represent a structured set of references to other signs or elementary components (in the case of an image, these are primary signs or data from sensors, in the case of personal meaning - operational components of actions). The causal matrix allows the encoding of information to represent both declarative and procedural knowledge. A set of sign components forms four types of causal networks, special types of semantic networks. Modeling of planning and reasoning functions is carried out by introducing the notion of activity (the set of active signs or causal matrices) and the rules of activity propagation on various types of networks [17]. In progress of a cognitive function, new causal matrices are formed, which can then be stored in the components of the new sign, similar to the experience preservation in systems based on precedents.

3 Dynamical Intelligent Systems

Let us introduce the basic concepts from the theory of intelligent rule-based dynamic systems following [5]. First of all, we will distinguish the main components of the system: working memory, in which a set of facts are stored (i.e. closed formulas of the first-order predicate calculus language), a set of rules and strategies for selecting rules.

The rule is the ordered triple of sets $r = \langle C, A, D \rangle$, where C is a condition of the rules, A is a set of facts added by the rule, D is a set of facts deleted by the rule. A special variable $t \in T$ is distinguished, where T is a discrete ordered set, is related to the discrete time. Thus, the concrete value of the variable t corresponds to a specific moment in time. The set of rules Π is divided into two subsets Π_{CL} and Π_{TR}. The set Π_{CL} consists of rules that do not correspond to any actions, their application only replenishes set of facts of the state (working memory). Such rules are called as the rules of communication, and the set Π_{CL} is called the set of rules for the closure of states. The set Π_{TR} includes rules defining actions, such rules are called transition rules, and the set itself is the set of transition rules. A distinctive feature of the transition rules is that the value of t changes at least by one for the conditions of the rule and the set of added and deleted rules:

$$\Pi_{CL} = \langle C(t), A(t), D(t) \rangle,$$
$$\Pi_{TR} = \langle C(t), A(t+1), D(t+1) \rangle.$$

Rules are applied to working memory, which, in turn, changes its state. The rule selection strategy determines which of the possible rules will be applied and terminates the application when the state of the working memory satisfies the target condition.

Let CL and TR be the strategies for applying the rules Π_{CL} and Π_{TR}, X be the set of facts, respectively. Then the strategy CL realizes a mapping $2^X \rightarrow 2^X$, and the strategy TR is a mapping $2^X \times T \rightarrow 2^X$. We introduce the functions

$$\Phi\left(\chi\left(t\right)\right) = \left(CL, \chi\left(t\right)\right) : 2^X \to 2^X,$$
$$\Psi\left(\chi\left(t\right), t\right) = \left(TR, \chi\left(t\right), t\right) : 2^X \times T \to 2^X.$$

Function Φ is called the closure function, it replenishes the description of the current state of the system. Function Ψ is called a transition function, it takes the system from one state to another.

Thus, a quadruple $D = \langle X, T, \Phi, \Psi \rangle$ is called an intelligent rules-based dynamic system.

Let us clarify the definitions introduced for the case of modeling reasoning and planning of relocation in the sign-based world model.

In the sign-based world model working memory and set of facts of the state will correspond to a set of active signs from which the description of the current situation is constructed (the causal matrix on the network of personal meanings), and the rules of the dynamic system correspond to rules for activity propagation in causal networks that change the set of active causal matrices and description of the current situation. Then the initial state of the working memory will correspond to the initial situation, and the state of the working memory that satisfies the target condition is the target situation.

The process of modeling is divided naturally into two stages: reasoning and planning actions, in our case, relocation. Moreover, while reasoning, obviously, only a change in the description of the current situation occurs (changing the world model of the agent) without modeling any actions in the environment. This process corresponds to the application of the closure function Φ to the working memory. In the process of relocation planning, the agent considers possible actions in the environment and the consequences of such actions, therefore, such a process can be associated with a function Ψ. Then the cognitive rule-based dynamic system defined in the sign-based world model is the quadruple $D_{SWM} = \langle X_{SWM}, T, \Phi_{SWM}, \Psi_{SWM} \rangle$, where X_{SWM} stays for the semiotic network and procedural matrices; T is discrete time; Φ_{SWM} are rules for activity propagation on causal networks in the implementation of the reasoning function; Ψ_{SWM} are rules for the activity propagation on causal networks in the implementation of the planning function.

4   Integration of Reasoning and Planning

The transition function Ψ is implemented due to the rules for activity propagation, designed as a MAP planning algorithm [17]. The MAP algorithm allows the cognitive agent with the sign-based world model to synthesize the optimal path to the required location on the map. The agent's sign-based world model for the relocation task includes elementary signs of objects, signs of actions, signs of spatial and quantitative relations modeling the relations of pseudo-physical logic [7], as well as signs of cells and regions (see Fig. 1) [18]. The process of map recognition by the agent begins with the stage of determining the regions. The map is divided into 9 identical segments that denote the "Region" sign. The regions do not have a fixed size and their area is calculated depending on the size of the map. The regions can contain objects, agents, obstacles, and cells. The cells are map segments obtained by dividing the larger segments (in the first step of recognition the segment is the region) by 9 equal parts until the central cell contains only the agent. As soon as such a cell is formed (its size cannot be less than the diameter of the agent), it is represented by the "Cell" sign. Further, around it, an additional 8 cells are built that describe the current situation. After that, the process of plan synthesis, presented in Algorithm 1, begins. It consists of two stages: the stage of replenishment of the agent's world model (step 1) and the stage of plan synthesis (steps 2–20).

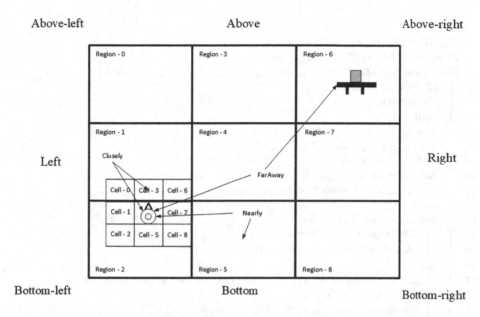

Fig. 1. Illustration of spatial relationships, cells and regions of the sign world model

The replenishment phase of the agent's world model begins with the creation of signs and causal matrices for objects (including cells and regions), their types, predicates and actions obtained from recognition of the map and the planning task, as well as the creation of the sign "I" [18]. Next, the agent creates causal matrices of the initial and final situations and locations on the map.

At the stage of plan synthesis, the agent recursively creates all possible plans to achieve the final situation, which describes the agent's target location on the map. To do this, in Step 7, the agent looks at all the signs that are included in the description of the current situation, and in Steps 8–9, using the activity propagation process over the network of significances, procedural action matrices are activated. Using the processes described in steps 10–12, action matrices are updated, replacing references to role signs and object types with references to specific task objects. Next, there is a step of choice $A_{checked}$ - actions that are heuristically evaluated, as the most appropriate in the situation $z_{sit-cur}$ to achieve the situation $z_{sit-goal}$. After that, from the effects of each action $A \in A_{checked}$ and the references to the signs that enter the current situation $z_{sit-cur+1}, z_{map-cur+1}$ is constructed, which describes the agent's state and the map after applying the action A. At the step 16, the action A and $z_{sit-cur}$ under consideration is added to the plan and, at the step 17, the entry $z_{sit-goal}$ in $z_{sit-cur+1}$, $z_{map-goal}$ in $z_{map-cur+1}$ is checked. If the matrices of the current state include the matrices of the target state, then the algorithm saves the found plan, as one of the possible ones, if not, then the plan search function is recursively repeated (step 20).

```
1   T_agent := GROUND(map, struct)
2   Plan := MAP_SEARCH(T_agent)
3   Function MAP_SEARCH(z_sit-cur, z_sit-goal, z_map-cur, z_map-goal, plan, i):
4       if i > i_max then
5           | return ∅
6       end
7       z_sit-cur, z_map-cur = Z^a_sit-start, Z^a_map-start
8       z_sit-goal, z_map-goal = Z^a_sit-goal, Z^a_map-goal
9       Act_chains = getsitsigns(z_sit-cur)
10      for chain in Act_chains do
11          | A_signif| = abstract_actions(chain)
12      end
13      for z_signif in A_signif do
14          Ch| = generate_actions(z_signif)
15          A_apl = activity(Ch, z_sit-cur)
16      end
17      A_checked = metacheck(A_apl, z_sit-cur, z_sit-goal, z_map-cur, z_map-goal)
18      for A in A_checked do
19          z_sit-cur+1, z_map-cur+1 = Sit(z_sit-cur, z_map-cur, A)
20          plan.append(A, z_sit-cur)
21          if z_sit-goal ∈ z_sit-cur+1 and z_map-goal ∈ z_map-cur+1 then
22              | F_plans.append(plan)
23          end
24          else
25              | Plans := MAP_SEARCH
                  (z_sit-cur+1, z_sit-goal, z_map-cur+1, z_map-goal, plan, i + 1)
26          end
27      end
```

Algorithm 1. Process of plan synthesis by cognitive agent

Thus, the agent forms an action plan using the rules for activity propagation Π_{TR}, changing the current state of the working memory (which consists in the formation and change of causal matrices $z_{sit-cur}$ and $z_{map-cur}$). When the state of working memory is reached, many facts of which include a set of facts that form the final state, the algorithm terminates.

Next, consider the process of reasoning in a sign-world model using elements of pseudo-physical logic.

To determine the location of an object relative to an agent, we define a set I_A^O such that the focus cell k_i^A, $i = 0 \ldots 8$ of the agent's attention belongs to the set I_A^O, if and only if it coincides with any focus cell k_j^O, $j = 0 \ldots 8$ of the object, i.e. focuses of the attention of the agent and the object are intersected in this cell: $I_A^O = \{k_i^A | k_i^A = k_j^O, i, j = 0 \ldots 8\}$ or $I_A^O = \{k_i^A | k_i^A \in F^A \cap F^O, i = 0 \ldots 8\}$,

where F^A, F^O are the focuses of the attention of the agent and the object, respectively. We apply exclusion *Exclude* and absorption *Absorb* operations to the set obtained. The operation *Exclude* checks whether the set I_A^O contains conflicting signs of cells, and if there are such cells, it excludes them. Also, this operation excludes the sign of the cell in which the agent is located, because it does not affect the location definition. The operation *Absorb* excludes a sign with a narrower significance if there is a sign with a wider significance. A set $I_O^A = \{k_j^O | k_j^O \in F^A \cap F^O, j = 0 \ldots 8\}$ is used to determine the location of the agent relative to the object.

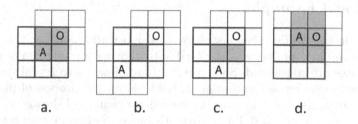

Fig. 2. Examples of the locations of the agent (A) and the object (O)

In Fig. 2 focuses of attention for the agent and the object intersect in four cells, then

$$I_A^O = \{\text{"Agent"}, \text{"Above"}, \text{"Right"}, \text{"Above-right"}\},$$
$$Exclude\left(I_A^O\right) = \{\text{"Above"}, \text{"Right"}, \text{"Above-right"}\},$$
$$Absorb\left(Exclude\left(I_A^O\right)\right) = \{\text{"Above-right"}\}.$$

This implies that the object O is on the right from above with respect to the agent A.

For the case presented in Fig. 2d, we determine the location of the agent A relative to the object O.

$$I_O^A = \{\text{"Object"}, \text{"Above"}, \text{"Left"}, \text{"Above-Left"}, \text{"Below"}, \text{"Below-Left"}\},$$
$$Exclude\left(I_O^A\right) = \{\text{"Left"}\},$$
$$Absorb\left(Exclude\left(I_O^A\right)\right) = \{\text{"Let"}\}.$$

Therefore, the agent A is to the left of the object O.

To determine the distance between the agent A and the object O, we will use the following rules:

1. If $\{$"Agent", "Object"$\} \in I_A^O \cup I_O^A$ then A "Closely" O;
2. If $\{$"Agent", "Object"$\} \notin I_A^O \cup I_O^A$ and $I_A^O \cup I_O^A \neq \{\emptyset\}$ then A "Close" O;
3. If $I_A^O = I_O^A = \{\emptyset\}$ or, equally, $I_A^A \cup I_O^A = \{\emptyset\}$ then A "Far" O.

The approach presented above implements the closure function Φ, which is described by the mechanism of activity propagation as follows. From the sign of the agent's focus of attention $k_i^A, i = 0 \ldots 8$, the activity, downward, spreads up to the actualization of the sign of the map cell. After that, the activity from the sign of the map cell spreads ascending and if the sign of the object's focus cell $k_j^O, j = 0 \ldots 8$ is updated, the sign of the cell $k_i^A, i = 0 \ldots 8$ is added to the set I_A^O. Procedures *Exclude*, *Absorb* and rules for determining the distance are implemented by the corresponding procedural matrices.

5   A Model Example

As part of the application demonstration of a sign world model to the problem of spatial planning, problems associated with moving an agent in a confined enclosed space are considered. Such a restriction allows us to reveal the advantages of a symbolic representation of spatial logic for the process of planning a route with obstacles and objects in the immediate vicinity of the agent. Here we present an example of scheduling an agent's move to an empty map in experiment 1, an example in which an agent plans to move away from an obstacle and, through logical inference, changes his view of the location of the obstacle in experiment 2 and an example in which the agent plans to bypass the obstacle in experiment 3 (see Fig. 3).

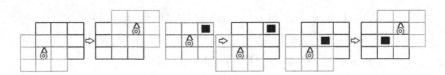

Fig. 3. Experiments 1, 2 and 3

Experiment 1 describes the process of constructing a plan with a length of 4 actions, the first iteration activated the matrix of the "Rotate" sign, which contained a reference to the "Closely" sign relative to the upper right cell and a reference to the sign mediating the direction to this cell. Next, the sign matrix "Move" was activated. At the next iteration, the matrices of signs "Move", "Right-from-top" and "Cell-6", which were referenced in the condition of "Location" matrix, are reactivated. In the final iteration, the matrix of the "Turn" sign was activated, which contained references to signs "Agent Direction" and "Top".

Experiment 2 describes the process of constructing a plan with a length of 3 actions, which includes 2 rotation actions and an action to move to the lower left region. The entire upper right area was occupied by an obstacle from which the agent retreated. At the first iteration of the planning process, the matrix of the "Rotate" sign, describing the change in the direction of the agent, as well as the matrix of the "Location" sign, which included a reference to the "Closely" sign with respect to the obstacle cell, was activated. Then, at the second iteration, the sign matrix "Move" was activated, and the matrix of the "Closely" sign with respect to the mentioned area ceased to be active. With the help of the reasoning process, the matrix of the sign "Close" (related to the area containing the obstacle) was activated. In the described task, matrices activated by the process of reasoning allow the agent not to repeat the process of finding objects that were included in the description of the previous situations of the plan.

In experiment 3, four possible plans were built to achieve the required location of the agent, of which a plan consisting of 6 actions was selected. At the first iteration of the planning, the sign matrix "Rotate" relative to the right upper region was not activated, because, through the heuristics used by the algorithm, the agent is not available actions that direct him to obstacles with which he can not interact. The matrix of the "Rotate" sign was activated, in the effects of which there was a reference to the "Agent Direction" sign, which mediates the direction to the adjacent to the target area. Further, the action plan according to the described heuristic was iteratively constructed.

6   Conclusion

In this paper we have presented an original approach to interacting mechanisms for the synthesis of the behavioral plan by the cognitive agent and reasoning procedures in its sign-based world model. A scheme of such interaction is proposed in the context of intelligent rule-based dynamic systems. The work of this approach in the problem of smart relocation in space is demonstrated.

Acknowledgements. This work was supported by Russian Foundation for Basic Research (Project No. 18-07-01011 and 17-29-07051).

References

1. Osipov, G.S., Panov, A.I., Chudova, N.V.: Behavior control as a function of consciousness. I. World model and goal setting. J. Comput. Syst. Sci. Int. **53**, 517–529 (2014)
2. Osipov, G.S., Panov, A.I., Chudova, N.V.: Behavior control as a function of consciousness. II. Synthesis of a behavior plan. J. Comput. Syst. Sci. Int. **54**, 882–896 (2015)

3. Alford, R., Shivashankar, V., Roberts, M., Frank, J., Aha, D.W.: Hierarchical planning: relating task and goal decomposition with task sharing. In: IJCAI International Joint Conference on Artificial Intelligence, pp. 3022–3028 (2016)
4. Stefanuk, V.L.: Dynamic expert systems. KYBERNETES Int. J. Syst. Cybern. **29**(5/6), 702–709 (2000)
5. Vinogradov, A.N., Osipov, G.S., Zhilyakova, L.Y.: Dynamic intelligent systems: I. Knowledge representation and basic algorithms. J. Comput. Syst. Sci. Int. **41**, 953–960 (2002)
6. Osipov, G.S.: Limit behaviour of dynamic rule-based systems. Inf. Theor. Appl. **15**, 115–119 (2008)
7. Pospelov, D.A., Osipov, G.S.: Knowledge in semiotic models. In: Proceedings of the Second Workshop on Applied Semiotics, Seventh International Conference on Artificial Intelligence and Information-Control Systems of Robots (AIICSR97), Bratislava, pp. 1–12 (1997)
8. Gemignani, G., Capobianco, R., Bastianelli, E., Bloisi, D.D., Iocchi, L., Nardi, D.: Living with robots: interactive environmental knowledge acquisition. Robot. Auton. Syst. **78**, 1–16 (2016). https://doi.org/10.1016/j.robot.2015.11.001
9. Galindo, C., Saffiotti, A., Coradeschi, S., Buschka, P., Fern, J.A., Gonz, J.: Multi-hierarchical semantic maps for mobile robotics. In Proceedings of the IEEE/RSJ International Conference on Intelligent Robots and Systems (2005)
10. Zender, H., Mozos, O.M., Jensfelt, P., Kruijff, G.M., Burgard, W.: Conceptual spatial representations for indoor mobile robots. Robot. Auton. Syst. **56**, 493–502 (2008). https://doi.org/10.1016/j.robot.2008.03.007
11. Kaelbling, L.P., Lozano-Prez, T.: Integrated task and motion planning in belief space. Int. J. Robot. Res. **32**(9–10), 1194–1227 (2013)
12. Garrett, C.R., Lozano-Prez, T., Kaelbling, L.P.: Backward-forward search for manipulation planning. In: IEEE International Conference on Intelligent Robots and Systems, (grant 1420927), pp. 6366–6373, December 2015
13. Erdem, U.M., Hasselmo, M.E.: A biologically inspired hierarchical goal directed navigation model. J. Physiol. Paris **108**(1), 28–37 (2014). https://doi.org/10.1016/j.jphysparis.2013.07.002
14. Milford, M., Wyeth, G.: Persistent navigation and mapping using a biologically inspired slam system. Int. J. Robot. Res. **29**(9), 1131–1153 (2010). https://doi.org/10.1177/0278364909340592
15. Milford, M., Schulz, R.: Principles of goal-directed spatial robot navigation in biomimetic models. Philos. Trans. Roy. Soc. B: Biol. Sci. **369**(1655), 20130484–20130484 (2014). https://doi.org/10.1098/rstb.2013.0484
16. Osipov, G.S.: Sign-based representation and word model of actor. In: Yager, R., Sgurev, V., Hadjiski, M., and Jotsov, V. (eds.) 2016 IEEE 8th International Conference on Intelligent Systems (IS), pp. 22–26. IEEE (2016)
17. Panov, A.I.: Behavior planning of intelligent agent with sign world model. Biol. Inspired Cogn. Archit. **19**, 21–31 (2017)
18. Kiselev, G.A., Panov, A.I.: Sign-based approach to the task of role distribution in the coalition of cognitive agents. SPIIRAS Proc. **57**, 161–187 (2018)

Data Mining for Automated Assessment of Home Loan Approval

Wanyok Atisattapong(✉), Chollatun Samaimai, Salinla Kaewdulduk, and Ronnagrit Duangdum

Department of Mathematics and Statistics, Faculty of Science and Technology, Thammasat University, Bangkok 12120, Thailand
awanyok@tu.ac.th

Abstract. Banks receive large numbers of home loan applications from their own customers and others each day. In this study, we investigated the use of data mining to decide whether or not to extend credit, based on two analytical approaches: Naïve Bayes and decision tree. Four independent factors were considered: the loan period, the net income of the applicant, the size of the loan, and other relevant characteristics of the potential borrower. Models were constructed that produce three outcomes: approval, conditional approval, and rejection. The predictive accuracy of the models was compared, to evaluate the effectiveness of the classifiers and a Kappa statistic was applied, to evaluate the degree of accuracy with which the models predicted the final outcome. The decision tree model performed better on both accuracy and the Kappa statistic. This model had an accuracy of 90% and a Kappa of 0.8140, whereas the Naïve Bayes had an accuracy of 65% and Kappa of 0.3694. We therefore recommend the use of decision tree-based models for home loan ranking. Data mining of the applicants history can support the decision-making of financial organizations, and can also help applicants realistically evaluate their own chances of securing a loan.

Keywords: Data mining · Naïve Bayes · Decision trees · Home loans

1 Introduction

Homebuyers need money to purchase a house and pay for decorating and other expenses. For most individuals, this involves taking out a loan. Banks and financial institutions are willing to offer loans for qualified borrowers. However, the credit risk assessment process takes at least two weeks to process customer data and approve the loan. Since the number of bank customers has significantly increased, the efficiency of credit granting methods must be improved for the benefit of both customers and the banking system.

Many techniques are used to support automatic decision making [1–3]. Approaches include techniques such as fuzzy logic [4,5], logistic regression [6,7], and artificial neural networks [8,9].

© Springer Nature Switzerland AG 2018
S. O. Kuznetsov et al. (Eds.): RCAI 2018, CCIS 934, pp. 11–21, 2018.
https://doi.org/10.1007/978-3-030-00617-4_2

Levy et al. [4] proposed the application of fuzzy logic to commercial loan analysis using discriminant analysis. Mammadli [5] used a fuzzy logic model for retail loan evaluation. The model comprised five input variables: income, credit history, employment, character, and collateral. The single output rated the credit standing as low, medium, or high. Dong et al. [6] proposed a logistic regression model with random coefficients for building credit scorecards. Majeske and Lauer [7] formulated bank loan approval as a Bayesian decision problem. The loan approval criteria were computed in terms of the probability of the customer repaying the loan.

Data mining is an emerging technology in the field of data analysis, and has a significant impact on the classification scheme. Alsultanny [10] proposed Naïve Bayes classifiers, decision tree, and decision rule techniques, for predicting labor market needs. The comparison between these three techniques showed the decision tree to have the highest accuracy. Hamid [11] proposed a model from data mining to classify information from the banking sector and to predict the status of loans. Recently, Bach et al. [12] compared the accuracy of credit default classification of banking clients through the analysis of different entrepreneurial variables, when using decision tree algorithms.

In this work, two data mining approaches, Naïve Bayes and decision tree, were investigated for evaluation of applications for a home loan. Instead of classifying only into two classes, loan approval and rejection, we added a third class in the middle, loan approval with conditions. These conditions are things like search for syndicate partners, adding of collateral assets, and anything else that the borrower needs to clear for loan approval. When the conditions are met, the bank will extend credit.

The remainder of the paper is structured as follows. In Sect. 2, the processes for constructing the models from Naïve Bayes and decision tree are described. In Sect. 3, the performance of both predictive models measured in terms of the accuracy and a Kappa statistic is presented. Finally, the conclusions are outlined in Sect. 4.

2   Proposed Models and Implementation

The data mining process can be divided into four steps, as follows.

1. Prepare the training set, using records that already have a known class label.
2. Build the model by applying the learning algorithm to the training set.
3. Apply the model to a test set containing unclassified data.
4. Evaluate the accuracy of the model.

The flow chart is shown as Fig. 1.

Many variables affect the customer evaluation. Obtaining comprehensive set of actual data from the banks was difficult as such information is considered confidential and thus should be hidden from unauthorized entities. To choose the appropriate variables, bank's lending policies, application forms, and loan review systems were considered.

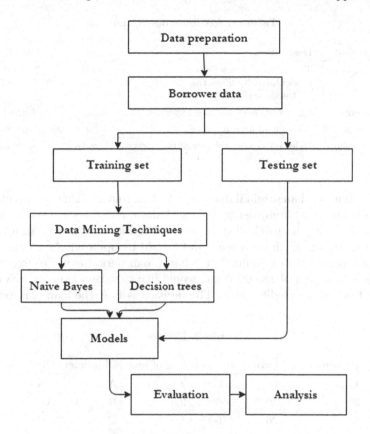

Fig. 1. Data mining process

In this work, the major variables were classified as independent or dependent. There were four independent variables: the loan period, net income, size of loan, and characteristics of the borrower. These are shown in Table 1.

For the period attribute, the timing was limited to a maximum of 35 to reflect bank policies. The 65 year age limit is based on the fact that the customer will usually cease earning at retirement. The net income represents the amount of money remaining after all expenses, interest, and taxes. From the bank's requirement, the net income per month must be greater than or equal to 15,000 Baht. The amount loaned ranged from one to ten million Baht. The relevant characteristics of the borrower were ranked into three levels: 'A', 'B', and 'C', with 'A' being the highest score and 'C' being the lowest. This was evaluated by scoring the customer's loan questionnaire, which collects data on education level, employment, any life insurance policies held, assets, and credit history.

The dependent variables were as follows: 'AP' indicated that the loan was approved. 'AC' indicated that the loan was conditionally approved. 'DN' indicated that the loan was rejected. The target class of the training set is based on the bank policy, which requires that a borrower taking a loan of one million Baht is able to make payments of not less than 7,000 Baht per month.

Table 1. Attribute description

No.	Attribute	Description	Unit	Data type
1	Period	Timing of disbursements (Calculated by 65 − the customer age ≤ 35)	Year	Numeric
2	Income	Excess of revenues over expenses	Baht	Numeric
3	Size of loan	Amount of loan required	Million Baht	Numeric
4	Character	Relevant characteristics of the borrower	Grade	Nominal

We first simulated an original dataset of 200 customers. This was divided into two sets, a training set comprising 80% of all data, and a testing set comprising 20%. Table 2 shows the original data set, which was designed to simulate the customer conditions. Each customer's data could be updated. As an example of a single data point, customer number 1 has a loan period equal to five years, a net income per month of 18,000 Baht, would like to take a loan for two million Baht, and has a good credit rating. The decision is that the loan is rejected.

Table 2. Dataset

Customer no.	Period	Income	Size of loan	Character	Class
1	5	18000	2	A	DN
2	20	50000	2.5	B	AP
3	26	15000	1	C	AC
4	16	25000	1.6	B	AP
5	10	20000	2.3	B	DN
6	32	45000	3	C	AP
7	33	55000	3.6	A	AP
8	35	65000	4.3	B	AP
9	35	75000	5	C	AP
10	35	85000	5.6	A	AP
...	...	...	...	...	...
200	24	27000	8.1	A	DN

Two models, Naïve Bayes and decision tree, were implemented using Weka Data Mining software version 3.9.2 [13] as shown in Fig. 2.

To investigate the appropriate size for the training set, we randomly chose data sets of ten different sizes, as shown in Table 3. Next, ten Naïve Bayes models were constructed and their accuracy computed automatically. After choosing the size that yielded the best accuracy, we used the Weka software to build a decision tree model of the same size to observe the nodes of the tree.

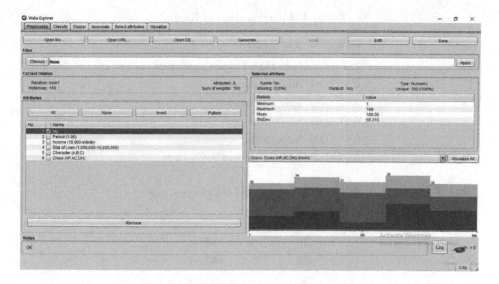

Fig. 2. Training set imported to Weka software

Table 3. Sizes of training sets

No.	Number of cases
1	16
2	32
3	48
4	64
5	80
6	96
7	112
8	128
9	144
10	160

3   Results

Classification accuracy is a standard metric used to evaluate classifiers [14]. Accuracy is simply the percentage of instances in which the method predicts the true outcome. The Kappa statistic [15] is used to measure the agreement of a prediction with the actual outcome. A Kappa of 1 indicates perfect agreement, whereas a kappa of 0 indicates agreement equivalent to chance. The scale of the Kappa is shown in Table 4.

Table 5 shows that the training set of 80 cases provided the best accuracy. The selected training dataset used to construct the decision tree is shown in

Table 4. Interpretation of Kappa

Kappa	Agreement
<0	Less than chance agreement
0.01–0.20	Slight agreement
0.21–0.40	Fair agreement
0.41–0.60	Moderate agreement
0.61–0.80	Substantial agreement
0.81–0.99	Almost perfect agreement

Table 5. Accuracy and training set size

No.	Number of cases	Accuracy
1	16	87.50
2	32	87.50
3	48	89.60
4	64	89.10
5	80	95.00
6	96	90.63
7	112	90.18
8	128	87.50
9	144	86.80
10	160	89.38

Tables 6 and 7 shows the accuracy and Kappa statistic of the Naïve Bayes and the decision tree models. The results showed the decision tree model to provide more precise results.

Figure 3 shows the structure of the tree obtained. As can be observed, the initial node of the tree was the net income of the applicant. This showed the first factor affecting the decision to be income. The other attributes were ranked as shown. Rectangular boxes show the final classification. The numbers in the box count the correctly and incorrectly categorized cases. In total, only four cases were incorrectly classified.

To find the appropriate class for a given customer, we start with the value at the root of the tree and keep following the branches until a leaf is reached. For example, the attribute values of customer number 1 are as follows: Income = 18,000, Size of Loan = 2, Character = A, and Period = 5. To classify this case, we start at the root of the tree in Fig. 3, which is labeled 'Income', and follow the true branch (Income ≤ 50,000) to the node 'Size of loan'. As a loan of 2 million Baht is less than the 3.3 million Baht threshold, Character is considered. Character 'A' leads finally to a leaf labeled 'AP', indicating that the bank can proceed with a loan approval.

Table 6. Example training set

No.	Customer no.	Period	Income	Size of loan	Character	Class
1	3	26	15000	1	C	AC
2	8	35	65000	4.3	B	AP
3	10	35	85000	5.6	A	AP
4	11	35	95000	6.3	B	AP
5	14	19	120000	8	B	AP
6	19	24	75000	6	A	AP
7	22	5	31000	5.2	A	DN
8	23	16	25000	7.3	A	DN
9	25	20	19000	1.9	B	AC
10	26	25	48000	3.2	C	AP
11	29	5	320000	9.8	B	AC
12	32	10	62000	1.9	C	AP
13	38	26	16000	3	C	DN
14	39	22	27000	2.5	A	AP
15	40	15	39000	6	A	DN
...	...	...	...	...	...	
71	175	16	55000	7	C	DN
72	176	24	65000	4.5	B	AP
73	185	20	23000	3	B	AC
74	187	21	64000	5.6	C	AP
75	188	19	29000	2.2	C	AC
76	189	30	15000	3.5	A	DN
77	192	20	46000	7.1	B	DN
78	194	18	25000	3	B	AC
79	195	19	35000	4.2	B	AC
80	196	19	25000	8	C	DN

Table 7. Accuracy and Kappa of two models from the 80 case training set

Technique	Accuracy	Kappa
Naïve Bayes	67.5%	0.4669
Decision tree	95.0%	0.9191

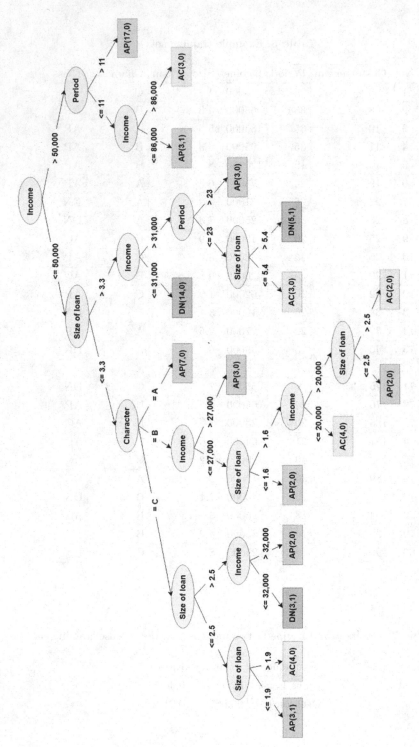

Fig. 3. Decision tree for the training set

Table 8. Test set

No.	Customer no.	Period	Income	Size of loan	Character	Target class
1	1	5	18000	2	A	AP
2	13	16	75000	8.8	A	AP
3	24	28	28000	6.4	A	DN
4	27	27	15000	1	C	AP
5	48	30	34000	3	A	AP
6	95	13	16000	3	A	AP
7	105	33	16000	1.4	A	AP
8	121	20	18000	4.6	C	DN
9	124	25	18000	2	C	AC
10	128	10	190000	8.7	B	AP
11	131	20	24000	3.3	A	AP
12	146	20	23000	5.2	C	DN
13	159	33	15000	1.5	C	AP
14	165	30	37000	4	C	AP
15	169	20	16000	2	B	AC
16	171	20	61000	9.2	C	AP
17	174	25	31000	3.3	B	AP
18	180	22	20000	4	C	DN
19	182	20	79000	4	B	AP
20	197	20	23000	3.3	C	AC

Table 9. Accuracy and Kappa of two models from the 20 case testing set

Technique	Accuracy	Kappa
Naïve Bayes	65%	0.8140
Decision tree	90%	0.3694

Next, the two models were applied to a testing set. Since the full dataset was split 80:20 into training and testing sets, and 80 case was determined to be the optimal size of the training set, 20 case was chosen for the testing set (Table 8). The accuracy and Kappa statistic from the two techniques are shown in Table 9. The decision tree model again yielded better accuracy.

4   Conclusions

In this study, two algorithms, Naïve Bayes and decision tree, were used to build predictive models and to classify applications for home loans. After adding a median approval class, the models were implemented in Weka and used to classify applicants base on simulated data. The results showed the decision tree model had both a higher accuracy and Kappa statistic than the Naïve Bayes model. The decision tree technique was efficient in predicting the values of instances that were not in the training set. It was able to deal with missing attribute values, because in some case the timing of disbursement is not critical. By applying this technique, banks can improve their predictions of which clients will have a higher chance of getting their loan application approved. This will expand access to home loans.

Future work should investigate the use of a wider range of data mining techniques. As the performance of an algorithm is known to be dependent on the domain and type of the dataset, it would be interesting to explore other classification algorithm such as those used in machine learning.

Acknowledgment. We would like to thank Mr. John Winward for comments and suggestions on the manuscript.

References

1. Chambers, M., Garriga, C., Schlagenhauf, D.: The loan structure and housing tenure decisions in an equilibrium model of mortgage choice. Rev. Econ. Dyn. **12**, 444–468 (2009)
2. Trönnberg, C.-C., Hemlin, S.: Lending decision making in banks: a critical incident study of loan officers. Eur. Manag. J. **147**, 362–372 (2014)
3. Lee, C.C., Ho, Y.M., Chiu, H.Y.: Role of personal conditions, housing properties, private loans, and housing tenure choice. Habitat Int. **53**, 301–311 (2016)
4. Levy, J., Mallach, E., Duchessi, P.: A fuzzy logic evaluation system for commercial loan analysis. OMEGA. Int. J. Manag. Sci. **19**, 651–669 (1991)
5. Mammadli, S.: Fuzzy logic based loan evaluation system. Proc. Comput. Sci. **102**, 495–499 (2016)
6. Dong, G., Lai, K.K., Yen, J.: Credit scorecard based on logistic regression with random coeggicients. Proc. Comput. Sci. **1**, 2463–2468 (2012)
7. Majeske, K.D., Lauer, T.W.: The bank loan approval decision from multiple perspectives. Expert Syst. Appl. **40**, 1591–1598 (2013)
8. Malhotra, R., Malhotra, D.K.: Evaluating consumer loans using neural networks. OMEGA. Int. J. Manag. Sci. **31**, 83–96 (2003)
9. Doori, M.A., Beyrouti, B.: Credit scoring model based on back propagation neural network using various activation and error function. Int. J. Comput. Sci. Netw. Secur. **14**, 16–24 (2014)
10. Alsultanny, Y.: Labor market forecasting by using data mining. Proc. Comput. Sci. **18**, 1700–1709 (2013)
11. Hamid, A.J., Ahmed, T.M.: Developing prediction model of loan risk in banking using data minding. Mach. Learn. Appl.: Int. J. **3**, 1–9 (2016)
12. Bach, M.P., Zoroja, J., Jakovi, B., and Sarlija, N.: Selection of variables for credit risk data mining models: preliminary research. In: 40th International Convention on Information and Communication Technology, Electronics and Microelectronics, pp. 1367–1372 (2017)
13. Witten, I.H., Frank, E., Hall, M., Pal, C.: Data Mining: Practical Machine Learning Tools and Techniques. Morgan Kaufmann, Burlington (2016)

14. Perlich, C., Provost, F., Simonoff, J.: Tree induction vs. logistic regression: a learning-curve analysis. J. Mach. Learn. Res. **4**, 211–255 (2003)
15. Landis, J.R., Koch, G.G.: The measurement of observer agreement for categorical data. Biometrics **33**, 159–174 (1977)

Feature Selection and Identification of Fuzzy Classifiers Based on the Cuckoo Search Algorithm

Konstantin Sarin, Ilya Hodashinsky[(⊠)], and Artyom Slezkin

Tomsk State University of Control Systems and Radioelectronics, Tomsk, Russia
hodashn@gmail.com

Abstract. Classification is an important problem of data mining. The main advantage of fuzzy methods for extracting classification rules from empirical data is that the user can easily understand and interpret these rules, which makes fuzzy classifiers a useful modeling tool. A fuzzy classifier uses IF-THEN rules, with fuzzy antecedents (IF-part of the rule) and class labels in consequents (THEN-part of the rule). A method to constructing fuzzy classifiers based on the cuckoo search metaheuristic is described. The proposed method to constructing fuzzy classifiers based on observations data involves three stages: (1) feature selection, (2) structure generation, and (3) parameter optimization. The contributions of this paper are: (i) proposal of Cuckoo Search based feature selection; (ii) proposal of Cuckoo Search based parameter optimization of fuzzy classifier; (iii) proposal of subtractive clustering algorithm for structure generation of fuzzy classifier; and (iv) experiments with well-known benchmark classification problems (wine, vehicle, hepatitis, segment, ring, twonorm, thyroid, spambase reproduction data sets).

Keywords: Classification · Feature selection · Cuckoo search

1 Introduction

Classification is an important problem of data mining. In contrast to other classification methods (neural network, support vector machine, and Bayesian classifier), fuzzy classification does not set rigid boundaries among neighbor classes; an object of classification can belong to several classes with different degrees of confidence.

The main advantage of fuzzy methods for extracting classification rules from empirical data is that the user can easily understand and interpret these rules, which makes fuzzy classifiers a useful modeling tool.

A fuzzy classifier uses IF-THEN rules, with fuzzy antecedents (IF-part of the rule) and class labels in consequents (THEN-part of the rule). Antecedent parts of fuzzy rules partition the input feature space into a set of fuzzy regions, while consequents determine the output of the classifier by marking these regions with proper class labels.

Identification of a classification system means constructing a model that predicts the class to which an object described by its feature vector belongs. To identify fuzzy classifiers, evolutionary computation and swarm intelligence methods are generally employed [1–4].

© Springer Nature Switzerland AG 2018
S. O. Kuznetsov et al. (Eds.): RCAI 2018, CCIS 934, pp. 22–34, 2018.
https://doi.org/10.1007/978-3-030-00617-4_3

The contributions of this paper are: (i) proposal of Cuckoo Search based feature selection; (ii) proposal of Cuckoo Search based parameter optimization of fuzzy classifier; (iii) proposal of subtractive clustering algorithm for structure generation of fuzzy classifier; and (iv) experiments with well-known benchmark classification problems (wine, vehicle, hepatitis, segment, ring, twonorm, thyroid, spambase reproduction data sets).

The rest of the paper is organized as follows: Sect. 2 presents a problem statement. An introduction of Cuckoo Search applications for search in a continuous and binary space is discussed in Sect. 3, where the methodology is described. Section 4 presents a brief description of fuzzy classifier construction. In Sect. 4.1, the application of Cuckoo Search for feature selection is introduced. In Sect. 4.2, the structure generation algorithm is proposed. Section 5 provides some experimentation to show the effectiveness of our method and finally Sect. 6 concludes the paper.

2 Problem Statement

Suppose that we have a universe $U = (A, G)$, where $A = \{x_1, x_2,\ldots, x_n\}$ is a set of input features and $G = \{1, 2,\ldots, m\}$ is a set of class labels. Suppose also that $\mathbf{X} = x_1 \times x_2 \times \ldots \times x_n \in \mathfrak{R}^n$ is an n-dimensional feature space. An object in this universe is characterized by a feature vector. The classification problem consists in predicting the class of the object based on its feature vector.

The feature selection problem is formulated as follows: on the given set of features $\mathbf{X}$, find a feature subset that does not cause a significant decrease in classification accuracy, or even increases it, when the number of features is decreased. The solution is represented as a vector $\mathbf{S} = (s_1, s_2,\ldots, s_n)^T$, where $s_i = 0$ means that the ith feature is excluded from classification and $s_i = 1$ means that the ith feature is used by the classifier. Classification accuracy is estimated for each feature subset. The traditional classifier can be represented as a function

$$f:\mathfrak{R}^n \to \{0, 1\}^m,$$

where $f(\mathbf{x}; \boldsymbol{\theta}) = (b_1, b_2, \ldots, b_m)^T$; here, $\beta_i = 1$ if the object given by the vector $\mathbf{x}$ belongs to the class i and $\beta_j = 0$ $(j = 1, 2\ldots, m, i \neq j)$ if it does not, and $\boldsymbol{\theta}$ is a parameter vector of the classifier.

A fuzzy classifier can be represented as a function that assigns a class label to a point in the feature space with a certain evaluable confidence:

$$f:\mathfrak{R}^n \to [0, g]^m,$$

where $g \in \mathfrak{R}$ is determined by individual characteristics of a classifier.

The fuzzy classifier uses production rules of the form

$$R_i: \text{IF } s_1 \wedge x_1 = A_{1i} \text{ AND } s_2 \wedge x_2 = A_{2i} \text{ AND} \ldots \text{AND } s_n \wedge x_n = A_{ni} \text{ THEN } y = L_i,$$

where $L_i \in \{1, 2,\ldots, m\}$ is the output of the ith rule ($i = 1, 2,\ldots, R$, where R is the number of rules); A_{ki} is a fuzzy term that characterizes the kth feature in the ith rule ($k = 1, 2,\ldots, n$); and $s_k \wedge x_k$ indicates the presence ($s_k = 1$) or absence ($s_k = 0$) of a feature in the classifier.

The fuzzy classifier yields a vector $(\beta_1, \beta_2,\ldots, \beta_m)^T$, where

$$\beta_j = \sum_{\substack{i=\overline{1,R} \\ L_i=j}} \prod_{k=1}^{n} \mu_{ki}(x_k), \ j = 1, 2, \ldots, m,$$

$\mu_{ki}(x_k)$ is a membership function for the fuzzy term A_{ki} at the point x_k, in this work, we use the Gaussian membership function:

$$\mu_{ki}(x_k) = e^{\dfrac{(x_k - s_{ki})^2}{\sigma_{ki}^2}},$$

where s_{ki} is the center and σ_{ki} is the deviation of the membership function. Using this vector, the class is assigned based on the winner-takes-all principle:

$$\text{class} = \arg \max_{1 \leq j \leq m} \beta_j.$$

On an observations table $\{(\mathbf{x}_p, c_p), p = 1, 2,\ldots, z\}$, the accuracy of the classifier can be expressed as follows:

$$E(\boldsymbol{\theta}, \mathbf{S}) = \dfrac{\displaystyle\sum_{p=1}^{z} \begin{cases} 1, & \text{if } c_p = \arg \max_{1 \leq j \leq m} f_j(\mathbf{x}_p; \boldsymbol{\theta}, \mathbf{S}) \\ 0, & \text{otherwise} \end{cases}}{z},$$

where $f(\mathbf{x}_p; \boldsymbol{\theta}, \mathbf{S})$ is the output of the fuzzy classifier with the parameters $\boldsymbol{\theta}$ and features $\mathbf{S}$ at the point $\mathbf{x}_p$. Thus, the problem of constructing the fuzzy classifier is reduced to finding the maximum of this function in the space of $\mathbf{S}$ and $\boldsymbol{\theta} = (\theta^1, \theta^2,\ldots, \theta^D)$:

$$\begin{cases} E(\boldsymbol{\theta}, \mathbf{S}) \rightarrow \max \\ \theta^i_{\min} \leq \theta^i \leq \theta^i_{\max}, \ i = \overline{1, D}, \\ s_j \in \{0, 1\}, \ j = \overline{1, n} \end{cases}$$

where $\theta^i_{\min}$ and $\theta^i_{\max}$ are the lower and upper boundaries of each parameter, respectively.

3   Cuckoo Search Optimization Algorithm

The cuckoo search is a metaheuristic optimization algorithm inspired by the cuckoo reproduction strategy whereby a cuckoo lays its eggs in the nests of other host birds (sometimes by removing host eggs); eventually, these alien eggs can be thrown away by the host [4, 5]. The cuckoo search algorithm is based on the following three rules:

(1) each cuckoo lays one egg at a time in a randomly chosen nest, which is a solution to the problem;
(2) the best nests with high-quality eggs (solutions) are carried over to the next generation;
(3) the number of available nests is fixed, and the host can discover an alien egg with a certain probability (in this case, the host bird can either throw the alien egg away or abandon the nest to build a completely new one in a new location).

An important feature of the cuckoo search algorithm is the use of Levy flights for local and global search [6]. A Levy flight is a random walk characterized by a series of sudden jumps described by a fat-tailed probability density function; as compared to the normal distribution, such jumps increase the probability of significant deviations from the mean.

3.1   Search in a Continuous Space

Essentially, search in a continuous space is an iterative procedure of finding new values for the solution vector $\mathbf{x} = [x_1, x_2, \ldots, x_n]^T$ through changing each coordinate of the vector by the size of the search step, $Levi_i$ $(i = 1, \ldots, n)$:

$$x_i = x_i + Levi_i. \tag{1}$$

The cuckoo search metaheuristic offers two stages to determine the step size: selection of a random direction and generation of steps. Direction is generated using uniform distribution, whereas the generation of the step size is quite complex. There are several ways of generation; one of the simplest and most effective ways is to use the algorithm proposed in [6] for symmetric stable Levy distribution. Here, "symmetric" means that steps can be positive or negative depending on the probability density of the Levy distribution. Using this algorithm, the step size is found as follows:

$$Levi_i = \gamma \cdot u_i / |v_i|^{1/\beta}. \tag{2}$$

Here, γ is a jump coefficient for Levy flights (in this work, it is 0.01 as recommended in [5]), while u_i and v_i are normally distributed parameters:

$$v_i \sim N(0; \sigma_v^2), \quad u_i \sim N(0; \sigma_u^2),$$

$$\sigma_v^2 = 1, \quad \sigma_u^2 = \left\{ \frac{\Gamma(1 + \beta) \sin(\pi\beta/2)}{\Gamma[(1 + \beta)/2] \cdot 2^{(\beta-1)/2}} \right\}^{1/\beta},$$

where $\Gamma(x)$ is a gamma-function and $\beta = 3/2$ as recommended in [5].

3.2   Search in a Binary Space

In a binary space, the components of the solution vector can take only two values, $x_i \in \{0, 1\}$. A new solution is derived from a continuous solution by using a transformation function that translates a real value into a value from the set $\{0, 1\}$.

In [7], a modification of the cuckoo search algorithm for the binary space was proposed. In this paper, we propose another modification. In contrast to [7], a binary value for the component of the solution vector is derived from a continuous value by transforming only the Levy jump, rather than the value of the vector with an added jump. In this case, the jump is transformed into the probability of obtaining the value inverse to the solution found at the previous step, rather than into the probability of obtaining 1.

For this transformation to reflect the cuckoo's behavior in the binary space, we should take the following into account. Longer jumps characterize a greater difference in solution than shorter jumps. In the binary domain, this means that, for long jumps, the probability that a solution component will invert its value should exceed that for short jumps. This probability should grow with increasing length of the jump. If the jump length tends to zero, then the inversion probability should also tend to zero. In other words, for $|Levi| \to \infty$, the probability of inverting the solution tends to 1; for $|Levi| \to 0$, this probability tends to 0. As a function for transforming the Levy jump into the inversion probability, we use the hyperbolic tangent modulus, $T(v) = |th(v)|$.

Taking this into account, a new solution is generated as follows:

$$x_i = x_i \oplus p_i, \tag{3}$$

where $p = 1$ if rand $< T(Levi_i)$, and $p = 0$ otherwise; $\oplus$ is the XOR operator.

4   Construction of the Fuzzy Classifier

The proposed method to constructing fuzzy classifiers based on observations data involves three stages: (1) feature selection, (2) structure generation, and (3) parameter optimization. At the first stage, a set of relevant features used to construct the classifier is found. Then, based on these features, the structure of the fuzzy classifier is generated and its parameters are optimized. We employ the cuckoo search metaheuristic for feature selection and parameter optimization [8], and the subtractive clustering algorithm for structure generation [9].

At a feature selection stage is determined features which are used at classification. At a structure generation stage, based on the training data, the structural characteristics of the classier (i.e., the number of fuzzy rules, the input variables, and the type of membership functions) are determined. This stage also determines the initial values of the membership function parameters and the labels for the fuzzy rules' outputs. Once the structure of the fuzzy classifier is generated, the membership function parameters and the output class labels are selected to maximize the accuracy E.

4.1 Feature Selection Algorithm

When constructing the classifier based on observations data, the procedure of searching for the optimum in the binary space can be employed to find the optimal set of features. Each coordinate of the solution vector is associated with a certain feature: 1 indicates the presence of the feature, while 0 indicates its absence. In this case, the objective function is the accuracy of classifying the training data objects on the features the solution vector of which contains 1. Each time, before evaluating the objective function, a classifier is constructed based on these features.

When constructing fuzzy classifiers, the cuckoo search metaheuristic can be used to generate the input feature space. In this case, the search vector is the binary vector $\mathbf{S}$, which indicates the presence or absence of features. The objective function is the classification accuracy on the features $\mathbf{S}$ that should tend to its maximum: $E(\mathbf{S}) \rightarrow$ max. Below is the step-by-step description of the feature selection algorithm.

Begin
Step 1. Initializing the population.
Specify the size of the population *popul* (population $\Theta = \{\mathbf{S}_i; \ i = 1,2,\ldots,popul\}$, where $\mathbf{S}_i$ are generated randomly), the probability of eliminating the worst nest p_a, and the number of iterations N. Initialize the iteration counter, $t=1$.
Step 2. Checking the termination condition.
Find the best solution $\mathbf{S}_{best}$ in the population Θ, where

$$best = \arg \max_{1 \le i \le popul} \left(E(\mathbf{S}_i) \right).$$

If $t = N$, then the algorithm terminates with the result $\mathbf{S}_{best}$.
Step 3. Eliminating the worst solution from the population Θ.
If $rand \le p_a$, then remove the solution $\mathbf{S}_k$ from the population Θ; here, *rand* is a random value uniformly distributed on the interval [0, 1] and

$$k = \arg \min_{1 \le i \le popul} \left(E(\mathbf{S}_i) \right).$$

Instead of the removed solution, randomly generate a new one.
Step 4. Generating a new solution.
For each element of the population, $\mathbf{S}_i$ ($i = 1,2,\ldots,popul$), generate a new solution $\mathbf{S}_i^{new}$ by using the Levy flight strategy with (3) and (2).
Step 5. Estimating the quality of new solutions.
IF $E(\mathbf{S}_i^{new}) > E(\mathbf{S}_i)$, THEN $\mathbf{S}_i = \mathbf{S}_i^{new}$.
Set $t = t+1$ and go to step 2.
End

4.2 Structure Generation

The proposed structure generation algorithm uses the assumption that the data of the same classes form compact regions (clusters) in the input space. To construct a rule base for the fuzzy classifier, we use a clustering algorithm that finds the data clusters and, then, attributes them to certain classes.

Data Clustering Algorithm. In this work, we employ the subtractive clustering algorithm [10]. This algorithm finds the clusters based on the potential of a data exemplar, which characterizes the closeness (proximity) of other data to the exemplar. In contrast to the k-means algorithms, the subtractive clustering algorithm has a fairly high efficiency and does not require specifying the number of clusters. Below is a brief description of the algorithm (for more details, see [10]).

Begin

Step1. Evaluate the potential of each point of the experimental data:

$$P_i = \sum_{j=1}^{z} e^{-\frac{4 \cdot d_{i,j}^2}{r_a^2}}.$$

Step 2. Find the point with the maximum potential:

$$new = \arg\max_{1 \le i \le z} P_i.$$

IF $P_{new} > \varepsilon_2 \cdot P_{C1}$ THEN *new* is a new cluster center, ELSE

IF $P_{new} < \varepsilon_1 \cdot P_{C1}$ THEN terminate the algorithm, ELSE

IF $\dfrac{d_{min}}{r_a} + \dfrac{P_{new}}{P_{C1}} \ge 1$ THEN *new* is a new cluster center, ELSE

set $P_{new} = 0$ and return to the beginning of Step 2.

Step 3. Reevaluate the potentials of all data points:

$$P_i = P_i - P_{new} \cdot e^{-\frac{4 \cdot d_{i,new}^2}{r_b^2}}, i = 1, ..., z,$$

and go to Step 2.

End

Here, $d_{i,j}$ is the distance (often, the Euclidean one) between the *i*th and the *j*th elements of the training data; d_{min} is the minimum distance between the candidate new and the current centers of the clusters; P_{C1} is the potential of the first cluster's center; $\varepsilon_1, \varepsilon_2$ are the lower and upper thresholds (respectively) that form the so-called grey zone (according to [10], these thresholds are set to be 0.15 and 0.5, respectively); and r_a, r_b are the radii determining the size of a cluster: the larger the radii, the less the number of clusters (according to the author of this algorithm, $r_b = 1.5 \cdot r_a$).

The clustering algorithm yields the cluster centers $\mathbf{v}_1, \mathbf{v}_2, ..., \mathbf{v}_r$ for the training data.

Structure Generation Algorithm. The input data for this algorithm are the training data from the observations table and the radius r_a of the clusters. The algorithm yields the fuzzy rule base $\boldsymbol{\theta}$.

Begin

Step 1. Using the subtractive clustering algorithm (see above), for the radius r_a, find the cluster centers $\mathbf{v}_1, \mathbf{v}_2, \ldots, \mathbf{v}_R$.

Step 2. For each cluster, determine the set of training data elements belonging to this cluster. These sets are denoted by C_i. The set C_i to which a training data exemplar $(\mathbf{x}_p, c_p)$ belongs is determined as follows:

$$i = \arg \min_{1 \leq i \leq R} \left\| \mathbf{v}_i - \mathbf{x}_p \right\|,$$

in this work, the metric $\|\cdot\|$ corresponds to the Euclidean distance.

Step 3. For each cluster i, find its generating class from the set of classes $\{1, 2, \ldots, K\}$. The label i-th cluster is determined as follows:

$$L_i = \arg \max_{1 \leq j \leq K} \left(\sum_{\substack{(\mathbf{x}_p, c_p) \in C_i \wedge \\ c_p = j}} \frac{\max\limits_{k, (\mathbf{x}_k, c_k) \in C_i} \left(\left\| \mathbf{v}_i - \mathbf{x}_k \right\| \right) - \left\| \mathbf{v}_i - \mathbf{x}_p \right\|}{\max\limits_{k, (\mathbf{x}_k, c_k) \in C_i} \left(\left\| \mathbf{v}_i - \mathbf{x}_k \right\| \right)} \right).$$

Step 4. Form the rule base with the number of fuzzy rules corresponding to the number of clusters. Set $y = L_i$ as an output of the ith rule. The parameters of the membership functions for the ith rule of the jth variable are determined as follows:

$$s_{ji} = v_{ji},$$

$$\sigma_{ji} = \sqrt{\frac{2}{|C_i|} \cdot \left(\sum_{(\mathbf{x}_p, c_p) \in C_i} \left(x_{jp} - s_{ji} \right)^2 \right)},$$

where $|C_i|$ is the cardinality of the set C_i.

End

Figure 1 shows the two-dimensional input space with three data clusters and the membership functions generated by these clusters.

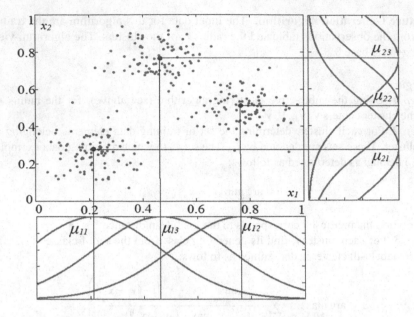

Fig. 1. Membership functions generated by three clusters.

Determining the Cluster Radius. The structural characteristics of the classifier, as well as its accuracy, depend on the parameter r_a of the structure generation algorithm. Since the structure generation is aimed at synthesizing high-accuracy classifiers with a reasonably small number of fuzzy rules, the problem of finding the optimal value of r_a can be reduced to the optimization problem with the following constraints:

$$\begin{cases} E(\theta_{r_a}, S) \to \min \\ 0.1 < r_a < 1.5 \\ R \le Rmax \end{cases},$$

where θr_a is the vector (formed by the structure generation algorithm with the parameter r_a) that describes the fuzzy rule base and $Rmax$ is the maximum number of fuzzy rules in the classifier. It should be noted that the value of r_a is bounded taking into account the normalization of the input data. This optimization problem is solved using the Hooke-Jeeves method [11].

4.3   Parameter Optimization Algorithm

At the stage of parameter optimization, the parameters of membership functions θ are tuned to maximize the classification accuracy: $E(\theta) \to \max$. For this purpose, the parameter optimization algorithm employs the cuckoo search metaheuristic in the continuous form. Below is the step-by-step description of the algorithm.

Begin

Step 1. Initializing the population.

Specify the size of the population *popul* (population $\Theta = \{S_i; i = 1,2,...,popul\}$), the probability of eliminating the worst nest p_a, and the number of iterations N. The value of θ_1 is given by the value of the classifier's parameters upon structure generation, while the values of θ_s (where $s = 2,...,popul$) are generated randomly. Initialize the iteration counter, $t=1$.

Step 2. Checking the termination condition.

Find the best solution θ_{best} in the population Θ, where

$$best = \arg \max_{1 \leq i \leq popul} \left(E\left(\theta_i\right) \right).$$

If $t = N$, then the algorithm terminates with the result θ_{best}.

Step 3. Eliminating the worst solution from the population Θ.

If $rand \leq p_a$, then remove the solution θ_k from the population Θ,

$$k = \arg \min_{1 \leq i \leq popul} \left(E\left(\theta_i\right) \right).$$

Instead of the removed solution, randomly generate a new one.

Step 4. Generating a new solution.

For each element of the population, θ_i ($i = 1,2,...,popul$), generate a new solution θ_i^{ne} by using the Levy flight strategy with (1) and (2).

Step 5. Estimating the quality of new solutions.

IF $E(\theta_i^{new}) > E(\theta_i)$, THEN $\theta_i = \theta_i^{new}$

Set $t = t+1$ and go to step 2.

End

5 Experiments on Real Data

The performance of the proposed algorithms for feature selection and parameter optimization was estimated on eight real datasets from the KEEL repository (http://keel.es). The experiments were carried out using 10-fold cross-validation.

Classifiers were constructed on the whole feature space and on the features selected by the corresponding algorithm. Table 1 shows the classification accuracies for the numbers of features used. Here, F_{all} is the total number of features in a dataset and F is the number of features selected by the algorithm; $Etra$ and $Etst$ are classification accuracies on the training and test data for selected features, while $Etra_{all}$ and $Etst_{all}$ are classification accuracies on the training and test data for all features. Table 2 shows Average results obtained by applying the D-MOFARC method and the FARC-HD algorithm [12].

Table 1. Average results obtained by applying the proposed method

Dataset	F_{all}	F	Etra	Etst	$Etra_{all}$	$Etst_{all}$
Wine	13	7	0.95	0.91	0.97	0.938
Vehicle	18	7.6	0.537	0.495	0.517	0.499
Hepatitis	19	9.5	0.957	0.836	0.972	0.938
Segment	19	17.7	0.634	0.629	0.575	0.568
Ring	20	3.1	0.729	0.733	0.826	0.825
Twonorm	20	8.3	0.885	0.883	0.937	0.933
Thyroid	21	16.6	0.94	0.94	0.929	0.928
Spambase	57	21.9	0.705	0.7	0.583	0.584

Table 2. Average results obtained by applying the D-MOFARC method and the FARC-HD algorithm.

Dataset	D-MORFAC			FARC-HD		
	F	Etra	Etst	F	Etra	Etst
Wine	13	1.000	0.958	13	1.000	0.955
Vehicle	18	0.845	0.706	18	0.772	0.680
Hepatitis	19	1.000	0.900	19	0.994	0.887
Segment	19	0.980	0.966	19	0.948	0.933
Ring	20	0.942	0.933	20	0.951	0.940
Twonorm	20	0.945	0.931	20	0.966	0.951
Thyroid	21	0.993	0.991	21	0.943	0.941
Spambase	57	0.917	0.905	57	0.924	0.916

The comparison analysis brings us at the following conclusions:

1. The Wilcoxon-Mann-Whitney test indicates that there is a statistically significant difference between the medians of the initial number of features and the number of features selected by the our algorithm (p-value < 0.015);
2. The Wilcoxon-Mann-Whitney test indicates that there is no statistically significant difference between the classification accuracies in the compared classifiers.

This, in turn, brings us to the following conclusion: given that the accuracies of the compared classifiers are statistically indistinguishable, our classifiers are preferable in view of the smaller number of rules, which eventually implies that they have less computational complexity and, possibly, are better interpretable.

6   Conclusions

In this paper, we have presented a method for constructing fuzzy classifiers based on the cuckoo search metaheuristic. We have also proposed an optimization algorithm which operates in the binary search space and is used to generate the input feature space. In addition to classification accuracy, the number of features is also an important characteristic of the classifier. This characteristic affects the computational resources used to construct the classifier and the time spent on evaluating irrelevant features, as well as the simplicity of the model. The continuous cuckoo search algorithm has been employed at the stage of parameter optimization, allowing the classification accuracy to be improved without modifying the structural properties of the classifier.

Using the proposed method, we have conducted a series of experiments on real datasets to construct fuzzy classifiers. On all datasets, we have managed to reduce the number of features, while the accuracy of classification has been improved only on three (out of eight) datasets. The results allow us to conclude that the reduction of the feature space does not necessarily improve the accuracy of classification. However, if the simplicity of the model, along with its accuracy, is of particular importance, then the feature space has to be reduced. That is why the proposed feature selection algorithm can be useful for constructing fuzzy classifiers.

Acknowledgements. The reported study was funded by RFBR according to the research project 16-07-00034.

References

1. Antonelli, M., Ducange, P., Marcelloni, F.: An experimental study on evolutionary fuzzy classifiers designed for managing imbalanced datasets. Neurocomputing **146**, 125–136 (2014)
2. Lahsasna, A., Seng, W.C.: An improved genetic-fuzzy system for classification and data analysis. Expert Syst. Appl. **83**, 49–62 (2017)
3. Jamalabadi, H., Nasrollahi, H., Alizadeh, S., Araabi, B.N., Ahamadabadi, M.N.: Competitive interaction reasoning: a bio-inspired reasoning method for fuzzy rule based classification systems. Inf. Sci. **352–353**, 35–47 (2016)
4. Yang, X.-S., Deb, S.: Engineering optimisation by cuckoo search. Int. J. Math. Model. Numer. Optim. **1**(4), 330–343 (2010)
5. Yang, X.-S., Deb, S.: Cuckoo search: recent advances and applications. Neural Comput. Appl. **24**, 169–174 (2014)
6. Mantegna, R.N.: Fast, accurate algorithm for numerical simulation of Levy stable stochastic processes. Phys. Rev. E **49**(5), 4677–4683 (1994)
7. Pereira, L.A.M., et al.: A binary cuckoo search and its application for feature selection. In: Yang, X.-S. (ed.) Cuckoo Search and Firefly Algorithm. SCI, vol. 516, pp. 141–154. Springer, Cham (2014). https://doi.org/10.1007/978-3-319-02141-6_7
8. Hodashinsky, I.A., Minina, D.Y., Sarin, K.S.: Identification of the parameters of fuzzy approximators and classifiers based on the cuckoo search algorithm. Optoelectron. Instrum. Data Process. **51**(3), 234–240 (2015)
9. Sarin, K.S., Hodashinsky, I.A.: Identification of fuzzy classifiers based on the mountain clustering and cuckoo search algorithms. In: International Siberian Conference on Control and Communications, SIBCON, pp. 1–6. IEEE, Astana (2017)
10. Chiu, S.L.: Fuzzy model identification based on cluster estimation. J. Intell. Fuzzy Syst. **3**(2), 267–278 (1994)

11. Hooke, R., Jeeves, T.A.: Direct search solution of numerical and statistical problems. J. Assoc. Comput. Mach. **8**, 212–229 (1961)
12. Fazzolari, F., Alcala, R., Herrera, F.: A multiobjective evolutionary method for learning granularities based on fuzzy discretization to improve the accuracy-complexity trade-off of fuzzy rule-based classification systems: D-MOFARC algorithm. Appl. Soft Comput. **24**, 470–481 (2014)

Knowledge Engineering in Construction of Expert Systems on Hereditary Diseases

Boris A. Kobrinskii[1(✉)], Nataliya S. Demikova[2], and Nikolay A. Blagosklonov[1]

[1] Federal Research Center "Computer Science and Control" of RAS, Vavilova str. 44, kor. 2, Moscow 119333, Russian Federation
kba_05@mail.ru
[2] Russian Medical Academy of Continuous Professional Education, Barrikadnaya str. 2/1, Moscow 125993, Russian Federation

Abstract. The paper considers medical knowledge extraction for the diagnosis of rare hereditary diseases. The specific feature of the proposed approach is the combination of expert estimates of the symptom presence probability with three complementary factors of confidence: time of manifestation, severity and frequency of symptom manifestation in different age ranges. Preliminary textological cards used by experts and cognitive scientists are formed based on the linguistic analysis of related literature.

Keywords: Knowledge engineering · Confidence factors · Linguistic analysis
Textological cards · Hereditary diseases · Orphan diseases
Holistic visual images

1 Introduction

Unclear pathological manifestation is a characteristic feature of clinical medicine. This is most clearly manifested in hereditary diseases of progredient nature. The difficulty is in the assessment of intermediate states which gradually increase with age of changes. Even a skillful physician experiences difficulty due to the rarity of genetically determined diseases in his/her practice. At the same time orphan (rare) hereditary diseases are an important social problem [1]. The necessity for early detection is determined by the fact that the progression of a number of them can be prevented. However, the regression of pathological changes is impossible. Therefore, the delay in treatment due to late diagnosis leads to remediless results. However, on the one hand, hereditary diseases are very rare but, on the other hand, their total number is near 6,000 nosological forms [2]. It is clear that no physician can remember the manifestations of even several hundreds of diseases. Moreover, the heterogeneity of clinically similar diseases (variants with different genetic changes) is combined with polymorphism (polyvariety) and at the same time partial overlapping of the spectrum of clinical features. These characteristics of hereditary pathology lead to a large number of erroneous and late diagnosed cases. Thus, more than 30 years ago the development of computer diagnostic systems in the field of clinical genetics started. Some of systems were later upgraded [3–9]. They allowed increasing the quality of genetically determined diseases recognition. However, up to

S. O. Kuznetsov et al. (Eds.): RCAI 2018, CCIS 934, pp. 35–45, 2018.
https://doi.org/10.1007/978-3-030-00617-4_4

date the hypotheses for differential diagnosis of many similar diseases, especially in the early stages, with poorly expressed pathological changes are practically not effective. Therefore, it is of vital importance to search for new approaches to the construction of intelligent diagnostic systems in the subject area of hereditary diseases that will take into account the dynamics of the processes.

The complexity of the subject area determines the search in the direction of knowledge engineering including the evaluation of confidence measures regarding the occurrence of signs characterized by unclear manifestations, an additional problem being the reflection of experts.

2   Features of the Clinic and Diagnosis of Hereditary Diseases

Hereditary diseases, which are related to orphan ones, present serious difficulties for the diagnostics. In most cases physicians have little personal experience of observing such patients. At the same time, one of the main tasks is early detection. This is especially important for lysosomal storage diseases occurring at the frequency of 1 per 7,000 to 8,000 newborns while each disease from this group occurs in only 1 in 100,000 newborns [10]. These diseases are characterized by a worsening progression, and it is extremely difficult to identify manifesting clinical signs. This is connected to the vagueness and at the same time to the low frequency of occurrence of individual pathological features. In addition, some signs may change to the opposite ones in the progression. For example, loose joints due to the accumulation of deposited macromolecules (glycosaminoglycans) at the mucopolysaccharidosis leads to an ever-increasing stiffness of joints.

Lysosomal diseases are characterized by a combination of numerous pathological changes associated with the accumulation of macromolecules in various organs and systems of the body. These diseases include mucopolysaccharidosis, mucolipidosis, gangliosidosis and others. Depending on the location of abnormal focuses and the speed of macromolecular deposition, a different clinical presentation or clinical variants ("portraits") of the disease takes place including the age of the manifestation, the severity of the symptoms, the degree of incidence at a particular age. In this regard, there are fuzzy intermediate states in the dynamics of the pathological process, various combinations of features, and varying degrees of their severity.

At the same time the need for the earliest possible detection of lysosomal accumulation diseases characterized by a deficiency of enzymes can be explained by the recent availability of enzyme replacement therapy.

The presented features of clinical manifestations of rare hereditary diseases create certain limitations in the application of computer decision support methods. At the same time, the requirements for knowledge about the diagnosis of orphan pathology are increasing.

Cumulative knowledge including its own verbalized and implicit (intuitive) knowledge in combination with well-known knowledge from various sources is an important factor for the knowledge base formation for a diagnostic system by hereditary diseases.

3   Fuzzy Sets of Transition States of Diseases

Each group of diseases can be formally represented by sets and subsets (multisets - for a higher level of associations of similar diseases). For orphan pathology, fuzzy sets (with inaccurately defined boundaries) are characteristic. This is particularly evident with progressive accumulation diseases. They are characterized by a constant process of increasing the severity of previously manifested pathological changes and the appearance of pathological changes in other organs and systems of the organism. In clinical practice, doctors constantly encounter situations that have minor differences. In addition, there can be fuzzy attributes, and fuzzy references to certain classes. In the course of chronic diseases, the signs tend to undergo changes, which are differently assessed and described by each of the physicians. A great diagnostic value in hereditary diseases can also have appearance (visual images) [11]. The idea of a visual series should be perceived as a visual quasi-continuum of close or relatively close images, which implies the vagueness of transitions between individual representatives (Fig. 1). However, an appearance is characterized by fuzzy transitions, which form a different visual row of patients of one subset, and sometimes of one set. This creates additional diagnostic difficulties and requires a verbal appreciation based on a fuzzy scale, similar to linguistic features. A fairly broad range of opinions of fuzzy verbal definitions to the views expressed by doctors is typical, which can be often combined with the modality "it seems to me".

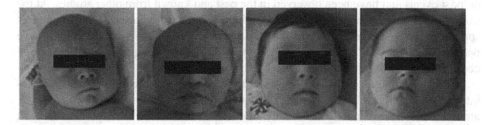

Fig. 1. Visual images of patients of the same type

To model the allowable level of inaccuracy in solving a particular task, Lotfi Zadeh proposed to consider degrees of granulation of measuring information [12], meaning by a granule a group of objects united by the relations of indistinguishability, similarity, proximity, or functionality. The granulation of information is based on the nonclassical representation of the set. Different types of fuzzy measures correspond to different points of view on the evaluation of the certainty of events. The theory of possibilities of Zadeh [13], allowing several (may be continuum set of) degrees of possibility, actually departs from modal semantics. In medicine, this corresponds to the existence of many variants of transitional states of manifestations of individual signs or health status. The problem of separating the continuum into diagnostically significant intervals is that the integrating estimates include numerous fuzziness. In the integral analysis of the organism's conditions, such concepts as trend, dynamics, which are almost always characterized by fuzzy marks, prevail. One of the variants of the subdivision of the health continuum into intervals meaningful for a particular subject area, corresponding to the range of values of some notion by the criterion of maximizing the "information gain" measure, is proposed by J.R. Quinlan [14]. The decomposition of fuzzy sets allows us to represent an arbitrary fuzzy set as a sum of fuzzy sets [15], corresponding to separate nosological forms or subclasses of diseases. These fuzzy sets can result from formalized expert evaluations in linguistic form or in the form of digital labels corresponding to the parameters at the linguistic scale. Combinations of features represented by fuzzy sets/ subsets allow us to characterize various variants of independent diseases or separate clinical forms. The method of ordering objects is based on an assessment of their proximity to some "ideal" object in a multivalued space. The method of ordering objects is based on an assessment of their proximity to some "ideal" object in a multivalued space.

The metric problem for measuring the distances between objects in this space can be solved by representing multisign objects based on the multiset formalism [16]. In medicine, as an "ideal" object, a case can be considered, characterized by a typical combination of symptoms most often repeated in a subclass of diseases.

4   Linguistic Analysis as the First Stage of Knowledge Extraction

The unclear representation of descriptions in the literature is determined by a number of factors: clinical picture of the disease, age characteristics of the patient, rate of the pathological process, knowledge of physicians and their subjective preferences based on past events that have been observed in the past and known from other studies. At the same time, various diseases may be hidden behind these differences. This fact requires special attention. Therefore, linguistic scales of medical parameters and fuzzy logic have particular importance for supporting effective differential diagnosis of diseases and conditions of the organism [17, 18].

Knowledge engineering assumes different approaches to the knowledge representation. At the first (preliminary) stage this can be the analysis of special (medical) literature sources. The data will serve as the additional material for the experts at the next stage.

Taking into account inhomogeneity and multilinguality of initial sources, an integrated approach combining the semantic, textual, and linguistic methods of text analysis was used to extract knowledge from literature.

Semantic analysis allows one to evaluate the meaning of the text. Based on results of the semantic analysis, it is possible to determine the frequency of various terms occurrence. In the problem area of orphan diseases diagnosis, it is necessary to identify and rank certain indicators. This is the age of manifestation of symptoms, the dynamics of their changes and the degree of incidence.

Textual analysis uses methods of studying the text based on the comparative research method. As applied to the problem under study, it should be a comparison of various publications on the diagnosis of hereditary diseases. Among others, a cognitive comparative analysis of manifestations of diseases in various ethnic groups is carried out.

Linguistic analysis involves the study of various aspects of diseases (etiology, pathogenesis, clinical manifestations), as well as the correlation of multilingual terms, the identification of terms-metaphors, etc. The result of a complex analysis of literature sources using the mentioned-above approaches is a structured representation of the extracted knowledge in the form of textological cards (see Table 1).

Table 1. Fragment of the textological card of the Hunter syndrome

Symptom	Presence	Level of manifestation	Age group	Degree of incidence	Source
Mental retardation	+	Strong	1 to 3 years	Frequently	*
	+	Very mild	over 6 years	Rare	**
	+	Strong	–	Frequently	***
Coarse facial features	+	Very strong	1 to 3 years	Very frequently	*
	+	Strong	1 to 3 years	Frequently	**
	+	Strong	–	Frequently	***
Macroglossia	+	Strong	1 to 3 years	Very frequently	*
	+	Strong	1 to 3 years	–	**
	+	Strong	–	Frequently	***

* Wraith, J.E., Scarpa, M., Beck, M., Bodamer, O.A., De Meirleir, L., Guffon, N., Meldgaard Lund, A., Malm, G., Van Der Ploeg, A.T., Zeman, J.: Mucopolysaccharidosis type II (Hunter syndrome): A clinical review and recommendations for treatment in the era of enzyme replacement therapy. Eur. J. Pediatr., **167**(3), 267–277 (2008)
** Namazova-Baranova, L.S., Vashakmadze, N.D., Gevorkyan, A.K., Altunin, V.V., Kuzenkova, L.M., Chernavina, E.G., Babaykina, M.A., Podkletnova, T.V., Kozhevnikova, O.V.: Obstructive Sleep Apnea Syndrome in Children with Type II Mucopolysaccharidosis (Hunter Syndrome). Pediatric pharmacology, **10**(6), 76–81 (2013) (In Russian)
*** Scarpa, M., Almássy, Z., Beck, M., Bodamer, O., Bruce, I.A., De Meirleir, L., Guffon, N., Guillén-Navarro, E., Hensman, P., Jones, S., Kamin, W., Kampmann, C., Lampe, C., Lavery, C.A., Leão Teles, E., Link, B., Lund, A.M., Malm, G., Pitz, S., Rothera, M., Stewart, C., Tylki-Szymaska, A., Van Der Ploeg, A., Walker, R., Zeman, J., Wraith, J.E.: Mucopolysaccharidosis type II: European recommendations for the diagnosis and multidisciplinary management of a rare disease. Orphanet J. Rare Dis., **6**(1), 1–18 (2011)

Knowledge engineer filling out a textological card uses the scales specially developed for this study.

The scale of symptom presence is presented in the form of " + " and "–" showing presence or absence of the indicator.

The rank scale of levels of symptom manifestation includes 5 linguistic estimates: very strongly expressed, strongly expressed, moderately expressed, mildly expressed, expressed very mildly.

The age scale contains five gradations: (1) the neonatal period i.e. the first 28 days of life, (2) the first year of the child life, (3) 1 to 3 years, (4) 4 to 6 years, (5) over 6 years. The linguistic scale of symptom incidence is the following: very frequent (more than 80%), frequent, (60–80%), relatively often (from more than 30% to 60%), rare (from more than 15 to 30%), very rare (less than 15%).

Textual cards are used by experts and knowledge engineer in the process of forming a "disease-signs" matrix.

5   Probability and Factors of Confidence in the Diagnosis of Hereditary Storage Diseases

In medicine, signs are in most cases stable concepts, but the attributes that characterize them indicate a significant variety of their manifestations and severity. Matrices for the formalized presentation of diseases with expert estimates for each nosological form at certain age intervals allow to be taken into account for the pathological process dynamics.

In this study the probability of developing a sign (attribute), called "the presence of a sign", was first indicated in each age group. This evaluation is based on the personal knowledge of the experts who took into account the views of other researchers presented in the textological cards. The scale for the expert evaluation of the probability of the sign presence includes 7 gradations in the interval from −1 to +5: −1 stays for the impossibility of the symptom manifestation in a certain age range, 0 stays for the normal range (the absence of a pathological feature being theoretically possible at this age), 1 stays for minimal possibility, 2 for remotely possible, 3 for possible, 4 for highly possible, 5 for maximum possibility. Such a linguistic scale corresponds to the accepted estimates of observed signs in medical practice. As an example of the gradation usage, −1 may be a sign of "tooth pathology in a newborn" because of the impossibility of its presence at a given age. When experts analyzed textological cards, the opinions of medical specialists (for example, ophthalmologists in assessing the pathological symptoms from the side of the vision organ) were of great importance.

The objectivity of the conclusion in each case was determined by a joint assessment of confidence measures by two experts who jointly make decisions. At first, one of the experts offered his assessment. If the second expert agreed the decision was taken. In case of disagreement, a dialogue of experts was carried out with the argumentation of their opinions. If necessary, additional literature sources were used in the discussion process. In the discussion, a cognitive scientist participated. He could comment on textological cards for these diseases.

Considerable part of the subjective knowledge is determined by the dispositions of the expert including those which are modified by the influence of their personal experience [19]. The result of personal experience is a measure of confidence in one's own and other people's knowledge.

Fuzzy logical conclusions create a model of rough reflection of a person, in particular a doctor, in the moment when a decision has to be made. A measure of confidence or an expert confidence factor is an informal assessment attached to the conclusions. In the Stanford model the certainty factor is a value from the interval [−1; 1], which numerically defines the confidence measure of the expert decision [20]. In linguistic form it can be presented, e.g., as follows: "almost impossible", "very doubtful", "doubtful", "can be suspected", "cannot be excluded", "most likely" or a large number of degrees of freedom.

In the medical diagnosis of diseases with progredient nature certainty factors by expert are specific for different age groups in terms of the severity of signs (symptoms) characterized by a fuzzy scale of changes, the timing of their manifestation and the degree of incidence.

In the clinical picture of diseases with progredient nature signs in most cases are stable concepts but the meanings of attributes that characterize the severity and time of manifestation are determined by the age-related dynamics of the pathological process.

"Sign presence" attributes are accompanied by three certainty factors: (a) the manifestation of the sign at a certain age, (b) the degree of incidence of the sign in the age group, (c) the severity of the sign. Estimates of the certainty factors lie in the area between 0.1 and 1.0. A confidence measure of 1.0 was used by experts only in cases where the presence of a sign is −1 or 0. The introduction of three complementary certainty factors allows us to obtain a complete ("three-dimensional") picture of the confidence measure of the analyzed sign including the time of its manifestation, the changing severity of manifestations and the frequency in each of the age groups. When deciding on the issue of the certainty factors the expert physicians turn to their intuitive knowledge, which expresses itself in the process of deciding on the choice of the estimates to be presented in verbally inexplicable changes.

Certainty factors significantly increase the likelihood of differentiation between certain diseases. It is especially important to introduce three confidence measures to the signs characterizing the clinical forms of diseases, for example, mucopolysaccharidosis type I, which according to the international classification of hereditary diseases [2] is represented by three types (IH − OMIM 607014, IH/S − OMIM 607015, IS − OMIM 607016). In fact, in this case there is a continuum of values of each characteristic, which was conditionally quantized by experts. The intermediate nosological form IH/S includes manifestations from severe (close to the subtype IH) to relatively light (close to the subtype IS). As another example, the "hepatomegaly (enlarged liver)" sign is considered in the case of mucopolysaccharidosis type II (Hunter syndrome − OMIM 309900), which includes two clinical forms, severe and light. In the severe form the sign presence in the first year of life corresponds to the "cannot be excluded" concept. Three certainty factors agreed by the experts were as follows: the sign manifestation at this age is 0.6, the degree of incidence is 0.4; level of manifestation is 0.3. The certainty factors in the next age range increase to 0.6; 0.5; 0.4, respectively. Another example concerns the "cuboidal vertebral bodies" sign at the age of 2 to 3 determined by radiological examination of the patients with mucopolysaccharidosis type IH (Hurler syndrome). Given the fact that the experience of the radiologist influences the detection of this sign, experts came to the conclusion that the manifestation of this sign can be characterized by the following certainty factors: manifestation − 0.4, the degree of incidence − 0.6, degree of severity − 0.7. An example of a table with signs (the top line for each sign-characteristic indicates presence in five age groups, respectively) and their confidence measure (three positions in the next line) for the Hunter syndrome (severe subtype) is presented in Table 2.

It should be noted that for Hurler syndrome and a number of other diseases, holistic signs (for example, rough facial features previously metaphorically named "gargoyle-like face") are of great importance in the diagnostic but their defragmentation reduces the ability to recognize the disease. Therefore, holistic visual images should be included in the knowledge base with expert confidence measures [11].

Table 2. Fragment of the table with expert confidence factors

Signs	Hunter syndrome (severe subtype)														
Macrocephaly	−1			0			2			3			4		
	1	1	1	1	1	1	0,9	0,8	0,6	1	0,9	0,8	1	0,9	0,9
Coarse facial features	0			0			3			4			4		
	1	1	1	1	1	1	0,4	0,4	0,3	0,6	0,6	0,5	0,8	0,8	0,9
Macroglossia	−1			0			3			4			4		
	1	1	1	1	1	1	0,3	0,4	0,4	0,5	0,6	0,6	0,8	0,8	0,7
Teeth widely spaced	−1			−1			3			3			3		
	1	1	1	1	1	1	0,3	0,3	0,3	0,5	0,5	0,5	0,7	0,8	0,8
Mental retardation	−1			0			2			4			4		
	1	1	1	1	1	1	0,3	0,3	0,3	0,5	0,5	0,5	0,2	0,8	0,9
Chest Funnel-shaped	−1			0			1			1			1		
	1	1	1	1	1	1	0,1	0,1	0,2	0,1	0,1	0,2	0,1	0,1	0,2
Joints hypermobility	0			0			0			0			0		
	1	1	1	1	1	1	1	1	1	1	1	1	1	1	1
Joint contractures	−1			−1			3			5			5		
	1	1	1	1	1	1	0,3	0,4	0,2	0,5	0,6	0,4	0,7	0,7	0,8
Carpal tunnel syndrome	−1			−1			0			4			5		
	1	1	1	1	1	1	1	1	1	0,5	0,5	0,6	0,7	0,7	0,8
Claw-like hand	−1			−1			0			3			4		
	1	1	1	1	1	1	1	1	1	0,5	0,5	0,6	0,7	0,8	0,8

Judgments about the plausibility of the signs accompanied by the certainty factors in the level and time of manifestation and degree of incidence can serve as a further basis for effective incensement of the suggested expert hypotheses system.

6   Reflection of Experts in the Evaluation of Fuzzy Signs

Personal features of cognitive style affect the quality of knowledge structures [21]. This should be taken into account, since the true expert never views his ideas about the subject domain as an axiom. He mentally analyzes the picture of the world he offers. In medicine, this is the significance of each of the signs, their specificity for this pathology, relevance for differential diagnosis, and much more. And cognitive scientists often observe elements of doubt among experts when they formulate their ideas about the subject domain. This is manifested in assessing the difference in situations in the proposed decisions based on hypotheses that are finally formed under the influence of the expert's reflexive system.

Reflection show at activity of expert-physicians in the process of extracting knowledge from him for creating an intelligent system for diagnostic and/or therapeutic decisions support. The greatest problems are estimates of indistinctly manifested states of the disease.

The results of the knowledge extraction sessions, including the reflexive composition of experts, can be presented in a formalized form [22]. The process of the intellectual activity of the expert as a whole, including the reflex expert, can be represented on the basis of a formula including a fuzzy variable, a confidence factor, relevance of the characteristics, and a temporal characteristic of the changes in the course of a disease.

7   Discussion

Nowadays computer systems are used in medical practice to diagnose hereditary diseases [5, 8, 23] based on the analysis of information databases, which does not allow to take into account the progression of the pathological process to the necessary extent. The basis of these systems is mainly databases containing various combinations of clinical signs (including post-diagnostic demonstration of photographs). Some systems are designed to demonstrate and analyze a large number of photographs, i.e. visual diagnosis of decision support system [24–28]. This is also a symbolic picture of the world. However, the presentation of photographs without a proper intellectual context does not allow the doctor to effectively solve the diagnostic problems of most orphan hereditary storage diseases, excluding certain congenital dysmorphisms. Inclusion of visual images (coarse facial features, a skull in the form of a shamrock, the shape of ears, and the like) at the stage of the formation of diagnostic hypotheses may allow for a new approach to the problem of recognizing orphan diseases characterized by specific external features. This was partially revealed in the DIAGEN [6] system, when it was discovered that doctors mistaken in determining, for example, the shape of the nose (beak-shaped, pear-shaped, plum-shaped, etc.).

Systems such as Human Phenotype Ontology allow one to focus on the phenotypic similarity between diseases [29] and on the semantic metrics to measure the phenotypic similarity between queries and hereditary diseases. However, the classification in the study on ontologies does not take into account the dynamic change of characteristics. To rank the expected diagnoses at ontology system used also a statistical model of diseases [30]. But the statistical approach allows diagnostics only for relatively common hereditary diseases.

This study is a continuation of the work on the creation of an intelligent decision support system for diagnostic rare diseases. The difficulties in identifying hereditary diseases, known from literary sources, imply that one can identify and assess the characteristics of clinical manifestations of orphan diseases in their age dynamics only with the help of the knowledge of medical experts. This is especially important, since early diagnosis is a prerequisite for the timely initiation of treatment and prevention of abnormalities from various organs and organism systems. But this is possible only if the clinical decision support system used at all stages of the development of children, beginning with the early neonatal period. The expert knowledge presented in the present work on the age-related manifestations of signs, accompanied by factors of confidence measure in their manifestation, severity and frequency of manifestations at different periods of life, is a condition for the formation of rules and logical conclusions of the knowledge base in the dynamics of the pathological process.

In contrast to other work the main attention of this study is paid to the formation of a field of knowledge taking into account the signs of the disease manifestation at different ages. The introduction of the three confidence factors allowed experts to reflect on their views, their confidence in the manifestation, and severity of signs at different ages and also in the frequency of symptom occurrence in each age group. A preliminary analysis of literary sources with the formation of textological cards provides a deeper subsequent analysis of clinical manifestations by experts. Thus, symptom assessments in combination with the triple certainty factors will help to increase the effectiveness of expert systems for different situations caused by the dynamics of hereditary accumulation diseases.

8 Conclusion

Due to the lack of statistically reliable samples of patients with orphan diseases, which are necessary for the statistical analysis, knowledge-based systems should be considered as means of creating support diagnostic solutions. The expert certainty factors include three aspects: (1) in relation to the probability of sign manifestation at a certain age ("time characteristic" including negative values when it is not possible to manifest at certain age limits), (2) in relation to the severity of signs ("characteristic on the depth of pathology") and (3) in relation to the degree of incidence of signs with a certain pathology ("frequency response characteristic"). This approach ensures the three-dimensionality of the observed sigh evaluation for each of the studied age periods. The proposed "triune" system of certainty factors opens up new opportunities for knowledge bases of the expert systems. In addition, the knowledge of engineering must take into account the vagueness of definitions in medicine, the presence of holistic images and the reflection of doctors, which creates serious difficulties in the diagnosis of hereditary orphan diseases. The created expert system will be implemented on ontologies.

References

1. Schieppati, A., Henter, J.I., Daiana, E., Aperia, A.: Why rare diseases are an important medical and social issue. Lancet **371**(9629), 2039–2041 (2008)
2. Online Mendelian Inheritance in Man® An Online Catalog of Human Genes and Genetic Disorders. https://www.omim.org/. Accessed 30 Mar 2018
3. Ayme, S., Caraboenf, M., Gouvernet, J.: GENDIAG: a computer assisted syndrome identification system. Clin. Genet. **28**(5), 410–411 (1985)
4. Pitt, D.B., Bankier, A., Haan, E.A.: A visual verbal computer assisted syndrome identification system. Aust. Paediat. J. **21**(4), 306–307 (1985)
5. POSSUM web 2018. Pictures of Standard Syndromes and Undiagnosed Malformations. https://www.possum.net.au/. Accessed 15 Apr 2018
6. Kobrinsky, B., Kazantseva, L., Feldman, A., Veltishchev, J.: Computer diagnosis of hereditary childhood diseases. Med. Audit News **4**(1), 52–53 (1991)
7. Guest, S.S., Evans, C.D., Winter, R.M.: The online London dysmorphology database. Genet. Med. **5**(1), 207–212 (1999)
8. Baraitser, M., Winter, R.M.: London Dysmorphology Database, London Neurogenetics Database and Dysmorphology Photo Library on CD-ROM, 3rd edn. Oxford University Press, Oxford (2001)
9. Ayme, S.: Orphanet, an information site on rare diseases. Soins **672**, 46–47 (2003)
10. La Marca, G.: Lysosomals. In: Blau, N., Duran, M., Gibson, K.M., Vici, C.D. (eds.) Physician's Guide to the Diagnosis, Treatment, and Follow-Up of Inherited Metabolic Diseases, pp. 785–793. Springer, Heidelberg. (2014). https://doi.org/10.1007/978-3-642-40337-8_52

11. Kobrinskii, B.A.: Approaches to the construction of cognitive linguistic–image models of knowledge representation for medical intelligent systems. Sci. Tech. Inf. Process. **43**(5–6), 289–295 (2016)
12. Zadeh, L.A.: Toward a theory of fuzzy information granulation and its centrality in human reasoning and fuzzy logic. Fuzzy Sets Syst. **90**(2), 111–127 (1997)
13. Zadeh, L.A.: Fuzzy sets as a basis for a theory of possibility. Fuzzy Sets Syst. **1**(1), 3–28 (1978)
14. Quinlan, J.R.: C4.5: Programs for Machine Learning. Morgan Kaufmann Publishers Inc., San Mateo (1993)
15. Malik, D.S., Mordeson, J.N.: Fuzzy Discrete Structures. Physica-Verlag, New York (2000)
16. Petrovsky, A.B.: Multi-attribute classification of credit cardholders: multiset approach. Int. J. Manage. Decis. Mak. **7**(2/3), 166–179 (2006)
17. Torres, A., Nieto, J.J.: Fuzzy logic in medicine and bioinformatics. J. Biomed. Biotechnol. **2006**(2), 1–7 (2006)
18. Prasath, V., Lakshmi, N., Nathiya, M., Bharathan, N., Neetha, N.P.: A survey on the applications of fuzzy logic in medical diagnosis. Int. J. Sci. Eng. Res. **4**(4), 1199–1203 (2013)
19. Popper, K.: Knowledge and the Body-Mind Problem: In Defence of Interaction. Routledge, Abingdon (2013)
20. Shortliffe, E.H., Buchanan, B.G.: A model of inexact reasoning in medicine. In: Buchanan, B.G., Shortliffe, E.H. (eds.) Rule-Based Expert Systems: The MYCIN Experiments of the Stanford Heuristic Programming Project, pp. 233–262. Addison-Wesley Publishing Company, London (1984)
21. Gavrilova, T., Leshcheva, I.: The interplay of knowledge engineering and cognitive psychology: learning ontologies creating. Int. J. Knowl. Learn. **10**(2), 182–197 (2015)
22. Kobrinskii, B.: Expert reflection in the process of diagnosis of diseases at the extraction of knowledge. In: Proceedings of the IV International research conference "Information technologies in Science, Management, Social sphere and Medicine" (ITSMSSM 2017), vol. 72, pp. 321–323. Atlantis Press, Paris (2017)
23. Rat, A., Olry, A., Dhombres, F., Brandt, M.M., Urbero, B., Ayme, S.: Representation of rare diseases in health information systems: the orphanet approach to serve a wide range of end users. Hum. Mutat.: Var. Inf. Dis. Special Issue: Deep Phenotyp. Precis. Med. **33**(5), 803–808 (2012)
24. Hammond, P., et al.: 3D analysis of facial morphology. Am. J. Med. Genet. **126A**(4), 339–348 (2004)
25. Hammond, P., et al.: Discriminating power of localized three-dimensional facial morphology. Am. J. Hum. Genet. **77**(6), 999–1010 (2005)
26. Vardell, E., Bou-Crick, C.: VisualDx: a visual diagnostic decision support tool. Med. Ref. Serv. Q. **31**(4), 414–424 (2012)
27. Kuru, K., Niranjan, M., Tunca, Y., Osvank, E., Azim, T.: Biomedical visual data analysis to build an intelligent diagnostic decision support system in medical genetics. Artif. Intell. Med. **62**(2), 105–118 (2014)
28. Ferry, Q., et al.: Diagnostically relevant facial gestalt information from ordinary photos. eLife **3**, e02020 (2014). https://doi.org/10.7554/eLife.02020
29. Robinson, P.N., Köhler, S., Bauer, S., Seelow, D., Horn, D., Mundlos, S.: The human phenotype ontology: a tool for annotating and analyzing human hereditary disease. Am. J. Hum. Genet. **83**, 610–615 (2008)
30. Köhler, S., et al.: Clinical diagnostics in human genetics with semantic similarity searches in ontologies. Am. J. Hum. Genet. **4**(85), 457–464 (2009)

Machine Learning Based on Similarity Operation

Dmitry V. Vinogradov[1,2](✉)

[1] FRC "Computer Science and Control" RAS, Moscow 119333, Russia
vinogradov.d.w@gmail.com
[2] Russian State University for Humanity, Moscow 125993, Russia

Abstract. The paper describes a machine-learning paradigm that uses binary semi-lattice operation for computing similarities between training examples, with Formal Concept Analysis (FCA) providing a technique for bitset encoding of the objects and similarities between them. Using this encoding, a coupling Markov chain algorithm can generate a random sample of similarities. We provide a technique to accelerate convergence of the main algorithm by truncating its runs that exceed sum of lengths of previous trajectories. The similarities are hypothetical causes (hypotheses) for the target property. The target property of test examples can be predicted using these hypotheses. We provide a lower bound on necessary number of hypotheses to predict all important test examples for a given confidence level.

Keywords: Semi-lattice · Bitset · Coupling Markov chain
Stopped Markov chain · Induction · Prediction

1 Introduction

Machine learning is to large extent concerned with discovering classification rules from empirical data. There are many different approaches to machine learning: metric, algebraic, linear-separable, committee-based, etc.

Our approach uses a binary operation that defines a low semi-lattice on a set of fragments. This set contains a training sample as a part. The term "fragment" is motivated by pharmacological applications, where chemical compounds can contain active fragments, causes of biological (target) properties. These fragments are found as common parts of similar compounds. Intersections of compound descriptions can be considered as hypothetical causes of biological activity.

We compute similarities between several objects by repeated application of binary operation of similarity between two fragments. The results will be independent of order of application if the similarity operation is idempotent, commutative, and associative, in other words, it makes a semi-lattice.

The Basic Theorem of FCA [6] implies a possibility of encoding descriptions by bitsets in such a way that bitwise multiplication corresponds to the similarity operation. At first, we describe an encoding algorithm, prove its correctness, and estimate its complexity.

© Springer Nature Switzerland AG 2018
S. O. Kuznetsov et al. (Eds.): RCAI 2018, CCIS 934, pp. 46–59, 2018.
https://doi.org/10.1007/978-3-030-00617-4_5

We use the bitset representation of the training set to compute a random subset of similarities by means of coupling Markov chain. This algorithm has many useful properties. We propose a means to accelerate computations by lazy evaluation scheme and stopping of the coupling Markov chain through the usage of information about lengths of previous trajectories.

The key parameter of the main algorithm is the number of hypotheses to generate. We provide a low bound on its value by a repeated sample technique of Vapnik and Chervonenkis [9]. However, our situation is dual to that described by Vapnik and Chervonenkis. We have random (independently similarly distributed) similarities (=hypotheses) and a fixed set of test examples, while hypotheses are fixed and examples are random in the classical situation.

2 Formal Concept Analysis and Bitset Encodings

Similarity is a binary operation $\cap : X \times X \to X$ on set X of *fragments* satisfying the following semi-lattice laws:

$$x \cap x = x \tag{1}$$

$$x \cap y = y \cap x \tag{2}$$

$$x \cap (y \cap z) = (x \cap y) \cap z \tag{3}$$

Usually, there is a unique *trivial fragment* $\varnothing$ with the bottom element property:

$$x \cap \varnothing = \varnothing \tag{4}$$

Semi-lattice operation defines the order:

$$x \leq y \equiv (x \cap y = x) \tag{5}$$

The most important example is low semi-lattice of bitsets (strings of bits of fixed length) with bitwise multiplication as similarity operation. The trivial fragment corresponds to the string of 0's. Every bit position is a binary attribute, and a bitset corresponds to a subset of attributes, where the bit is equal to 1.

The value of this example is due to the efficient implementation of the bitwise multiplication on modern computers supported by special C++-class boost::dynamic_bitset<>. Moreover, modern compilers provide additional optimizations through the vector parallelism.

Basic Theorem of FCA [6] states maximal generality of these examples, since every finite lattice corresponds to some bitset semi-lattice with bitwise multiplication, if we add the string of all 1's as the top element. We use the modern approach to the proof of the theorem to provide an efficient encoding of objects represented by features whose values form arbitrary semi-lattice.

Context is a binary relation between set O of objects and set F of attributes. If string $o \in O$ contains 1 in position $f \in F$, then *object o possesses attribute f*, denoted by oIf. Otherwise, *object o does not possess attribute f*.

Similarity between subset $A \subseteq O$ of objects is $A' = \{f \in F : \forall o \in A[oIf]\} \subseteq F$. By definition $\varnothing' = F$. This definition coincides with bitwise multiplications of all strings corresponding to objects from subset A. *Similarity* between subset $B \subseteq F$ of attributes is $B' = \{o \in O : \forall f \in B[oIf]\} \subseteq O$. Here $\varnothing' = O$.

Lemma 1. *For $A_1 \subseteq O$ and $A_2 \subseteq O$ $(A_1 \cup A_2)' = A_1' \cap A_2'$ holds. For $B_1 \subseteq F$ and $B_2 \subseteq F$ $(B_1 \cup B_2)' = B_1' \cap B_2'$ holds.*

Lemma 2. *Similarity operations define Galois connection:*

$$\forall A \subseteq O\left[A \subseteq A''\right] \qquad \forall B \subseteq F\left[F \subseteq F''\right] \tag{6}$$

$$\forall A_1 \forall A_2 \left[A_1 \subseteq A_2 \Rightarrow A_1' \supseteq A_2'\right] \quad \forall B_1 \forall B_2 \left[B_1 \subseteq B_2 \Rightarrow B_1' \supseteq B_1'\right] \tag{7}$$

$$\forall A \subseteq O\left[A' = A'''\right] \qquad \forall B \subseteq F\left[B' = B''\right] \tag{8}$$

Definition 1. A pair $\langle A, B \rangle$ such that $A = B' \subseteq O$ and $B = A' \subseteq F$ will be called here a *candidate*.

FCA uses term "formal concept" instead, but in machine learning "concept" corresponds to a different notion.

Definition 2. The *order* between candidates is defined as follows: $\langle A_1, B_1 \rangle \leq \langle A_2, B_2 \rangle$ *iff* $B_1 \subseteq B_2$..

FCA uses the dual definition; our variant corresponds to Russian school tradition. The easier part of Basic Theorem of FCA is:

Proposition 1. *For any context $I \subseteq O \times F$ the set of all candidates forms a lattice with respect to*

$$\langle A_1, B_1 \rangle \cap \langle A_2, B_2 \rangle = \left\langle (B_1 \cap B_2)', B_1 \cap B_2 \right\rangle \tag{9}$$

$$\langle A_1, B_1 \rangle \cup \langle A_2, B_2 \rangle = \left\langle A_1 \cap A_2, (A_1 \cap A_2)' \right\rangle. \tag{10}$$

Definition 3. Subset $S \subseteq L$ of lattice L is called $\cap$-**dense** if any element $x \in L$ admits representation $x = \cap X$ for some subset $X \subseteq S$. A subset $S \subseteq L$ is called $\cup$-**dense**, if any $x \in L$ has representation $x = \cup X$ for some subset $X \subseteq S$.

Lemma 3. *In lattice L of all candidates generated by context $I \subseteq O \times F$, image $g(O) =$* $\left\{ \langle \{o\}'', \{o\}' \rangle : o \in O \right\}$ *of map $g : O \to L$ is a $\cap$-dense subset, and image $h(F) =$* $\left\{ \langle \{f\}', \{f\}'' \rangle : f \in F \right\}$ *of map $h : F \to L$ is a $\cup$-dense subset.*

The harder part of the basic theorem of FCA is the reverse of Lemma 3 statement, for the finite case it was proved in [7].

Proposition 2. *Let a finite lattice L have maps $g : O \to L$ and $h : F \to L$ with $\cap$-dense image $g(O) \subseteq L$ and $\cup$-dense image $h(F) \subseteq L$. Then $oIf \Leftrightarrow g(o) \geq h(f)$ defines a context with the lattice of all candidates isomorphic to L.*

The proof of the Basic Theorem of FCA in [6] uses the identity maps $g = id :$ $L \to L$ и $h = id : L \to L$. This approach generates unnecessarily large context. Moreover, the set of training objects is given a priori. Therefore, our goal is to select the minimal set of attributes $F \subseteq L$. To this end we need the following

Definition 4. Element $x \in L$ is called $\cup$-*irreducible* if $x \neq \varnothing$ and for any $y, z \in$ $Ly < x$ and $z < x$ imply $y \cup z < x$.

Lemma 4. *Any superset of all $\cup$-irreducible elements of a finite lattice forms $\cup$-dense subset.*

Restrict ourselves to bitset encoding of set V of values of a single feature, where some similarity operation exists. We concatenate the bitset encodings of values of single features into bitset representation of the whole object.

Since similarity operation defines an order (see Eq. (5)), the covering relation $(x \prec y \equiv x < y \& \neg \exists z [x < z < y])$ gives a directed acyclic graph (DAG).

The encoding algorithm consists of four parts. At first, we make a topological sorting. We use some initial enumeration of vertices of DAG $G = \langle V, E \rangle$.

Definition 5. Linear ordering (permutation) $V[0] < V[1] < \ldots < V[n-1]$ is a *topological sorting* if $\forall i, j [V[i] \prec V[j] \Rightarrow i < j]$.

Topological sorting requires $O\left(|V| + |V|^2\right)$ steps (see, for example [3]). We omit details in the representation below. The rest parts use the list of columns as data structure, and we give computational details to estimate time complexity.

The second part of the algorithm computes matrix T as the reflexive and transitive closure of the covering relation. Computational complexity of this part is $O(|V|^3)$.

The main third part of the algorithm consists in detecting unnecessary columns. Computational complexity of this part is $O\left(|V|^4\right)$. If the storage structure of matrix T is *std::list<boost::dynamic_bitset<>>*(columns list), then internal cycle operation is efficiently parallelized by modern compilers to decrease the computation time.

The last part of the algorithm constructs encoder/decoder matrix B from necessary columns (=binary attributes). The computational complexity of this part is $O\left(|V|^2\right)$.

Data: set $V = [0,1,\ldots,n-1]$ of values of the selected feature
Result: Matrix B, where $B[j]$ is a bitset encoding of value j
$V = topological\_sort(V)$; // topological sorting
$\forall i \forall j[T[i][j] = false]$; // T is a reflexive and transitive closure of $V[i] \prec V[j]$
for ($index$=0; $index$<n; ++$index$) **do**
   $T[V[index]][V[index]] = true$;
   for ($indx$=0; $indx$<$index$; ++$indx$) **do**
      if ($V[indx] \prec V[index]$) **then**
         for (ndx=0; ndx<n; ++ndx) **do**
            $T[V[index]][ndx] \mathrel{|=} T[V[indx]][ndx]$;
         end
      end
   end
end
$\forall i[Del[i] = false]$; // columns to delete
for ($index$=2; $index$<n; ++$index$) **do**
   for ($indx$=1; $indx$<$index$; ++$indx$) **do**
      for (ndx=0; ndx<$indx$; ++ndx) **do**
         if ($T[][V[index]] == T[][V[indx]] \& T[][V[ndx]]$) **then**
            $Del[V[index]] = true$;
         end
      end
   end
end
for (ndx=$index$=0; ndx<n; ++$index$) **do**
   $ndx = index$;
   while ($Del[ndx]$ && ndx<$index$) **do**
      ++ndx;
   end
   if ($ndx < n$) **then**
      for ($indx$=0; $indx$<n; ++$indx$) **do**
         $B[index][indx] = T[ndx][indx]$;
      end
   end
end

Algorithm 1. Bitset encoding

The correctness of Algorithm 1 follows from

Theorem 1. *In lattice L of all candidates of the context $(L, L, \leq)$ the attribute image* $h(f) = \langle \{f\}', \{f\}'' \rangle$ *is $\cup$-reducible if and only if there are attributes $f_1 < f$ and $f_2 < f$ with $\{f\}' = \{f_1\}' \cap \{f_2\}'$.*

Proof. For $\cup$-reducible $\langle \{f\}', \{f\}'' \rangle$ Definition 4 implies existence of pairs $\langle A_1, B_1 \rangle$ and $\langle A_2, B_2 \rangle$ such that $\langle \{f\}', \{f\}'' \rangle = \langle A_1, B_1 \rangle \cup \langle A_2, B_2 \rangle$. Let $f_1 = B_1$ and $f_2 = B_2$. Then $\{f_1\}' = A_1$ and $\{f_2\}' = A_2$. Proposition 1 and Lemma 1 imply $\langle \{f\}', \{f\}'' \rangle = \langle A_1 \cap A_2, (B_1 \cup B_2)'' \rangle = \langle \{f_1\}' \cap \{f_2\}', (f_1 \cup f_2)'' \rangle$. Definitions 2, 4, and Lemma 2 imply $f_1 \in \{f_1\}'' \subset \{f\}''$ (i.e. $f_1 < f$). Similarly, $f_2 < f$ holds.

Definition 6. Operation **CbO (Close-by-One)** replaces a candidate $\langle A, B \rangle$ with respect to an object $o \in O$ by pair $CbODown(\langle A, B \rangle, o) = \left\langle \left(B \cap \{o\}' \right)', B \cap \{o\}' \right\rangle$. It also replaces a candidate $\langle A, B \rangle$ with respect to an attribute $f \in F$ by pair $CbOUp(\langle A, B \rangle, f) = \left\langle A \cap \{f\}', \left(A \cap \{f\}' \right)' \right\rangle$.

Lemma 5. *For any candidate $\langle A, B \rangle$ and any object $o \in O$ pair $CbODown (\langle A, B \rangle, o) = \left\langle \left(B \cap \{o\}' \right)', B \cap \{o\}' \right\rangle$ is a candidate too. For any candidate $\langle A, B \rangle$ and any attribute $f \in F$ pair $CbOUp(\langle A, B \rangle, f) = \left\langle A \cap \{f\}', \left(A \cap \{f\}' \right)' \right\rangle$ is a candidate too.*

Lemma 6. *For any ordered pair of candidates $\langle A_1, B_1 \rangle \leq \langle A_2, B_2 \rangle$ and any object $o \in O$ the order $CbODown(\langle A_1, B_1 \rangle, o) \leq CbODown(\langle A_2, B_2 \rangle, o)$ holds. For any ordered pair of candidates $\langle A_1, B_1 \rangle \leq \langle A_2, B_2 \rangle$ and any attribute $f \in F$ the order $CbOUp(\langle A_1, B_1 \rangle, f) \leq CbOUp(\langle A_2, B_2 \rangle, f)$ holds.*

3   Coupling Markov Chains

In 1980's Victor K. Finn proposed JSM-method of automatic generation of hypotheses [1, 2] based on Boolean algebra representation of objects and set-theoretic intersection as a similarity operation. 'Induction' step of JSM-method computes all the candidates and then rejects a subset of them by testing additional logical conditions. However, this approach has exponentially high computational complexity. From machine learning point of view this approach results in overfitting, where a part of computed hypotheses corresponds to accidently appeared subsets of unessential features common to several training examples each of which has a different 'real' cause, see paper [11] for details.

Here we propose to use a coupling Markov chain to find a random hypothesis. A state of the coupling Markov chain is an ordered pair of candidates $\langle A_1, B_1 \rangle \leq \langle A_2, B_2 \rangle$. Initially, the smallest candidate is the bottom element $Min := \langle O, O' \rangle$ and the largest one is the top element $Max := \langle F', F \rangle$. The man algorithm cycle applies to them either as $CbODown$ w.r.t. an object or as $CbOUp$ w.r.t. an attribute. The process terminates when the candidates coincide.

Paper [10] describes the properties of the coupling Markov chain algorithm. The most important one is that on the termination of the algorithm almost surely.

Data: Training sample; external $CbOUp(\ ,\ )$ and $CbODown(\ ,\ )$ operations
Result: some candidate $\langle A, B \rangle$
$O :=$ training examples, $F :=$ attributes; $I \subseteq O \times F$ is a corresponding context;
$R := O \cup F$; $Min := \langle O, O' \rangle$; $Max := \langle F', F \rangle$;
while $(Min \neq Max)$ **do**
    Choose a random element $r \in R$;
    if $(r \in O)$ **then**
        $Min := CbODown(Min, r)$; $Max := CbODown(Max, r)$;
    else
        $Min := CbOUp(Min, r)$; $Max := CbOUp(Max, r)$;
end
return Max;

Algorithm 2. Coupling Markov chain

Components of operation $CbODown(\langle A, B \rangle, o) = \langle (A \cup \{o\})'', B \cap \{o\}' \rangle$ has incomparable computational complexity. While intersection $B \cap \{o\}'$ corresponds to a bitwise multiplication of two bitsets B and $\{o\}'$, the closure $(A \cup \{o\})'' = (B \cap \{o\}')'$ requires to check at most all objects to add $(A \cup \{o\})''$ to the parents list.

Similarly, operation $CbOUp$ has the computationally difficult component $(B \cup \{f\})'' = (A \cap \{f\}')'$ and the easy part $A \cap \{f\}'$.

Hence, lazy coupling Markov chain accelerates the computations:

Data: Training sample
Result: some candidate $\langle A, B \rangle$
$O :=$ training examples, $F :=$ attributes; $I \subseteq O \times F$ is a corresponding context;
$R := O \cup F$; $\langle A_1, B_1 \rangle := \langle O, O' \rangle$; $\langle A_2, B_2 \rangle := \langle F', F \rangle$; $moveUp :=$ **true**
while $(\langle A_1, B_1 \rangle \neq \langle A_2, B_2 \rangle)$ **do**
    Choose a random element $r \in R$;
    if $(r \in O \ \&\& \ moveUp)$ **then**
        $B_1 := A_1'$; $B_1 := A_1'$;
    end
    if $(r \in O)$ **then**
        $B_1 := B_1 \cap r'$; $B_2 := B_2 \cap r'$;
    if $(r \in F \ \&\& \ !moveUp)$ **then**
        $A_1 := B_1'$; $A_1 := B_1'$;
    if $(r \in F)$ **then**
        $A_1 := A_1 \cap r'$; $A_2 := A_2 \cap r'$;
end
if $(moveUp)$ **then**
    $B_1 := A_1'$; $B_1 := A_1'$;
else
    $A_1 := B_1'$; $A_1 := B_1'$;
return $\langle A_2, B_2 \rangle$;

Algorithm 3. Lazy coupling Markov chain

Now we estimate an acceleration by the lazy scheme of computing. Consider a sequence of elements types (objects or attributes) selected by a run of Algorithm 2. It forms a (possible infinite) sequence of Bernoulli trials $\langle \sigma_1, \ldots, \sigma_j, \ldots \rangle$ with success probability $p = \frac{n}{n+k}$ of choosing an attribute, where n denotes the number of attributes and k is the number of training examples.

Fix two events $\sigma_1 = 0$ и $\sigma_1 = 1$. Consider random variables R^σ of the first moment of selecting an attribute (for $\sigma = 0$) and the first moment of selecting an object (for $\sigma = 1$). Then lazy coupling Markov chain computes only two closures instead of $1 + R^0 + R^1$ closures computed by ordinary coupling Markov chain.

Theorem 2. $\mathbf{E}[R^0 + R^1] = 1 + \frac{p}{1-p} + \frac{1-p}{p}$.

Proof. The law of total probability implies $\mathbf{E}[R^0 + R^1] = \mathbf{E}[R^0 + R^1 | \sigma_1 = 0] \cdot \mathbf{P}[\sigma_1 = 0] + \mathbf{E}[R^0 + R^1 | \sigma_1 = 1] \cdot \mathbf{P}[\sigma_1 = 1]$.

First summand corresponds to event $\{\sigma_1 = \cdots = \sigma_{R^1-1} = 0, \sigma_{R^1} = \cdots = \sigma_{R^1+R^2-1} = 1, \sigma_{R^1+R^0} = 0\}$ and the second term corresponds to the event $\{\sigma_1 = \cdots = \sigma_{R^0-1} = 1, \sigma_{R^0} = \cdots = \sigma_{R^0+R^1-1} = 0, \sigma_{R^0+R^1} = 1\}$, respectively. It implies that $\mathbf{E}[R^0 + R^1 | \sigma_1 = 0] = (\mathbf{E}[R^0 | \sigma_1 = 1] + 1) + \mathbf{E}[R^1 | \sigma_1 = 0]$ and $\mathbf{E}[R^0 + R^1 | \sigma_1 = 1] = (\mathbf{E}[R^1 | \sigma_1 = 0] + 1) + \mathbf{E}[R^0 | \sigma_1 = 1]$. Then $\mathbf{E}[R^0 + R^1] = 1 + \mathbf{E}[R^0 | \sigma_1 = 1] + \mathbf{E}[R^1 | \sigma_1 = 0]$.

Conditional distribution $\mathbf{P}[R^1 = k | \sigma_1 = 0] = p \cdot (1-p)^{k-1}$ $(k = 2, 3, \ldots)$ has a mean $\mathbf{E}[R^1 | \sigma_1 = 0] = \frac{1-p}{p}$. Similarly, distribution $\mathbf{P}[R^0 = k | \sigma_1 = 1] = (1-p) \cdot p^{k-1}$ $(k = 2, 3, \ldots)$ has mean $\mathbf{E}[R^0 | \sigma_1 = 1] = \frac{p}{1-p}$.

Theorem 2 states that mean acceleration of lazy evaluation scheme is $2 + \frac{p}{1-p} + \frac{1-p}{p} = 2 + \frac{n}{k} + \frac{k}{n}$ times of closure computations replaced by 2 times.

Note, that $\frac{n}{k} + \frac{k}{n} \geq 2$, and minimum occurs at $n = k$, where we have 4:2 ratio of mean numbers of closure operations between Algorithm 2 and Algorithm 3, respectively.

The more different the number k of training examples from the number n of attributes, the greater the win. Since majority of dataset has number of training examples exceeding the number of attributes, the lazy coupling Markov chain is more preferable than the ordinary one. Another way to accelerate coupling Markov chains is to stop it on very long runs. Coupling Markov chains admit acceleration if trajectory terminates when its length exceeds a sum of lengths of previous runs. Algorithm begins a new trajectory in this case.

Definition 7. Let $T_1, \ldots, T_r$ be independent integer-valued random variables with common distribution of time T of coupling. Then *upper bound* (based on r previous runs) is $\hat{T} = T_1 + \ldots + T_r$.

In practice algorithm makes r runs of coupling Markov chain of lengths $t_1, \ldots, t_r$, respectively, and compute an estimation $t_1 + \ldots + t_r$ of the upper bound.

Definition 8. For random variable $\hat{T}$, independent to random variable T, *conditional distribution* of terminated Markov chain state i w.r.t. $B = \{T \leq \hat{T}\}$ is

$$\mu(\hat{T})_i = \frac{\mathbf{P}[X_T=i, T \le \hat{T}]}{\mathbf{P}[T \le \hat{T}]}.$$

Definition 9. *Total variation* distance between distributions $\mu = (\mu_i)_{i \in U}$ and $\nu = (\nu_i)_{i \in U}$ on finite states space U is $\|\mu - \nu\|_{TV} = \frac{1}{2} \cdot \sum_{i \in U} |\mu_i - \nu_i|$.

This distance is equal to the half of metric l_1, hence it is a metric (for instance, it is a symmetric relation).

Lemma 7. $\|\mu - \nu\|_{TV} = \max_{R \subseteq U} |\mu(R) - \nu(R)|$.

Lemma 7 holds for $R = \{i \in U | \mu_i > \nu_i\}$.

Lemma 8. $\|\mu - \mu(\hat{T})\|_{TV} \le \frac{\mathbf{P}[T > \hat{T}]}{1 - \mathbf{P}[T > \hat{T}]}$, *where $\mu(\hat{T})$ is the distribution of states of Markov chain terminated by upper bound $\hat{T}$ based on $r > 1$ previous runs, and μ is the distribution of states of original (non-terminated) coupling Markov chain.*

Proof. Definition 6 gives $\mu(\hat{T})_i = \frac{\mathbf{P}[X_T=i, T \le \hat{T}]}{\mathbf{P}[T \le \hat{T}]}$. Then

$$\mathbf{P}[T \le \hat{T}] \cdot (\mu(\hat{T})_i - \mu_i) = \mathbf{P}[X_T = i, T \le \hat{T}] - \mathbf{P}[T \le \hat{T}] \cdot \mu_i =$$
$$= \mathbf{P}[T > \hat{T}] \cdot \mu_i - \mathbf{P}[X_T = i, T > \hat{T}] \le \mathbf{P}[T > \hat{T}] \cdot \mu_i$$

Summing up to $R = \{i \in U | \mu_i > \mu(\hat{T})_i\}$, we obtain
$$\mathbf{P}[T \le \hat{T}] \cdot \|\mu - \mu(\hat{T})\|_{TV} \le \mathbf{P}[T > \hat{T}].$$

Lemma 9. $\|\mu - \mu(\hat{T})\|_{TV} \le \frac{1}{2^r - 1}$, *where $\mu(\hat{T})$ is the distribution of states of Markov chains terminated by upper bound $\hat{T}$ based on $r > 1$ previous runs, and μ is the distribution of states of original (non-stopped) coupling Markov chain.*

Proof. Lemma 8 implies the result if we prove $\mathbf{P}[T > \hat{T}] \le 2^{-r}$. $\mathbf{P}[T > T_j] \le \frac{1}{2}$ for all $1 \le j \le r$ by definition of $T, T_1, \ldots, T_r$ as independent identically distributed random variables.

It suffices to prove submultiplicativity $\mathbf{P}[T > \sum_{j=1}^k T_j] \le \mathbf{P}[T > \sum_{j=1}^{k-1} T_j] \cdot \mathbf{P}[T > T_k]$ for all $1 < k \le r$. It follows from the definition of conditional probability.

If $T > \sum_{j=1}^{k-1} T_j$ then $\left\langle A_1\left(\sum_{j=1}^{k-1} T_j\right), B_1\left(\sum_{j=1}^{k-1} T_j\right)\right\rangle < \left\langle A_2\left(\sum_{j=1}^{k-1} T_j\right), B_2\left(\sum_{j=1}^{k-1} T_j\right)\right\rangle$.

Apply operations *CbODown* and *CbOUp* to four candidates

$$Min \le \left\langle A_1\left(0 + \sum_{j=1}^{k-1} T_j\right), B_1\left(0 + \sum_{j=1}^{k-1} T_j\right)\right\rangle < \left\langle A_2\left(0 + \sum_{j=1}^{k-1} T_j\right), B_2\left(0 + \sum_{j=1}^{k-1} T_j\right)\right\rangle \le Max$$

simultaneously. If internal pair

$$\left\langle A_1\left(t+\sum_{j=1}^{k-1}T_j\right),B_1\left(t+\sum_{j=1}^{k-1}T_j\right)\right\rangle < \left\langle A_2\left(t+\sum_{j=1}^{k-1}T_j\right),B_2\left(t+\sum_{j=1}^{k-1}T_j\right)\right\rangle$$

has coupled after $T_k + \sum_{j=1}^{k-1}T_j = \sum_{j=1}^{k}T_j$, then identity of $Min < Max$ occurs after T_k, and we have $\mathbf{P}\left(T > \sum_{j=1}^{k}T_j | T > \sum_{j=1}^{k-1}T_j\right) \leq \mathbf{P}[T > T_k]$. Induction concludes the proof.

Lemmas 7 and 9 imply

Theorem 3. *For any subset R with $\mu(R) = \rho$ and any $r > \log_2\left(1 - \frac{1}{\rho}\right)\mu(\hat{T})(R) \geq \rho - \frac{1}{2^r - 1}$ holds for upper bound $\hat{T}$ based on $r > 1$ previous runs.*

Proof. $\rho - \frac{1}{2^r-1} \leq \mu(R) - \left\|\mu - \mu(\hat{T})\right\|_{TV} = \mu(R) - \mathbf{max}_{Q\subseteq U}\left|\mu(Q) - \mu(\hat{T})(Q)\right| \leq \mu(R) - \left|\mu(R) - \mu(\hat{T})(R)\right| \leq \mu(\hat{T})(R)$

4 Machine Learning

The general scheme of many machine learning models consists of two parts: inductive generalization of training examples and prediction of the target property of testing examples. We use '*counterexample forbidding*' condition (from JSM method) to block some candidates to become a hypothetical cause. A counterexample is a negative (positive) example with a set of attributes which is a superset of an intersection of positive (negative) examples.

> **Data:** training sample; number N of hypotheses to generate
> **Result:** N random hypotheses (without counterexamples)
> $O :=$ training examples, $F :=$ attributes; $I \subseteq O \times F$ is a context for training examples;
> $C :=$ counter examples, $S := \emptyset$; $i := 0$;
> **while** $(i < N)$ **do**
>   Generate a candidate $\langle A, B \rangle$ by coupling Markov chain;
> $hasObstacle :=$ **false**;
>   **for** $(c \in C)$ **do**
>   **if** $(B \subseteq \{c\}')$ **then**
> $hasObstacle :=$ **true**;
>   **end**
>   **end**
>   **if** $(hasObstacle =$ **false**$)$ **then**
> $S := S \cup \{\langle A, B \rangle\}$; $i := i+1$;
>   **end**
> **end**

Algorithm 4. Inductive generalization

Testing condition $B \subseteq \{c\}'$ means testing inclusion of fragment B of candidate $\langle A, B \rangle$ into fragment (=set of attributes) of counterexample c. This inclusion means negation of 'counterexample forbidding' condition. If a candidate avoids all such inclusions, it becomes a hypothesis about causes of the target property.

The target property prediction corresponds to the following procedure

> **Data:** sample of hypotheses, test examples
> **Result:** predictions of the target property
> $X :=$ test examples;
> **for** $(o \in X)$ **do**
>   *PredictPositively(o)* := **false**;
>   **for** $(\langle A, B \rangle \in S^+)$ **do**
>   **if** $(B \subseteq o')$ **then**
>   *PredictPositively(o)* := **true**;
>   **end**
>   **end**
> **end**

Algorithm 5. Prediction (by analogy)

This procedure looks for an inclusion $B \subseteq \{o\}'$ of fragment of some inductively generated hypothesis $\langle A, B \rangle$ into description $\{o\}' \subseteq F$ of test example o. If this inclusion holds, then the target property of test example o is predicted positively by analogy with training examples $A \subseteq O$ of hypothesis $\langle A, B \rangle$, whose fragment B is included in $\{o\}'$. Otherwise, the target property is not predicted for test example o.

Number N of hypotheses is a key parameter of Algorithm 4. We apply a technique of repeated sample of Vapnik and Chervonenkis [9] to obtain a lower bound on it, in order to predict positively all important test examples with given confidence level.

If the number of attributes is $n = |F|$, then fragment B corresponds to a vertex of n-dimensional hypercube $B \in \{0, 1\}^n$.

Definition 10. *Low half-space* $H^{\downarrow}(o)$ defined by test object o with fragment $\{o\}' \subseteq F$ is a set of solutions of linear inequality $x_{j_1} + \ldots + x_{j_k} < \frac{1}{2}$, where $F \setminus \{o\}' = \{f_{j_1}, \ldots, f_{j_k}\}$. The class of low half-spaces $Sub^{\downarrow}$ contains also half-space $\{0, 1\}^n$ defined by $0 < \frac{1}{2}$ and corresponding to the test object o with fragment $\{o\}' = F$.

Lemma 10. *Test example o is predicted positively if and only if the low halfspace $H^{\downarrow}(o)$ contains a fragment of some hypothesis.*

Proof. Algorithm 5 predicts the target property of test example o positively if there exists a hypothesis $\langle A, B \rangle$ with $B \subseteq \{o\}'$, i.e. $B \cap (F \setminus \{o\}') = \emptyset$, or in terms of Definition 10, $\forall i [f_{j_i} \notin B]$. Therefore, $0 = f_{j_1} + \ldots + f_{j_k} < \frac{1}{2}$ and we use Definition 10 to conclude the proof.

Definition 11. Test object o is an ε- *important* if probability of all hypotheses $\langle A, B \rangle$ with $B \in H^{\downarrow}(o)$ exceed $\varepsilon > 0$.

Definition 12. *Growth function* $m^{Sub^{\downarrow}}(l)$ *is a maximal number of subsets shattered by different* $H \in Sub^{\downarrow}$ *from a set of cardinality* l.

Lemma 11. $m^{Sub^{\downarrow}}(l) = 2^n$ *holds for any* $l \geq n$.

Proof. On the one hand, $Sub^{\downarrow}$ contains 2^n elements. On the other hand, $Sub^{\downarrow}$ shatters all 2^n subsets of $\{(1, 0, \ldots, 0), (0, 1, \ldots, 0), \ldots, (0, 0, \ldots, 1)\}$.

The next lemma is a core of repeated sample method [9] of V.N. Vapnik and A.Ya. Chervonenkis. There $|S_2 \cap H|$ denotes multiset cardinality, since elements of second sample S_2 can be equal.

Lemma 12. *For any* ε *and* $l > \frac{2}{\varepsilon}$ *for independent samples* S_1 *and* S_2 *of cardinalities* l

$$\mathbf{P}^l\{S_1 : \exists H \in Sub^{\downarrow}[S_1 \cap H = \varnothing, PH > \varepsilon]\} \leq$$
$$2 \cdot \mathbf{P}^{2l}\{S_1 S_2 : \exists H \in Sub^{\downarrow}[S_1 \cap H = \varnothing, |S_2 \cap H| > \tfrac{\varepsilon \cdot l}{2}]\} \textbf{\textit{holds}}.$$

Proof. For sample S_1 consider low half-space $H \in Sub^{\downarrow}$ such that $S_1 \cap H = \varnothing$ and $PH > \varepsilon$. Triangle inequality implies

$$\mathbf{P}^l\left\{S_2 : l \cdot PH - |S_2 \cap H| \leq \frac{\varepsilon \cdot l}{2}\right\} \leq \mathbf{P}^l\left\{S_2 : |S_2 \cap H| > \frac{\varepsilon \cdot l}{2}\right\}.$$

For $l > \frac{2}{\varepsilon}$ it suffices to prove

$$\mathbf{P}^l\left\{S_2 : l \cdot PH - |S_2 \cap H| \leq \frac{\varepsilon \cdot l}{2}\right\} = \mathbf{P}^l\left\{S_2 : |S_2 \cap H| \geq l \cdot PH - \frac{\varepsilon \cdot l}{2}\right\} \geq \frac{1}{2}.$$

It is a probability of Binomial random variable $|S_2 \cap H|$ to exceed its mean $l \cdot PH$ decreased by $\frac{\varepsilon \cdot l}{2} > 1$ (for $l > \frac{2}{\varepsilon}$). It is a well-known fact that the mean of binomial distribution differs from the median less than 1. Therefore, the inequality holds.

Independence of S_1 and S_2 implies

$$\mathbf{P}^l\{S_1 : S_1 \cap H = \varnothing, PH > \varepsilon\} \cdot \mathbf{P}^l\{S_2 : |S_2 \cap H| > \tfrac{\varepsilon \cdot l}{2}\}$$
$$= \mathbf{P}^{2l}\{S_1 S_2 : S_1 \cap H = \varnothing, |S_2 \cap H| > \tfrac{\varepsilon \cdot l}{2}\}.$$

Therefore,

$$\tfrac{1}{2} \cdot \mathbf{P}^l\{S_1 : S_1 \cap H = \varnothing, PH > \varepsilon\}$$
$$\leq \mathbf{P}^l\{S_2 : l \cdot PH - |S_2 \cap H| \leq \tfrac{\varepsilon \cdot l}{2}\} \cdot \mathbf{P}^l\{S_1 : S_1 \cap H = \varnothing, PH > \varepsilon\}$$
$$\leq \mathbf{P}^{2l}\{S_1 S_2 : S_1 \cap H = \varnothing, |S_2 \cap H| > \tfrac{\varepsilon \cdot l}{2}\}.$$

The unions of events on both extreme sides over half-space $H \in Sub^{\downarrow}$ satisfying conditions $S_1 \cap H = \varnothing$ and $PH > \varepsilon$ conclude the proof.

Lemma 13. *For any ε and two independent samples S_1 and S_2 of hypotheses of cardinalities l*

$$\mathbf{P}^{2l}\{S_1 S_2 : \exists H \in Sub^{\downarrow}[S_1 \cap H = \varnothing, |S_2 \cap H| > \varepsilon \cdot l]\} \leq m^{Sub^{\downarrow}}(2l) \cdot 2^{-\varepsilon \cdot l} holds.$$

Proof. Define mapping g of united sample $S_1 S_2$ to a multiset over $\{0, 1\}^n$ to describe a content. Probability $\mathbf{P}^{2l}$ on $(\{0, 1\}^n)^{2l}$ determines probability measure $g(\mathbf{P})$ on multisets of cardinality $2l$. Independence and identical distributions of hypotheses computed in different runs of coupling Markov chain imply equal probabilities of samples of same contents. Fix sample content v. For low half-space $H \in Sub^{\downarrow}$ conditional probability bounds by hypergeometric distribution we have:

$$\mathbf{P}\{S_1 S_2 : S_1 \cap H = \varnothing, |S_2 \cap H| > \varepsilon \cdot l | g(S_1 S_2) = v\} \leq \frac{\binom{l}{\varepsilon \cdot l}}{\binom{2l}{\varepsilon \cdot l}} = \frac{l! \cdot (2l - \varepsilon \cdot l)!}{(l - \varepsilon \cdot l)! \cdot (2l)!}$$

$$= \frac{l \cdot (l-1) \cdot \ldots \cdot (l - \varepsilon \cdot l + 1)}{2l \cdot (2l-1) \cdot \ldots \cdot (2l - \varepsilon \cdot l + 1)} \leq 2^{-\varepsilon \cdot l}.$$

Then conditional probability on samples of given content is

$$\mathbf{P}\{S_1 S_2 : \exists H \in Sub^{\downarrow}[S_1 \cap H = \varnothing, |S_2 \cap H| > \varepsilon \cdot l] | g(S_1 S_2) = v\} \leq m^{Sub^{\downarrow}}(2l) \cdot 2^{-\varepsilon \cdot l}.$$

The left-side expression does not depend on content v, hence integration over $g(\mathbf{P})$ concludes the proof.

Theorem 4. *All ε-important test examples will be predicted positively with probability $> 1 - \delta$ if Algorithm 3 generates $N \geq \frac{2 \cdot (n+1) - 2 \cdot \log_2 \delta}{\varepsilon}$ hypotheses for any $\varepsilon > 0$ and $1 > \delta > 0$, where n is the number of attributes.*

Proof. Lemmas 12 and 13 for $N > \frac{2}{\varepsilon}$ imply

$$\mathbf{P}^N\{S \subseteq \{0, 1\}^n : \exists H \in Sub^{\downarrow}[S \cap H = \varnothing, PH > \varepsilon]\} \leq 2 \cdot m^{Sub^{\downarrow}}(2N) \cdot 2^{-\frac{\varepsilon N}{2}}.$$

Lemma 11 and inequality $2 \cdot 2^n \cdot 2^{-\frac{\varepsilon N}{2}} \leq \delta$ resolved with respect to N conclude the proof .

5   Conclusion

Formal Concept Analysis together with randomized algorithms forms a new approach to machine learning [10] based on binary similarity operation. We described all essential ingredients of this approach: encoding technique, various coupling Markov chain algorithms, estimation of acceleration by lazy evaluation and stopping mechanisms, inductive generalization of training examples and prediction of the target property of test examples. The main result is an analogue of Vapnik-Chervonenkis theorem [9] that converts our approach into machine learning procedure.

There are precursors of our approach: JSM method [1, 2] and its description in terms of FCA [4, 5]. However described in these works procedures use only semilattice operation of intersection. The approach proposed here uses the full power of FCA. Another feature of the approaches known so far is their exponential computational complexity, hence these approaches cannot be called efficient methods of machine learning (with respect to the viewpoint initiated by Valiant [8]).

More results on the approach described here can be found in the author's Doctor of Science thesis [12] (in Russian).

Acknowledgments. Author would like to thank his colleagues from Federal Research Center for Computer Science and Control and colleagues from Russian State University for Humanities for support and scientific discussions.

References

1. Finn, V.K.: Plausible inferences and plausible reasoning. J. Sov. Math. **56**(1), 2201–2248 (1991)
2. Anshakov, O.M., Finn, V.K. (eds.): JSM Method of Automatic Generation of Hypotheses (Logical and Epistemological Foundations). URSS, Moscow (2009). (in Russian)
3. Cormen, T.H., Leiserson, C.E., Rivest, R.L., Stein, C.: Introduction to Algorithms, 3rd edn. The MIT Press, Cambridge (2009)
4. Kuznetsov, S.O.: Mathematical aspects of concept analysis. J. Math. Sci. **80**(2), 1654–1698 (1996)
5. Ganter, B., Kuznetsov, Sergei O.: Hypotheses and version spaces. In: Ganter, B., de Moor, A., Lex, W. (eds.) ICCS-ConceptStruct 2003. LNCS (LNAI), vol. 2746, pp. 83–95. Springer, Heidelberg (2003). https://doi.org/10.1007/978-3-540-45091-7_6
6. Ganter, B., Wille, R.: Formal Concept Analysis. Springer, Berlin (1999). https://doi.org/10.1007/978-3-642-59830-2
7. Barbut, M., Monjardet, B.: Ordre et classification. Hachette, Paris (1970)
8. Valiant, L.G.: A theory of the learnable. Commun. ACM **27**, 1134–1142 (1984)
9. Vapnik, V.N., Chervonenkis, A.Ya.: Theory of Pattern Recognition: Statistical Problems of Learning. Nauka, Moscow (1974, in Russian)
10. Vinogradov, D.V.: VKF-method of hypotheses generation. In: Ignatov, D.I., Khachay, M.Y., Panchenko, A., Konstantinova, N., Yavorskiy, Rostislav E. (eds.) AIST 2014. CCIS, vol. 436, pp. 237–248. Springer, Cham (2014). https://doi.org/10.1007/978-3-319-12580-0_25
11. Vinogradov, D.V.: Accidental formal concepts in the presence of counterexamples. In: Kuznetsov, S.O., Watson, B.W. (eds.) Proceedings of International Workshop "Formal Concept Analysis for Knowledge Discovery" (FCA4KD 2017), CEUR Workshop Proceedings, vol. 1921, pp. 104–112. HSE, Moscow(2017)
12. Vinogradov, D.V.: Doctor of Science thesis. http://frccsc.ru/sites/default/files/docs/ds/002-073-05/diss/17-vinogradov/17-vinogradov_main.pdf?587. Accessed 05 July 2018. (in Russian)

A Method of Dynamic Visual Scene Analysis Based on Convolutional Neural Network

Vadim V. Borisov and Oleg I. Garanin[(✉)]

NRU "MPEI", Moscow, Russia
{vbor67, hedgehog91}@mail. ru

Abstract. In this paper, we analyze the existing methods of Multiple Object Tracking (MOT), point out their advantages and disadvantages. It is noted that the MOT task must be solved together with the detection of these objects, thus developing a method of the analysis of the dynamic visual scene. We propose a method of dynamic visual scene analysis based on the appearance object model. This method allows one to detect images and to get the "deep features" of detection in one Convolutional Neural Network forward pass, as well as to improve the accuracy of tracking objects construction compared to other online methods and perform processing in real time, at the speed of 24 FPS, which is shown experimentally. In addition, the method works both in the conditions of uncertainty and in the conditions of noise detection data.

Keywords: Tracking · Tracking-by-detection · Single Shot Multibox Detector Convolutional Neural Network

1 Introduction

The task of dynamic visual scene analysis is to build the tracks of objects on the input sequence of frames. Thus, the tasks of detection and tracking objects on multiple frames (Multiple Object Tracking) are solved.

To develop a method of dynamic visual scene analysis, it is necessary to solve the problem of object detection, and then perform tracking-by-detection task. The task of tracking objects is most often solved separately from the detection and the joint use of the detector and the tracking method requires first to detect the object, and then to get the features separately for each object, which increases the complexity of processing. For example, in [9, 11], each detection is fed separately to the input of the Convolutional Neural Network (CNN) to get its "deep" features. This approach can improve accuracy, but requires significant computing resources.

In addition, there are quite accurate methods of tracking objects, such as MHT, ELP [5, 8], but they do not allow processing in real time, because they solve the problem of global optimization and require obtaining the entire sequence of frames (off-line methods). Other methods perform real-time processing, but are not enough accurate, for example SORT [1] (online methods).

At the same time, to study the developed methods of tracking, the most often existing datasets offer [6] to work with a large number of false detections and build algorithms that allow to get data from noise. However, while using CNN, it is

© Springer Nature Switzerland AG 2018
S. O. Kuznetsov et al. (Eds.): RCAI 2018, CCIS 934, pp. 60–69, 2018.
https://doi.org/10.1007/978-3-030-00617-4_6

sometimes necessary to deal not with a large number of false positive detections, but with false negative ones.

Thus, the most promising methods for solving problems of recognition and tracking objects today are the methods based on "deep" learning and CNN, because they allow to obtain high accuracy of recognition and tracking compared with other methods.

Therefore, while solving the complex problem of detection and tracking objects, it is advisable to propose a method that allows performing the detection and construction of an appearance model of the object in one pass CNN, under conditions of uncertainty of their detection. The proposed method will improve the tracking accuracy and perform processing in real time.

2 Appearance Object Model and Algorithm for Obtaining "Deep Features" of the Objects Detection

2.1 Model Description

In this paper, the detection of an object in the image is understood as the area of the image, on which the object is selected with the help of a bounding box. While building a method of Multiple Object Tracking Appearance, models that use a certain set of features to describe the object are often used.

Figure 1 shows the structure of the appearance object model using a multiscale detection model. Unlike the well-known solution of [7] Single Shot MultiBox Detector (SSD), this structure is characterized by the addition of a layer of ROI-Pooling to form the "deep features" of detection. The method of tuning a multiscale model for detecting visual objects in CNN is described in [2, 3].

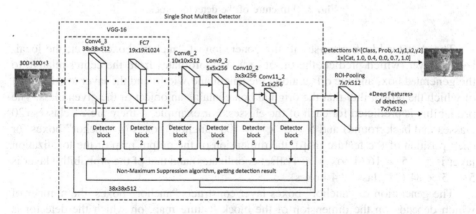

Fig. 1. Structure of the appearance object model

We give a detailed description of the structure of SSD model. This model is based on CNN VGG-16 [10], which contains 5 blocks of convolution layers with a max-pool layer after each block, 3 fully-connected layers and a classifier layer. The first two blocks contain two layers, the other three have three 3 × 3 convolution layers.

To use the VGG-16 as a detector, it is necessary to exclude the last fully connected layer and the classifier one, as well as to include several detection blocks and the layers used to change the scale of the analyzed image. So, one detector block is added to both the feature map of the fourth block (conv4_3) and the fully connected layer (FC7) of model VGG-16. In addition, the network is supplemented by four additional blocks, used to change the scale (conv8_2, conv9_2, conv10_2, conv11_2) with detectors blocks, as shown in Fig. 1. At the network output, the detection results of each block (6 detectors in all) are analyzed with the help of non-Maximum Suppression algorithm and the result of detection is formed.

The structure of the detector block of the Conv9_2 layer is shown in Fig. 2. The rest of the detector blocks are constructed in a similar way.

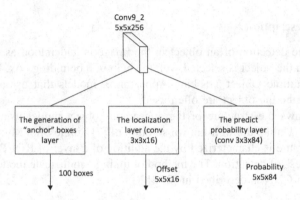

Fig. 2. Structure of the detector block

The detector block consists of: the generation of "anchor" boxes layer, the localization layer (which predicts the offset of X/Y coordinates from the center relative to the generated box, and the offset along the length and width) and the layer at the output of which the probability that the original box contains an object of the given class. The probability is predicted for each of the classes. For example, if there are 21 classes (20 classes and background) and 5 × 5 × 4 = 100 "anchor" boxes (4 "anchor" boxes for each position of the feature map), the dimension of the output map of the localization layer is 5 × 5 × 16 (4 boxes × 4 offset coordinates) and that of the probability layer is 5 × 5 × 84 (21 class × 4 boxes).

The generation of "anchor" boxes layer constructs "anchor" boxes, the number of which depends on the dimension of the block feature map, on which the detector is built. Such rectangles cover the entire image with a grid, as shown in Fig. 3.

It is important to consider the inputs and outputs of this model. The input gets an input RGB image (frame), which is converted to 300 × 300 resolution. At the output, object detections (class, detection confidence, object coordinates) and their "deep features" are formed.

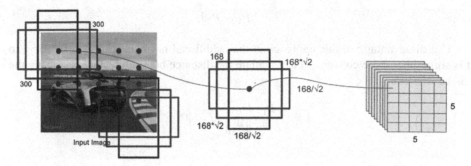

Fig. 3. Covering the original image with a grid of "anchor" boxes for a 5×5 feature map

Thus, applying the proposed model, it becomes possible to detect the image and get the "deep features" of detection in one pass of the CNN.

2.2 Algorithm for Obtaining "Deep Features" of the Objects Detection

Input Data: a feature map of the m-layer CNN $n \times h_m \times w_m$, coordinates of the object in the image $o = (x_1, y_1, x_2, y_2)$, image dimensions $h \times w$, dimension of the ROI-pooling layer $n \times h_r \times w_r$.

Output Data: a set of features P of the object o.

Step 1. Calculation of the scaling factors:

$$K_h = \frac{h_m}{h}, K_w = \frac{w_m}{w}.$$

Step 2. Calculation of the coordinates of object projections on a feature map:

$$x_i^m = x_i \cdot K_h, y_i^m = y_i \cdot K_w, i = 1, 2.$$

Step 3. Getting features of the object P_{proj} with dimension $n \times (y_2^m - y_1^m) \times (x_2^m - x_1^m)$ and coordinates $(x_1^m, y_1^m, x_2^m, y_2^m)$ on m-layer CNN feature map.

Step 4. Scaling of projections of the features P_{proj} object o on feature map with coordinates $(x_1^m, y_1^m, x_2^m, y_2^m)$ and dimension $n \times (y_2^m - y_1^m) \times (x_2^m - x_1^m)$ to dimension $n \times h_r \times w_r$ using max-pooling (ROI-pooling) $Ceil\left(\frac{y_2^m - y_1^m}{h_r}\right) \times Ceil\left(\frac{x_2^m - x_1^m}{w_r}\right)$, where $Ceil$ – ceiling operator. As a result of this operation, a set of P features is formed. In the same way feature sets are computed for all the detections which are sent to the input of the ROI-Poling.

The distance between two sets of features P_1 and P_2 of two detections can be calculated as the norm of the difference between two vectors of dimension $n \times h_r \times w_r$ in Euclidean space:

$$d = \|P_1 - P_2\|.$$

The disadvantage of this approach is that additional normalization is necessary, so it is sometimes more convenient to calculate the distance between feature sets using the cosine distance:

$$d = \frac{P_1 \cdot P_2}{\|P_1\| \cdot \|P_2\|}, d = [0, 1]$$

3   Method of Dynamic Visual Scene Analysis

3.1   Formulation of the Problem

Let $K = \{k_n\}$ be a set of frames (a dynamic visual scene) $n = 1, \ldots, N$, $k_n : O_{k_n} = \{o_{m_n}^{(k_n)}\}$ be a set of object in the frame k_n, $m_n = 1, \ldots, M_n$, $n = 1, \ldots, N$.

Each of the objects $o_{m_n}^{(k_n)}$ is represented by a set of features $P_{m_n} \forall o_{m_n}^{(k_n)} : P_{m_n} = \{p_l^{(m_n)}\}, l = 1, \ldots, L$ and coordinates $\left\{ \left(x_1^{(m_n)}, y_1^{(m_n)} \right), \left(x_2^{(m_n)}, y_2^{(m_n)} \right) \right\}$, $TR = \{tr_z\}$ is a set of tracks $z = 1, \ldots, Z$.

Each of the tracks tr_z is represented by a sequential representation of a single object $o_z^{(k_i)}$ in a sequence of frames k_i:

$$tr_z = \left\{ o_z^{(k_i)} \right\}, i = i_{in}, i_{in+1}, \ldots, i_{out}, z = 1, \ldots, Z, i_{in}, i_{out} \in K, i_{in} \leq i_{out}, o_z^{(k_i)} \in O_{k_i}$$

moreover, the same object is not included in different tracks. Let θ be the degree of similarity between the sets, S be a method of dynamic visual scene analysis.

It is required to find $TR^* = \{tr_z^*\}$, a set of tracks, each of which is represented by a sequential representation of a single object $o_z^{*(k_i)}$ on a sequence of frames k_i, tracked using S methods, $z = 1, \ldots, Z$ such that

$$\theta(TR, TR^*) \xrightarrow{S} \max, \text{ with the set of constraints T.}$$

3.2   Description of the Method

The developed method differs from the existing ones in detection of objects and calculation of their "deep" features in one pass of the detector based on the CNN work in conditions of uncertainty or noise detection data.

This method allows one to select objects on each frame using a detector and compare them with existing tracks (sequences of object coordinates on previous frames) or delete a track if the sequence of frames associated with the track is missing.

Input Data: $K = \{k_n\}$, ρ_{thr} is a maximum allowed threshold of difference between a track and an object, $Count_{max}$ is a maximum allowed number of consecutive frames, for which the degree of difference ρ between an object and a track is greater than ρ_{thr}, n_{av}

is the number of frames on the basis of which the averaging features P_{m_n} of the object $o_{m_n}^{(k_n)}$ with the coordinates $\left\{ \left(x_1^{(m_n)}, y_1^{(m_n)} \right), \left(x_2^{(m_n)}, y_2^{(m_n)} \right) \right\}$ is performed.

Output Data: $TR = \{tr_z\}$ is a set of tracks, each of which is represented by the coordinates $\left\{ \left(x_1^{(m_n)}, y_1^{(m_n)} \right), \left(x_2^{(m_n)}, y_2^{(m_n)} \right) \right\}$ of an individual object $o_z^{(k_n)}$ in a sequence of frames $k_n, o_z^{(k_n)} \in O_{k_n}$.

Initialization of a set of tracks, the number of which on the given frame initially equals zero.

Stage 1. The definition of the detector operating conditions is the uncertainty or noise in data detection.

If the operating conditions of the detector are unknown, any of the conditions can be selected, and then the correctness of the selection based on the detection accuracy indicators is evaluated. Such indicators can be represented as the number of false positives and the number of false negatives. If there are a lot of false positives of the detector, we can talk about noise detection data, otherwise the uncertainty of detection data is implied. The following sequence of stages is performed for each k_n:

Stage 2. A search of the coordinates $\left\{ \left(x_1^{(m_n)}, y_1^{(m_n)} \right), \left(x_2^{(m_n)}, y_2^{(m_n)} \right) \right\}$ for each object $o_{m_n}^{(k_n)}$, the features P_{m_n} of this object using an appearance object model and an algorithm for obtaining "deep features" of the objects detection. This step may also include a filtering procedure in case of noisy detection data.

Thus, the stage is designed to search for the objects on the current frame and their features, which can then be assigned to the appropriate tracks on the basis of the comparison of the attributes of these objects.

Stage 3. Matching of the objects $\left\{ o_{m_n}^{(k_n)} \right\}$ found in stage 1 with the existing tracks $\left\{ tr_{z_{n-1}} \right\}$, if $TR \neq \varnothing$:

Step 1. Addition to the circular list $P_{m_{n-1}}$ and displacement $P_{m_{n-1-n_{av}}}$.
Step 2. Calculation of the average-object features P_z^C of the track tr_z as the arithmetic mean of the object features in the circular list:

$$P_z^C = \frac{\sum_{j=1}^{n_{av}} P_{m_{n-j}}}{n_{av}}.$$

At the same time, if the number of features in the list is less than n_{av}, the average-object features are calculated on the base of the existing number of features.

Step 3. Construction of the cost-matrix $m_n \times z_{n-1}$:
Each element of the matrix is the distance between features P_{m_n} (the total number of objects – m_n) and the features of the track average-object P_z^C (total number of tracks – z_{n-1}) based on the appearance model.

Step 4. Solution of the problem of assigning m_n objects to z_{n-1} tracks using the Hungarian algorithm [4].
As a result of the stage, some of the found objects will be assigned to the existing tracks.

Stage 4. Initialization of new tracks for objects from stage 3 that have not been assigned to tracks:

$$tr_{z_n} = tr_{z_{n-1}} \cup \left\{ o_{m_n}^{k_n} \notin tr_{z_n} \right\}.$$

When Step 3 is completed, the objects that could not be assigned to the tracks remain. This step is designed to create new tracks for such objects.

Stage 5. Check of the validity of detection assignment to a track. The assignment is considered valid in case the degree of difference between the track and the designated object does not exceed the allowable threshold $\rho\left(tr_{z_n}, o_z^{(k_n)}\right) \leq \rho_{thr}$.

The stage allows to exclude those detection assignments on the track, which are not allowable and represent false detector triggering.

Stage 6. Deletion of tracks from $\{tr_{z_n}\}$ if the number of frames in the sequence is greater than the threshold number of frames $k_{n-m}, \ldots, k_n, m \geq Count_{\max}$ at which the degree of difference between the track and the assigned object exceeds the allowed threshold $\rho\left(tr_{z_n}, o_z^{(k_n)}\right) \geq \rho_{thr}$.

The stage allows to delete those tracks on which the object was missing (perhaps, it was occluded by another object) several frames in a row.

Stage 7. Selection of tracks from $\{tr_{z_n}\}$: $\rho\left(tr_{z_n}, o_z^{(k_n)}\right) \leq \rho_{thr}$.

The stage allows one to select only those tracks on which the object is present. The tracks on which an object is overlapped by other objects or is missing are not displayed.

Thus, as a result of the stages of the developed method, a set of tracks $\{tr_{z_n}\}$ is formed on each frame k_n.

4 Effectiveness Evaluation of the Developed Method

Effectiveness evaluation of the developed method is carried out with the help of a training and test dataset 2D MOT15, so this technique corresponds to the recommendations given in [6]. The effectiveness comparison of the developed method with other existing methods (SORT, ELP, MHT, DPM) was carried out taking into account the following conditions: uncertainty of detection and noise detection data. To simulate the conditions of uncertainty, detection that was obtained using an appearance model of the object was taken. To simulate the noise conditions of the detection data, the detections from the 2D MOT 15 dataset was taken. Then the gained detection was combined into tracks using the method of dynamic visual scene analysis.

Furthermore, the proposed method was further implemented with the procedure of filtration in case of noise in the data: from the dataset 2DMOT15 in each frame the detection with detection confidence at least 30% was chosen. For each of such detection with the help of the algorithm for getting "deep features" of object detection, its "deep features" were formed. Then these features were compared in pairs and the detection with a difference degree less than 0.2 (selected experimentally) was selected. The detection, which has a lower detection confidence threshold, was excluded from each of such pair.

In addition, for the tracks that have a minimum frames number of successful track detections greater than 3, this detection confidence threshold was reduced to 20%.

The degree of difference ρ in this method between an object and a track was calculated on the basis of the cosine distance with the maximum difference threshold $\rho_{thr} = 0.5$ selected for the following reasons: the object can be overlapped by another object or changed between adjacent frames by no more than the half.

The remaining parameters are selected as follows: $n_{av} = 10$, $Count_{max} = 3$.

Table 1 presents the obtained results. The table shows that the proposed method in case of noise detection data exceeds the methods that work online (SORT, DPM) and works online, but at the same time worse than the methods that work offline (ELP, MHT). With the uncertainty of detection, the proposed method is strongly superior to SORT.

Table 1. Evaluation results of the method effectiveness

Method	$MOTA$, % (noise detection data)	$MOTA$, % (uncertainty of detection data)	Online
SORT	26.0	20.2	Yes
ELP	30.3	28.1	No
MHT	34.2	In these conditions, the test fails	No
DPM	26.8	In these conditions, the test fails	Yes
Proposed method	27.3	26.3	Yes

The time spent on processing one frame of the proposed CNN was experimentally estimated using the appearance object model on GPU TITAN BLACK. These experiments do not take into account the time of reading the image from the hard disk, its transformation and loading in the CNN.

t_1 is one frame time processing using the proposed appearance object model;

t_2 is one frame time processing using SSD model detection without a ROI-Pooling layer;

t_3 – time for obtaining features on the frame based on CNN VGG 16.

The results of the comparison are presented in Table 2.

Table 2. Execution time comparison of different parts of the algorithm

Time, ms	t_1	t_2	t_3
	42	41	18

Thus, the use of the developed algorithm allows one to reduce frame processing time by 18 ms for each detection.

5 Conclusion

In this article we proposed a method of dynamic visual scene analysis based on the appearance model. This method allows one to detect images on the offered model and obtain "deep features" of detection for one CNN pass, to increase accuracy of tracks construction, and to execute processing in real time at the 24 FPS. In addition, this method can work both in conditions of uncertainty and in conditions of noise detection data.

Time spent on one frame processing using the appearance object model on GPU TITAN BLACK was experimentally estimated. The assessments results in conclusion that the application of the developed model allows one to reduce the processing time of the frame by 18 MS for each detection.

Acknowledgments. The work was supported by grant RFBR № 18-07-00928_a "Methods and technologies of intelligent support for research of complex hydro-mechanical processes in conditions of uncertainty on the convoluted neuro-fuzzy networks".

References

1. Bewley, A., Ge, Z., Ott, L., Ramos, F., Upcroft, B.: Simple online and realtime tracking. arXiv preprint arXiv: 1602.00763 (2016)
2. Garanin, O.I.: Method for selecting receptive field of convolutional neural network. Neurocomputers (3) 63–69 (2017)
3. Garanin, O.I.: Tuning method of multiscale model for detecting visual objects in a convolutional neural network. Neurocomputers (2) 50–56 (2018)
4. Kuhn, H.W.: The Hungarian method for the assignment problem. Nav. Res. Logist. Q. **2**, 83–97 (1955)
5. Kim, C., Li., F.: Multiple hypothesis tracking revisited. In: Proceedings of the IEEE International Conference on Computer Vision, pp. 4696–4704 (2015)
6. Leal-Taixe, L., Milan, A., Reid, I., Roth, S., Schindler, K.: MOTChallenge 2015: towards a benchmark for multi-target tracking. arXiv preprint arXiv: 1504.01942 (2015)
7. Liu, W., Anguelov, D., Erhan, D., Szegedy, C., Reed, S. E.: SSD: single shot multibox detector. arXiv preprint arXiv: 1512.02325 (2015)
8. McLaughlin, N., Rincon, J. M. D., Miller, P.: Enhancing linear programming with motion modeling for multi-target tracking. In: Proceedings of the IEEE Winter Conference on Applications of Computer Vision, pp. 271–350 (2015)

9. Sadeghian, A., Alahi, A., Savarese, S.: Tracking the untrackable: learning to track multiple cues with long-term dependencies. arXiv preprint arXiv: 1701.01909 (2017)
10. Simonyan, K., Zisserman A.: Very deep convolutional networks for large-scale image recognition. arXiv preprint arXiv: 1409.1556 (2015)
11. Wojke, N., Bewley, A., Paulus, D.: Simple online and realtime tracking with a deep association metric. arXiv preprint arXiv: 1703.07402 (2017)

Extended Stepping Theories of Active Logic: Paraconsistent Semantics

Michael Vinkov[1] and Igor Fominykh[2(✉)]

[1] Bauman Moscow State Technical University, Moscow, Russia
vinkovmm@mail.ru
[2] Moscow Power Engineering Institute, Moscow, Russia
igborfomin@mail.ru

Abstract. Active Logic is a conceptual system principles of which are satisfied by reasoning formalisms that allow correlation of their results with specific points in time and that are contradiction-tolerant. Today, tolerance for inconsistencies is theoretically justified in the works of the authors and is attributed to the so-called formalism of stepping theories, integrating the principles of Active Logic and logic programming. This paper represents further development and refinement of this concept for stepping theories with two kinds of negation. In this respect, more expressive and complete technique enables satisfaction of the principles of logic programming to the maximum extent as compared to other versions of formalism of stepping theories. The obtained results illustrate that the proposed argumentation semantics of stepping theories with two kinds of negation is paraconsistent in the sense that logical inconsistencies existing in these theories does not lead to their destruction.

Keywords: Active Logic · Reasoning situated in time · Stepping theories
Paraconsistent semantics

1 Introduction

Active Logic [1, 2] is a conceptual system that combines a number of formalisms (reasoning situated in time) that are not viewed as a static sequence of beliefs, but as a process that flows through time. The principles established in Active Logic systems are relevant for solving the tasks of operating complex objects in hard real-time environment. Such task solving is attributive for situations when the excess of the admissible amount of time allocated for their solution is fraught with catastrophic consequences. Furthermore, information coming in the process of solving such problems often proves to be contradictory and requires special processing techniques. Reasonings that fit the Active Logic concept have two important properties: (1) their results can be linked with the moments of time, which a reasoning agent is able to perceive (temporal sensitivity) and (2) these reasonings are tolerant to inconsistencies that occur during their construction (reasoning characterized by such feature are called paraconsistent). Although assuming a temporal sensitivity in active logic systems was not associated with serious theoretical problems, until recently the attempts to create declarative paraconsistent semantics for such logic systems failed, as shown in [2]. This state of things lasted until

© Springer Nature Switzerland AG 2018
S. O. Kuznetsov et al. (Eds.): RCAI 2018, CCIS 934, pp. 70–78, 2018.
https://doi.org/10.1007/978-3-030-00617-4_7

paraconsistency of argumentation semantics was introduced [3] into a version of stepping theory formalism that combines the principles of Active Logic and logic programming.

This paper contributes to the issues of paraconsistency of Active Logic Systems in relation to another, more general form of stepping theory formalism, the stepping theory formalism with two kinds of negation [4] (further, formalism of extended stepping theories), for which paraconsistency of its argumentative semantics is illustrated.

2 Paraconsistent Logics

Paraconsistent logics is a subdomain of modern non-classical logic, where the principle of entailing an arbitrary sentence from a logical contradiction is not supported. In classical logic, a certain theory is called contradictory, when it is possible to simultaneously prove both a definite statement and denial of the same statement, i.e., its negation. If – based on this fact – an arbitrary sentence can also be proved in a theory, the latter is called trivial. In standard logical systems the notions of contradiction notion and triviality do not differ, i.e., a contradiction in such theory means its triviality. Paraconsistent logics treats the contradiction differently from the classical logics. It excludes the possibility to derive from the contradictions any kind of statement, thus the contradiction ceases to be a threat of destruction of the theory. However, this does not remove the fundamental need to get rid of contradictions in the course of further development of the theory. Hence it implies another definition of paraconsistent logic: a logic is called paraconsistent if it can be used as a basis of inconsistent but non-trivial theories.

The strict definition of paraconsistent logic is related to the characteristic of logical sequence relationship. It can be called explosive if it satisfies the condition that for any formulas A and B, A and non-A entails an arbitrary formula B (symbolically: $\{A, \neg A\} \vdash B$). Classical logic, intuitionistic logic, multi-valued logics, and most of other standard logics are explosive. A logical consequence relation is said to be paraconsistent iff it is not explosive.

3 Extended Stepping Theories of Active Logic

Compared with other stepping theories, the syntax of extended stepping theories has the following features: an operator of subjective negation not^t is added to the alphabet of the rule language. The rule is as follows:

$$N : a_1 \wedge a_2 \ldots \wedge a_m \wedge not^t c_1 \wedge not^t c_2 \ldots \wedge not^t c_n \Rightarrow b, \tag{1}$$

where N is a character string that indicates the name of the rule, b is a propositional literal, and $a_1 \ldots a_m$ are propositional literals or first-order literals of the form later(j) or $\neg$later(j), where j is a natural number. Propositional literals can be objective or subjective, it will be discussed below.

The rules describe the principle of negative introspection in the following interpretation: if the formula $a_1 \wedge a_2 ... \wedge a_m$ is feasible and at this inference step it is unknown whether the formula $c_1 \wedge c_2 ... \wedge c_n$ is feasible, then it is admissible to assume that the formula b is feasible. In this case, $-b$ means (always objective) literal that is a supplement to the contrary pair for the literal b. In the situations where it is convenient, rule antecedents are viewed as sets of literals. Therefore, the system of stepping theories can be considered as a variant of the Active Logic system, which is completely rule-based and satisfying the principle of logic programming, according to which the formula models are sets of literals rather than more complex structures like in Kripke-style semantics. It is noteworthy that the rules of extended stepping theories in outward appearance are similar to default rules of R. Reiter's theories. The issue of correlation between the extended stepping theories of Active Logic and extended logic programs under answer set semantics (which are known to be special cases of default theories) is presented in the case study [5].

An *extended stepping theory is a* pair T = (R, Ck), where R is a finite set of rules of the form (2.1), Ck is a so-called *clock* of a stepping theory, which represents a finite subsequence of the sequence of natural numbers, e.g., Ck = (0, 1, 2, 4, 7, 11). Members of this subsequence identify duration of successively performed deductive cycles determining a reasoning process in all systems of Active Logic. A more detailed discussion of this issue can be found in [6]. Here, to ensure the generality of derived results, for simplicity's sake we assume that $Ck = Ck^1 = (1, 2, 3, 4, ...)$. We'll use the symbol Ck^* to designate the set of all members of the given sequence.

Further, for any extended stepping theory T = (R, Ck) designation R[q] denotes the set of all rules whose consequent is q. The set of literals forming the antecedent of rule r will be denoted by A(r). Let Lit_T be the set of all literals occurring in the rules of stepping theory T. *The belief set of* stepping theory T = (R, Ck) is a set of the form $\{now(t)\} \cup L_T^t$, where t is a natural number representing a point of time on the clock ($t \in Ck^*$) or 0, $L_T^t \subset Lit_T$. Let's consider operator ϑ_T that transforms the belief sets into belief sets in such a way that if B is a belief set such that $now(t) \in B$, then now $(t + 1) \in \vartheta_T(B)$. The sufficient conditions under which the literals will belong to belief set $\vartheta_T(B)$ will be different for different step theories, see, e.g., [7]. Below we will review these conditions as they apply to a system of stepping theories in the context of their argumentation semantics.

Now let B be the belief set of theory T such that literal $now(t) \in B$. We take B *as a quasi-fixed point* of operator ϑ_T iff for any B_1, such that $now(t_1) \in B_1$, where $t_1 > t$, the sequence includes $B_1 \backslash \{now(t_1)\} = \vartheta_T(B_1) \backslash \{now(t_1 + 1)\}) = B \backslash \{now(t)\}$. *The history* in stepping theory T is a finite sequence of belief sets $\boldsymbol{B}$. $\boldsymbol{B}(i)$ is the i-th member in the history, $\boldsymbol{B}(0) = \{now(0)\}$, for any t $\boldsymbol{B}(t + 1) = \vartheta_T(\boldsymbol{B}(t))$. The last element in the history is a belief set denoted by B_{fin}, *(final)*. It is the minimal quasi-fixed point of operator ϑ_T in the sense defined above. *An inference step* in stepping theory T = (R, Ck) is any pair of the form $(\boldsymbol{B}(i), \boldsymbol{B}(i + 1))$, and *the inference step number* is the number equal to (i + 1). The consequent (t-*consequent*) of extended stepping theory T is a literal belonging to belief set $B_{fin}(\boldsymbol{B}(t), t \in Ck^*)$.

One of the two kinds of negations in the extended stepping theories of the type under review will be called *subjective* negation and denoted by not^t. While in logic programming the expression not q means that the attempt to infer the literal q using the given logical program failed, the expression not^{t}q in the antecedent of the rule stepping theory means that a reasoning agent did not manage to infer the literal q to the current point of time (= at the given inference step). Another kind of negation, used in step-type theories of a given type, is denoted by the unary logical connection ¬, we shall henceforth call it a *strong negation*, like it is called in logic programming.

Hereafter, any propositional literal of type a_i that is not preceded by a subjective negation operator not^t will be called an objective literal. Any literal of type not^tc_j will be hereafter called a subjective literal. In this regard it should be noted that any further reasoning will remain valid even when the first order literals without functional symbols, i.e., with the finite Herbrand universe, are used instead of the propositional literals.

4 Explosive Argumentation Semantics for Extended Stepping Theories of Active Logic

The argumentation theory [8] has proved to be quite fertile in presenting of non-monotonic reasoning [9]. Argumentation semantics for formalisms of stepping theories of Active Logic was suggested in some our previous works [7, 10]. A variant of such kind of argumentation semantics with the consideration of specifics of extended stepping theories and representing an example of explosive semantics will be given below. Further we will refer to it as to E-semantics (E – explosive).

Definition 1. We assume that T = (R, Ck) is an extended stepping theory. Let an argument for T be defined as follows:

(1) any literal (of the first order logic) of type now(t), later(t) or ¬later(t), where t > 0, there exists rule r ∈ R, such that l ∈ A(r);
(2) sequence of rules Arg = [r_1,..., r_n], where r_1,..., r_n ∈ R, such that for any 1 ≥ i ≥ n, if p ∈ A(r_i), where p is the objective propositional literal, then there might be such j < i that r_j ∈ R[p];
(3) any subjective literal m of type not^{t}q, such that there exists a rule r ∈ R and m ∈ A (r);
(4) any two-element set of objective literals {b. −b}, where b, −b ∈ Lit$_T$.

For this extended stepping theory T = (R, Ck), *a set of all its arguments* is denoted by Args$_T$. If an argument is a first-order literal of type later(t) (¬later(t)), then we'll call such argument *limiting* (the function of the other argument after and before in time). An argument of type Arg = [r_1,..., r_n] is called *supporting argument*. Propositional literal b is called a *conclusion* of supporting argument Arg = [r_1,..., r_n] iff r_n ∈ R[b]. An argument of type not^{t}q is called a *subjective argument*. An argument of type {b. −b} is called an *absolute argument*. Any subsequence of [r_1,..., r_n] sequence, meeting Definition 1 is called *supporting subargument* of argument Arg = [r_1,..., r_n]. Limiting or subjective argument is a *subargument* of the argument Arg = [r_1,..., r_n] if a corresponding first-order literal or a subjective literal is included in the antecedent of any of the rules r_1,..., r_n.

Any supporting subargument of the argument Arg = $[r_1,..., r_n]$ is called its maximum subargument if a literal which is the conclusion of the said subargument is included in the antecedent of r_n rule. Limiting, or subjective argument is a maximum subargument of argument Arg = $[r_1,..., r_n]$ if a corresponding literal is included in the antecedent of r_n rule.

Example 1. Let R1 set of stepping theory T1 = (R1, Ck) consist of the following elements:

{N1: $\Rightarrow$ p,
N2: p $\Rightarrow$ q,
N3: q $\Rightarrow$ r,
N4: later(4) $\wedge$ not^t r $\Rightarrow$ ¬tusk_is_solved,
N5: ¬later(4) $\wedge$ r $\Rightarrow$ tusk_is_solved,
N6: tusk_is_solved $\Rightarrow$ tusk_is_solved}.

Supporting arguments of T_1 theory are Arg_1 = [N1, N2, N3, N5, N6] and all of its supporting subarguments, as well as Arg_2 = [N4]. Arg_3 = [N1, N2, N3, N5] supporting subargument is the maximum supporting subargument of argument Arg_1. Arg_4 = ¬later(4) and Arg_5 = later(4) are limiting arguments of T_1 theory (and corresponding subarguments of arguments Arg_1 and Arg_2). Arg_6 = not^{t}r is a subjective subargument of argument Arg_2.

Going over to definition of an argument status and a conflict between arguments of extended stepping theories, we should take into consideration that unlike other systems of argumentation, where the interrelations of various arguments are studied "in statics", here we discuss the development of these interrelations over time, i.e. in steps of inference of the stepping theory. On a certain step of inference, a specific argument may not have been constructed, i.e. it failed to become active, while after putting into action it may be in action until the moment of time (= step of inference) of its withdrawal from action. The latter means that the given argument (on the given step) is denied (= disposed of) due to the consequences of a conflict with other arguments. Thereby, the notion of "putting an argument into action" plays the key role in determining the status of the arguments and a conflict between them. For simplicity we will refer to the inference step numbered i as to simply step i.

Definition 2. Putting arguments of extended stepping theories in action is performed according to the following rules:

1. Any limiting argument of type *now(t)* is put into action on step i, so that clock (i − 1) < t ≤ i (at time point t). Any limiting argument of type ¬later(t) is activated on step 1 (at time point 0).
2. If the supporting argument does not have any subarguments, then it will be activated on step 1.
3. Any supporting argument is activated on step i iff all of its subarguments are put into action on the previous steps and there is a maximum subargument activated on step (i − 1) (at time point (i − 2)).
4. Any subjective argument is put into action on step 1(at time point 0).

5. Any absolute argument of type $\{b, -b\}$ is activated at step i, iff b, $-b \in B$ (i $-$ 1), where B(i $-$ 1) is a set of beliefs obtained at step i $-$ 1.

Definition 3. Withdrawal of limiting arguments from action is performed by the following rule:

Any limiting argument of type $\neg$later(t) is withdrawn from action on step i + 1, where i is such that (i $-$ 1) $\leq$ t.

All other arguments, including limiting arguments of the type later(t) after they are put into action on subsequent steps of inference, they have the status *active*.

The notion of arguments attacking each other, which is present in practically all argumentation systems, has its specific features in stepping theory of Active Logic.

Definition 4 (attacking arguments)

1. Arg_1 supporting argument *attacks* the other supporting arguments with conclusion of q, or subjective argument $Arg_2 = not^t q$ on step i iff the following conditions are simultaneously fulfilled:
   (1) the conclusion of Arg_1 is literal $-q$;
   (2) Arg_1 (like all of its subarguments) is active on step i;
   (3) none of the Arg_1 subarguments is attacked on step i by none of the other supporting or subjective arguments.
2. The subjective argument $Arg_1 = not^t q$ *attacks* subjective argument $Arg_2 = not^t -q$ on step i iff it is not attacked on step i by any supporting argument with the conclusion q.

Any set of beliefs in the stepping theory, in addition to a meta-literal of type now (i), consists of objective literals. The definition given below establishes necessary and sufficient condition of attributing of the objective literal to a belief set in the extended stepping theory under E-semantics.

Definition 5. Let B(i) be a belief set of a certain extended stepping theory. The objective literal q $\in B$(i) iff at least one of the following conditions are satisfied:

(1) there is the supporting argument Arg_1 of the given extended stepping theory, which conclusion is literal q, such that all its subarguments are active on step i, it is put into action no later than on step i $-$ 1, and Arg_1 or any of its subjective subarguments is not attacked on step i by any other supporting or subjective arguments;
(2) there is an absolute argument introduced into action not later than step i $-$ 1.

It should be noted that the fulfillment of the second condition is only possible if the first condition on step i $-$ 2 is satisfied for two supporting arguments with the conclusions forming a contrary pair. Such extended stepping theories will be called *inconsistent theories*. Also, when the second condition is satisfied, B(i)/$\{$now (i)$\}$ = Lit_T. The latter means that E-semantics for extended stepping theories introduced in this section is explosive in the sense defined in Sect. 2.

Example 2. Let belief set R_2 of stepping theory $T_2 = (R_2, Ck)$ consist of the following elements:

{N1: ⇒ p,
N2: later(1) ∧ not^{t}r ⇒ ¬d,
N3: p ⇒ d,
N4: ¬later(4) ⇒ r}.

The stepping theory T_2 is trivial in the sense defined in Sect. 1, its smallest quasi-fixed point $B_{fin\_} = Lit_{T2} \cup \{now(3)\} = \{now(3), p, r, d, ¬d\}$.

5 Explosive Argumentation Semantics for Extended Stepping Theories of Active Logic

Below, we will define the argumentation semantics for stepping theories with two kinds of negation, which we henceforth will refer to as P-semantics (P – paraconsistency). All definitions introduced for E-semantics remain in effect, except for 4 and 5, which respectively determine the attack by arguments of each other and the conditions for attributing objective literals to sets of beliefs. Instead of the said definitions we introduce the following.

Definition 6

1. The supporting argument Arg_1 *attacks* other supporting arguments with the conclusion q or the subjective argument $Arg_2 = not^t q$ on step i iff the following conditions are simultaneously fulfilled:
   (1) the conclusion of Arg_1 is the literal −q;
   (2) Arg_1 (like all its subarguments) is active on step i;
   (3) none of the Arg_1 subarguments is attacked on step i by none of the other supporting or subjective arguments.
2. The subjective argument $Arg_1 = not^t q$ *attacks* subjective argument $Arg_2 = not^t −q$ on step i iff it is not attacked on step i by any supporting argument with the conclusion q.

Definition 7. Let $B(i)$ be the set of beliefs of certain extended stepping theory. The objective literal $q \in B(i)$ iff the following condition is satisfied:
   There is a supporting argument Arg_1 of the extended stepping theory, which conclusion is literal q, such that all its subarguments are valid on step i, it is put into action no later than on step $i − 1$, and Arg_1 or any of its subjective subarguments is not attacked on step i by any other supporting or subjective arguments.
   Let T = <R, Ck> be certain stepping theory with two kinds of negation. We denote the transformation operators of belief sets of the theory T as ϑ_T^P и ϑ_T^E, which refer to P-semantics and E-semantics, respectively.

Theorem 1. Let $B = B(i)$ be a set of beliefs of certain extended stepping theory T = <R, Ck>. Then, $\vartheta_T^P(B) \subseteq \vartheta_T^E(B)$, and if the respective stepping theory T is not inconsistent, then the equality $\vartheta_T^P(B) = \vartheta_T^E(B)$ is valid.

Theorem 2. Let the extended stepping theory, which is inconsistent in terms of E-semantics, be T = <R, Ck>, let $B = <B(0),...B(n)>$ be its history, such that for any step i, $0 \le i \le n$ $B(i+1) = \vartheta_T^P(B(i))$. Then for any i, $0 \le i \le n$, there exists at least one literal $l \in \mathrm{Lit}_T$ such that $l \notin B(i)$.

The proofs of both theorems are based on the fact that until an absolute argument under E-semantics is introduced into action, any supporting argument of the extended stepping theory T, acting under P-semantics at a certain inference step, is active at the same step and also under E-semantics. If an absolute argument of type $\{b, -b\}$ under E-semantics is put into action at a certain inference step i, then all subsequent belief sets under E-semantics will have the form $B(j)/\mathrm{now}(j) = \mathrm{Lit}_T$, where $j > i$. Along with this, it is easily understand that under P-semantics, the objective literals b and $-b$ according to Definition 6 will not appear simultaneously in any belief sets of the theory of T.

Theorem 2 states that the extended stepping theories – though inconsistent – are non-trivial theories under P-semantics in the sense defined in Sect. 2.

6   Conclusion

The article sets out that the argumentation semantics introduced for the extended stepping theories is paraconsistent. The stepping theories of this type can contain inconsistencies (in terms of P-semantics, any stepping theory with supporting arguments attacking each other at certain points of time will be paraconsistent), but this does not lead to their destruction as it happens with E-semantics.

Acknowledgment. This work was supported by Russian Foundation for Basic Research (projects 17-07-00696, 15-07-02320, 18-51-00007).

References

1. Elgot-Drapkin, J., Perlis, D.: Reasoning situated in time I: basic concepts. J. Exp. Theoret. Artif. Intell. **2**(1), 75–98 (1990)
2. Hovold, J.: On a semantics for active logic. MA thesis. Department of Computer Science, Lund University (2011)
3. Fominykh, I., Vinkov, M.: Paraconsistency of argumentation semantics for stepping theories of active logic. In: Abraham, A., Kovalev, S., Tarassov, V., Snášel, V. (eds.) IITI 2016. AISC, vol. 450, pp. 171–180. Springer, Cham (2016). https://doi.org/10.1007/978-3-319-33609-1_15
4. Vinkov, M.M., Fominykh, I.B.: Stepping theories of active logic with two kinds of negation. Adv. Electr. Electron. Eng. **15**(1), 84–92 (2017)
5. Fominykh, I., Vinkov, M.: Step theories of active logic and extended logical programs. In: Abraham, A., Kovalev, S., Tarassov, V., Snasel, V., Vasileva, M., Sukhanov, A. (eds.) IITI 2017. AISC, vol. 679, pp. 192–201. Springer, Cham (2018). https://doi.org/10.1007/978-3-319-68321-8_20
6. Vinkov, M.M.: Time as an external entity in modeling reasoning of a rational agent with limited resources. In: Proceedings of 11th National Conference on AI, CAI 2008. Fizmatlit. Publ. (2008). (in Russian)

7. Vinkov, M.M., Fominykh, I.B.: Argumentation semantics for active logic step theories with granulation time. In: Artificial Intelligence and Decision-Making, no. 3, pp. 3–9. URSS Publ., Moscow (2015). (in Russian)
8. Dung, P.M.: Negation as hypothesis: an abduction foundation for logic programming. In: Proceedings of the 8th International on Logic Programming. MIT Press, Paris (1991)
9. Vagin, V.N., Zagoryanskaya, A.A.: Argumentation in plausible reasoning. In: Proceedings of 9th National Conference on AI, CAI 2010, vol. 1, pp. 28–34. Fizmatlit. Publ., Moscow (2000). (in Russian)
10. Vinkov, M.M.: Argumentation semantics for theories stepper active logic. In: Proceedings of 10th National Conference on AI, CAI 2006, Obninsk, vol. 1, pp. 64–72 (2006). (in Russian)

Processing Heterogeneous Diagnostic Information on the Basis of a Hybrid Neural Model of Dempster-Shafer

Alexander I. Dolgiy[1(✉)], Sergey M. Kovalev[1(✉)],
and Anna E. Kolodenkova[2(✉)]

[1] Rostov State Transport University, Rostov-on-Don, Russia
adolgy@list.ru, ksm@rfniias.ru
[2] Samara State Technical University, Samara, Russia
anna82_42@mail.ru

Abstract. In the this work it is emphasized that fusion of the diverse data obtained from sources of primary information (sensors, the measuring equipment, systems, subsystems) for adoption of diagnostic decisions at a research of faults of devices, is one of the main problems in information processing. A generalized scheme of fusion of diverse data reflecting features of this process is considered. Classification of levels, modern methods of fusion of diverse data in the conditions of incomplete, indistinct basic data is also considered. The article develops a new hybrid approach to the diagnosis of technical objects based on multisensory data in terms of heterogeneity of the original information. We consider a new class of adaptive network models focused on the implementation of the procedures of logical-probabilistic inference using the Dempster-Shafer methodology and fuzzy logic. The adaptive Dempster-Shafer model (*DS* model) is a multilayered network of neurons mapped to the elements of the hypothesis space together with the current values of their base probabilities, on the basis of which the confidence probabilities of hypotheses are calculated. The original training algorithm for the neural network model with the attraction of experimental data is based on the principle of the distribution of the total error in the neural network in proportion to their confidence probabilities. The network model of Dempster-Shafer functions is trained together with the neural network model, which simulates the process of forming empirical estimates of hypotheses on the basis of subjective preferences of experts for the influence of various factors on diagnostic solutions. The principal advantage of the hybrid system is the ability to jointly adapt the parameters of both models in the learning process, which increases the reliability of the results of calculations due to the diversity of the used expert statistical information. The adaptability of the hybrid system also makes it possible to implement a new approach to the calculation of probability estimates of hypotheses based on a combination of several evidence by training a hybrid system based on data from several sources.

Keywords: Hybrid neural model · Heterogeneous information
Data processing

© Springer Nature Switzerland AG 2018
S. O. Kuznetsov et al. (Eds.): RCAI 2018, CCIS 934, pp. 79–90, 2018.
https://doi.org/10.1007/978-3-030-00617-4_8

1 Introduction

Currently, in the field of technical diagnostics at the peak of application, there are hybrid approaches to the detection and prediction of faults based on the combination of heterogeneous data obtained from different sensor systems [1–7]. For the development of hybrid diagnostic technologies, the key methods are the combination (fusion) of heterogeneous data and, in particular, the approach based on the combination of probabilistic and intellectual methods [8–12].

The *Dempster-Shafer* (*D-Sh*) methodology [13], which is used in conjunction with adaptive network models, is an effective and promising means of modeling and processing heterogeneous data used together with network models. Neural network models are able to give adaptability to the *D-Sh* methodology and improve its "accuracy" due to the possibility of training on experimental data, and network models of the *D-Sh* methodology will make it possible to process quantitative information on fuzzy judgments and confidence ratings on the basis of strict mathematical methods.

The main objects and parameters of the *D-Sh* theory are confidence functions, basic probabilities and probability estimates of hypotheses. However, in practice, there are serious problems with the quantitative evaluation of these parameters, due to the combinatorial "explosion" of the space of hypotheses, on the set of which the expert should evaluate these parameters. However, in reality, it is always possible to obtain estimates for some of the hypotheses for which expert or statistical information is available. To extend partial estimates to the entire area of hypotheses, methods are needed to calculate these parameters based on the attraction of additional expert information.

The approach discussed below is based on the identification of the most important parameters of the *D-Sh* methodology, called basic probability hypotheses or masses, with the involvement of statistical and empirical information obtained from experts. The proposed approach also makes it possible to implement procedures for combining information in the process of obtaining new certificates.

2 The Generalized Scheme of Fusion of Diverse Data

At the heart of the offered approach to fusion of the diverse data obtained from a set of various sensors is the use of the generalized scheme presented in Fig. 1.

From Fig. 1 we can see that fusion of data can be classified on three levels [14]:

1. Low level of fusion. This level is often called the level of raw data (raw data level) or level of signals (signals level). The raw data are considered as entrance data, then unite. As a result of association it is expected to obtain new more exact and informative data, than the raw entrance data. For example, in work [15] the example of low-level fusion to use of the filter of moving average is given.
2. Medium level of fusion. This level is called the level of signs (attributes, characteristics) (feature level). There is a fusion of signs (a form, texture, edges, corners, lines, situation) as a result of which new objects, or cards of objects which can be used for other problems, for example, of segmentation and recognitions turn out.

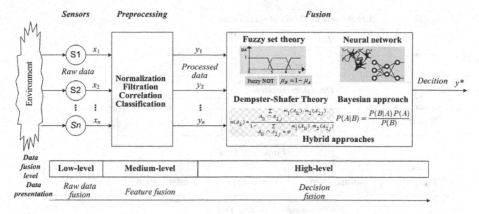

Fig. 1. The generalized scheme of fusion of diverse data.

Also at this level there is a data processing, namely a filtration (fight against noisy data), normalization (transformations to one type of data), correlation, classification of data, to use of methods of "soft calculations" and methods of data mining. Examples of medium-level fusion are given in [16, 17].

3. High level of fusion. This level is called the level of decision (decision level) or level of symbols (symbol level). There is a fusion at the level of the decision as a result of which the global decision turns out. The most traditional and known methods of fusion of data are probabilistic methods (Bayesian networks, the theory of proofs); computational intelligent methods (*Dempster-Shafer*, theory of indistinct sets and neural networks). These methods allow to present the coordinated and uniform opinion on diagnostic process to the person making the decision. High level fusion is considered, e.g. in [18].

3 Classification of Levels and Methods of Diverse Data Fusion

Now rather large number of methods is developed for fusion of data. However, at the choice of this or that method it is necessary to consider some aspects (what are the best fusion methods for the available data?; what is preliminary processing necessary?; how to choose from a data set those which fusion will give the best effect?, etc.).

On the basis of systematization of the review of references in Fig. 2, classification of levels [19–22] and modern methods of diverse data fusion [23–28] in the conditions of incomplete, indistinct basic data is presented.

Let us note that the given division of methods of fusion of the diagnostic decisions given for acceptance as conditional character as in practice they are crossed and interact among themselves.

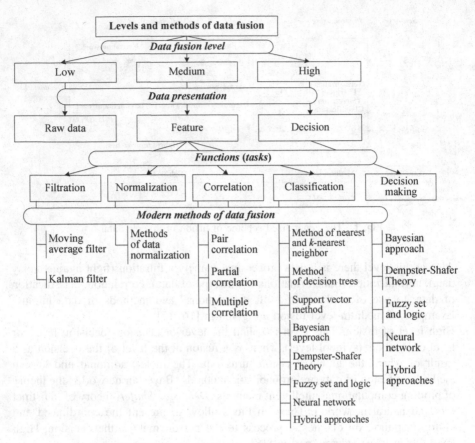

Fig. 2. Classification of levels and methods of data fusion.

4   Elements of the Dempster-Shafer Methodology

The main objects of the *D-Sh* methodology are confidence-building measures and plausibility, which are calculated on the basis of the basic probabilities of hypotheses. A set of hypotheses describing the decision-making situation is compared with the confidence interval, which should belong to the degree of confidence in each hypothesis. The confidence measure is denoted by $Bel(P)$ changing from zero, indicating full confidence in this hypothesis. A measure of the plausibility of the hypothesis $Pl(P)$ is determined using a measure of credibility:

$$Pl(P) = 1 - Bel(not(P)).$$

Let X be a universal set of conjectures and 2^X be a set of all subsets of X called an exponential set. The key in the *D-Sh* methodology is the notion of mass $m(A)$ of the element of the exponential set. It expresses the ratio of all relevant and available evidence that supports the claim that a certain conjecture X belongs to A, but does not belong to any subset of A. The value of $m(A)$ refers only to set A and carries no additional information about other subsets of A, each of which has its own mass.

Based on the assigned masses, the upper and lower limits of the range of possibilities can be determined. This interval contains the exact probability of the considered subset of hypotheses and is bounded by two continuous measures called trust (*belief*) and plausibility. The confidence of $Bel(A)$ to set A is defined as the sum of all masses of the eigenvectors of the set in question

$$Bel(A) = \sum_{B \subseteq A} m(B),$$

and plausibility $Pl(A)$ is the sum of the masses of all subsets corresponding to at least one element with A:

$$Pl(A) = \sum_{B \cap A \neq \varnothing} m(B).$$

In practice, the base probability function is often defined as a frequency function based on statistics: $m(A_i) = c_i/N$, where N – the total number of observations; c_i – the number of observations of the event A_i.

The most important element of the theory of $D\text{-}Sh$ is the rule of evidence combination. The original Union rule, known as the *Dempster* rule combination [13], is a generalization of the Bayes rule. In fact, the Union (called the connected mass) is calculated from two sets of masses m_1 and m_2 as follows:

$$m_{1,2}(\varnothing) = 0, \quad m_{1,2}(A) = \frac{1}{1-K} \sum_{B \cap C = A \neq \varnothing} m_1(B) m_2(C),$$

$$K = \sum_{B \cap C = \varnothing} m_1(B) m_2(C).$$

The K factor is a normalizing factor and, at the same time, characterizes the measure of conflict between the two sets of masses.

5 The Network Dempster-Shafer Model

In the practical application of the $D\text{-}Sh$ methodology, an important problem is to determine the underlying probabilities of $m(X)$ hypotheses on the basis of which confidence and plausibility estimates are calculated. As a rule, the user a priori does not have such information in full.

To solve this problem, we use an approach based on the identification of the *Dempster-Shafer* model (*DS*-model) parameters with the involvement of additional expert statistics. Identification of parameters is based on the unification of the two network models: adaptive network DS models, performing the calculation and adjustment of the basic and trust probabilities of the hypotheses, and neuro-fuzzy model to obtain probabilistic assessments of hypotheses based on empirical views of experts.

Consider the organization of the *DS*-model on the example of one of the diagnostic subsystems of railway automation [1]. Suppose that the state of a controlled technical object (*TO*) is characterized by a set of parameters (numeric attributes), one of which is *X*. Based on the analysis of the values of the parameter *X*, hypotheses are put forward about the technical state of the object being diagnosed.

For this interval of values of the attribute is under divided into several fuzzy intervals $\alpha_1, \ldots, \alpha_n$ characterizing different degrees of efficiency of *TO*, including: the valid values for the parameter *X* under which *TO* is considered fully functional; invalid values at which the *TO* is considered unhealthy; intermediate values characterizing the pre-failure states of the *TO*.

Each *TO* state is associated with an elementary hypothesis α_i, which for each specific value $x \in X$ has a certain base probability (mass) $m(\alpha_i) = \mu_{\alpha_i}(x)$ where $\mu_{\alpha_i}(x)$ is membership function (*MF*) of the fuzzy interval α_i. All possible combinations of elementary hypotheses form an exponential set 2^L ($L = \{\alpha_i\}$) of compound hypotheses. The mass of composite hypotheses $\{\alpha_{i_1}, \ldots, \alpha_{i_k}\}$ is calculated through a conjunction of MF constituent elementary hypotheses:

$$m(\alpha_{1_1}, \ldots, \alpha_{i_k}) = \bigwedge_{j=1}^{k} \mu_{\alpha_{i_j}}(x),$$

where $\wedge$ is fuzzy conjunction calculated on the basis of the *T*-norm operator.

The decision-making situation is characterized by a specific value of the diagnostic feature $x \in X$, on the basis of which the probabilistic estimates of all hypotheses $\{\alpha_{i_1}, \ldots, \alpha_{i_k}\}$ ($k = 1, 2, \ldots, n$) on the technical condition of the controlled object can be calculated.

The network *DS* model is focused on the calculation and adaptation of the basic and trust probabilities of the hypotheses $\{\alpha_{i_1}, \ldots, \alpha_{i_k}\}$ and contains *n* layers of neurons $s_k^i = s_k^i(\{\alpha_{i_1}, \ldots, \alpha_{i_k}\})$, corresponding to the hypotheses of the 2^L, so that each *k*-th layer contains neurons corresponding to all possible combinations of *k* elementary hypotheses, the index *k* indicates the layer number of the network, and the index *j* is the number of the neuron in the layer. Interlayer connections between neurons is organized in such a way that the output of the *j*-th neuron $s_k^j(\{\alpha_{j_1}, \ldots, \alpha_{j_k}\})$ in the *k*-th layer is connected to the input *l*-th neuron $s_{k+1}^l(\{\alpha_{l_1}, \ldots, \alpha_{l_{k+1}}\})$ in the subsequent $(k + 1)$-th layer if and only if $\{\alpha_{j_1}, \ldots, \alpha_{j_k}\} \subset \{\alpha_{l_1}, \ldots, \alpha_{l_{k+1}}\}$.

The first (input) layer contains *n* neurons s_1^i ($i = 1, \ldots, n$) corresponding to elementary hypotheses $\alpha_1, \ldots, \alpha_n$, the inputs of which are the value of the parameter *x*. The outputs stay for the basic probabilities of the elementary hypotheses $m(\alpha_i)$ and at the same time, of the confidence probabilities $Bel(\alpha_i)$ calculated on the basis of the *MF*:

$$m(s_1^i) = Bel(\alpha_i) = \mu_{\alpha_i}(x, \overrightarrow{p_i}) \cdot K,$$

where $K = \left(\sum\limits_{(\alpha_{i_1},\ldots\alpha_{i_k}) \in 2^L} m(\alpha_{i_1},\ldots,\alpha_{i_k}) \right)^{-1}$ is a normalized coefficient, $\vec{p_i}$ is a vector

of *MF* parameters.

The neurons of the subsequent hidden layers of the *DS-model* calculate the masses and confidence probabilities of compound hypotheses by formulas:

$$ m(s_k^i) = \bigwedge_{s_{k-1}^i \subset s_k^i} m(s_{k-1}^i) \cdot K, \quad Bel(s_k^i) = \sum_{s_{k-l}^i \subseteq s_k^i} m(s_{k-l}^i), \quad (l = 1, \ldots, k-1), $$

where $\wedge$ is fuzzy conjunction defined on the basis of the T-norm operator. Thus, the DS-model has n inputs and 2n outputs, on which the values of the confidence probabilities of the hypotheses of the exponential set 2L are formed. In Fig. 3 the structure of a *DS* model is shown for the four hypotheses $\alpha_1,\ldots,\alpha_4$ and distribution error signal when adapting the network to the received probability estimation P($\{\alpha_1, \alpha_2, \alpha_3\}$) for the hypothesis $\{\alpha_1, \alpha_2, \alpha_3\}$.

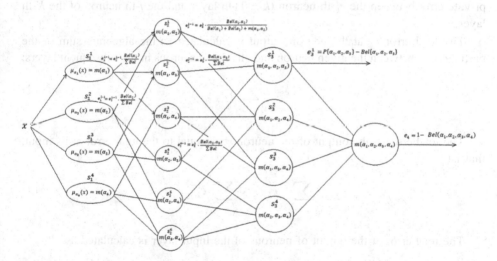

Fig. 3. The structure of the *DS*-model for the four-element set of hypotheses $\{\alpha_1, \alpha_2, \alpha_3, \alpha_4\}$ and the propagation of the error signal from the neuron s_3^1 to the input neurons $s_1^1 \& s_1^2$.

Training and adaptation of the network model is based on the error back propagation algorithm, a generalized description of which is given below.

Suppose that with respect to some particular decision-making situation characterized by the value of the controlled parameter $x \in X$ and a hypothesis $\{\alpha_{i_1},\ldots,\alpha_{i_k}\}$, there was a statistical or expert information about the likelihood of this hypothesis $P(\alpha_{i_1},\ldots,\alpha_{i_k})$, that is, information about the likelihood of finding something in suitable condition. The output of the corresponding neuron $s_k^j = s(\{\alpha_{i_1},\ldots,\alpha_{i_k}\})$ of the network model generates a value of confidence probability of the hypothesis, $Bel(\{\alpha_{i_1}, \ldots, \alpha_{i_k}\})$.

The difference $e = P(\alpha_{i_1}, \ldots, \alpha_{i_k}) - Bel(\{\alpha_{i_1}, \ldots, \alpha_{i_k}\}) - Bel(\alpha_{i_1}, \ldots, \alpha_{i_k})$ between the expert rating and the confidence value of the hypothesis is considered as an error of the network model on the given input parameter value $x \in X$. The adaptation algorithm aims at minimizing the deviation of e by adjusting the baseline probabilities of hypotheses $Bel(\alpha_{i_1}, \ldots, \alpha_{i_k})$ network model and adjusting the MF of the masses of the elementary hypotheses $m(\alpha_i)$.

Adaptation of the parameters of DS-model is reduced to the error distribution at the output of the j-th neuron of the current layer k for all neurons of the preceding $(k - 1)$-th layer according to the following law:

$$e_{k-1}^{i \to j} = e_k^j : \left[\sum_{s_{k-1} \subset s_k^j} Bel(s_{k-1}) \right] \cdot Bel(s_{k-1}^i), \tag{1}$$

where e_k^j is the total error at the output of the j-th neuron of the K-th layer; $e_{k-1}^{i \to j}$ is a private error between the e_k^i-th neuron $(k - 1)$-th layer and the j-th neuron of the K-th layer.

The total error e_k^j at the neuron output is calculated as the algebraic sum of the partial errors between the given neuron and all related neurons in the subsequent layers:

$$e_k^j = \sum_{s_{k+1}^l \supset s_k^j} e_k^{j \to l}. \tag{2}$$

The total error at the output of the neuron e_k^i is equal to the total error at its input, that is:

$$\sum_{s_{k+1}^i \supset s_{k+1}^j} e_k^{i \to j} = \sum_{s_{k-1}^j \supset s_k^i} e_{k-1}^{j \to i}. \tag{3}$$

The total error at the output of neurons of the input layer is calculated as:

$$e_1^i = \mu_{\alpha_i}(x_i, \overrightarrow{p_i}) - \sum_{s_2^j \supset s_1^i} e_1^{i \to j}. \tag{4}$$

Minimization of the error e is reduced to the adjustment of masses of neurons $m(\{\alpha_{i_1}, \ldots, \alpha_{i_k}\})$ MF parameters $p \in \overrightarrow{p_i}$ in accordance with the law:

$$\Delta e_1^i = \frac{de_1^i}{dp_j} \cdot e_1^i \cdot \eta, \tag{5}$$

where η is the coefficient of the speed of learning.

Correction of neuronal masses is carried out by distributing the total error at the output of neurons of the k-th layer on all related neurons of the previous $k - 1$ layer in proportion to the confidence probabilities of these neurons.

If at the beginning of training for some hypotheses the values of the basic probabilities are a priori unknown for these hypotheses, the same starting masses are assigned and normalized to one.

6 Hybrid Neural-Fuzzy Dempster-Shafer Model

In real-world decision-making situations, the user often does not have complete statistical or expert information about the probability of finding a particular state with this combination of features. In such situations, the classical probabilistic inference procedures developed in the Framework of the D-Sh theory are not applicable.

To solve this problem, the paper proposes a hybrid approach based on the combination of the network DS-model and adaptive fuzzy model that simulates the empirical process of experts' formation of probability estimates based on subjective preferences about the impact of certain sensor readings on diagnostic solutions. As a fuzzy model, it is proposed to use a neural network model of the first order *Takagi-Sugeno* (*TS*-model) [29].

TS-model is a neural network that describes the relationship between the input values of the diagnostic feature $x \in X$ and probabilistic estimates of hypotheses $P(\alpha_{i_1}, \ldots, \alpha_{i_k})$. *TS*-model has one input, which receives the value of the controlled parameter x, and 2^n outputs, which form the probability estimates of hypotheses. The *TS*-model knowledge base includes $m \leq 2^n$ fuzzy rules of the form:

$$R_i : (x^* = \alpha_{i_1}) \vee \ldots \vee (x^* = \alpha_{i_r}) \Rightarrow P(\alpha_{i_1}, \ldots, \alpha_{i_r}) = c_r, \tag{6}$$

where x^* are the values of parameter X; α_{i_j} are fuzzy terms corresponding to the conjectures of the exponential set 2^L; c_r are output adaptable parameters of fuzzy rules.

The distinguishing feature of the fuzzy system (6) from the traditional *TS*-model is the presence of not one, but a set of outputs corresponding to the hypotheses of the exponential set, relative to which there is expert information about the validity of hypotheses. Another feature is the aggregation of fuzzy terms in the antecedents of fuzzy rules based on the disjunction operator implemented in the class of S-norms. This is due to the fact that the probability of a composite hypothesis about the technical condition is obviously a non-decreasing function of the probabilities of its elementary hypotheses.

The output values from *TS*-model when entering the input parameter values $x^* \in X$ is computed in the standard way:

$$P(\alpha_{i_1}, \ldots, \alpha_{i_r}) = S(\mu_{\alpha_{i_1}}(x^*), \ldots, \mu_{\alpha_{i_r}}(x^*)) \cdot c_r,$$

where $S(*)$ is S-standard operator.

TS-model is an adaptive system, in which, in addition to calculating the probabilistic estimates of hypotheses $(\alpha_{i_1}, \ldots, \alpha_{i_r})$, the adaptation of the parameters MF

$\mu_{\alpha_{ij}}(x_{ij}, \overrightarrow{p_{ij}})$ and the output parameters c_r is used. The adaptation properties of the *TS*-model are used in the hybrid *D-Sh* system to replenish information about the probability of estimates of hypotheses when identifying the parameters of the *DS*-model, as well as to adjust the parameters of the *TS*-model upon receipt of additional expert information about the probabilities of hypotheses. The overall structure of the hybrid system is shown in Fig. 4 below.

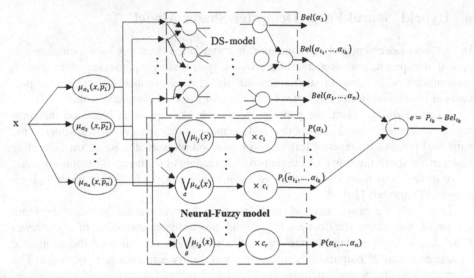

Fig. 4. Structure of a hybrid adaptive network based on the TS model and the Dempster-Shafer network model.

The principle feature of the developed hybrid *DS*-system is the possibility of joint adaptation of the parameters $\overrightarrow{p_i}$ *MF* $\mu_{\alpha_i}(x_{ij}, \overrightarrow{p_i})$, representing the basic probabilities of elementary hypotheses $\{\alpha_1\},...\{\alpha_n\}$, in both models, which, firstly, increases the flexibility of the system due to its adaptation to both objective statistical and subjective expert data, and, secondly, increases the speed of learning.

7   Conclusion

A new class of neural network models focused on the implementation of the probabilistic inference methodology developed in the framework of the *Dempster–Shafer* theory is developed in the article. A new hybrid approach to the diagnosis of technical objects using multisensory data, based on the combination of neural network model *Dempster-Shafer* and neural network, whose principal advantages are:

1. The possibility of realizing probabilistic inference in the absence of full information on masses of hypotheses due to the attraction of expert information on probabilistic

estimates of hypotheses received from a neural network on the basis of the analysis of subjective preferences of an expert.

2. The possibility of implementation of a fundamentally new approach to combining multiple evidence in the methodology of *Dempster-Shafer* adaptation of parameters of neural network models as new evidence from multiple sources and adjusting the results of the calculations simultaneously for all certificates.

3. Joint adaptation of the parameters of both models in the learning process, which increases the reliability of the results of computation due to the diversity of the expert statistical information in the process of computing the confidence.

Acknowledgments. The work was supported by RFBR grants No. 16-07-00032-a, 16-07-00086-a and 17-08-00402-a.

References

1. Kovalev, S.M., Tarassov, V.B., Dolgiy, A.I., Dolgiy, I.D., Koroleva, M.N., Khatlamadzhiyan, A.E.: Towards intelligent measurement in railcar on-line monitoring: from measurement ontologies to hybrid information granulation system. In: Abraham, A., Kovalev, S., Tarassov, V., Snasel, V., Vasileva, M., Sukhanov, A. (eds.) IITI 2017. AISC, vol. 679, pp. 169–181. Springer, Cham (2018). https://doi.org/10.1007/978-3-319-68321-8_18

2. Kovalev, S.M., Kolodenkova, A.E.: Creation of the knowledge base of the intelligent control system and preventions of risk situations for the design stage complex technical systems. Des. Ontol. **7**(4), 398–409 (2017). (26)

3. Kirchner, D., Geihs, K.: Qualitative Bayesian failure diagnosis for robot systems. http://people.csail.mit.edu/gdk/iros-airob14/papers/QBFD_CameraReady.pdf. Accessed 09 May 2018

4. Yi, Y.: Fault diagnosis based on Bayesian Petri Nets. Sens. Transducers **179**(9), 114–120 (2014)

5. Abdulaiev, P.Sh., Mirzoev, A.G.: Application of the theory of Dempster-Shafer in the synthesis of information on the technical state of GTE. Aerosp. Tech. Technol. **9**(116), 128–132 (2014)

6. Gholamshahi, R., Poshtan, J., Poshtan, M.: Improving stator winding fault diagnosis in induction motor using Dempster-Shafer theory. Electr. Electron. Eng.: Int. J. **3**(2), 161–173 (2014)

7. Hui, K.H., Lim, M.H., Leong, M.S., Al-Obaidi, S.M.A.: Dempster-Shafer evidence theory for automated bearing fault diagnosis. Int. J. Mech. Eng. Technol. **8**(4), 277–288 (2017)

8. Hall, D.L., Llinas, J.: Handbook of Multisensor Data Fusion. CRC Press LLC, Boca Raton (2001)

9. Vovchenko, A.E., Kalinichenko, L.A., Kovalev, D.Yu.: Methods of entity resolution and data fusion in the ETL-process and their implementation in the environment Hadoop. Inf. Appl. **4**(8), 94–109 (2014)

10. Gorodezkiy, V.D., Karsaiev, O.V., Samoilov, V.V.: Multi-agent technology for decision-making in the task of combining data, **1**(2), 12–37 (2002)

11. Khaleghi, B., Khamis, A., Karray, F.O., Razavi, S.N.: Multisensor data fusion: a review of the state-of-the-art. Inf. Fus. **14**(1), 28–44 (2013)

12. Bumystriuk, G.: Sensor fusion technologies for management in critical situation. Control Eng. Rus. (5), 49–53 (2014)

13. Dempster, D., Shafer, G.: Upper and lower probabilities induced by a multi-valued mapping. Ann. Math. Stat. **38**, 325–339 (1967)
14. Brahim, A.B.: Solving data fusion problems using ensemble approaches (2010)
15. Polastre, J., Hill, J., Culler, D.: Versatile low power media access for wireless sensor networks, pp. 95–107 (2004)
16. Nowak, R., Mitra, U., Willett, R.: Estimating inhomogeneous fields using wireless sensor networks. IEEE J. Sel. Areas Commun. **22**, 999–1006 (2004)
17. Zhao, J., Govindan, R., Estrin, D.: Residual energy scans for monitoring wireless sensor networks. In: IEEE Wireless Communications and Networking Conference, Orlando, FL, vol. 1, pp. 356–362 (2002)
18. Krishnamachari, B., Iyengar, S.: Distributed Bayesian algorithms for fault-tolerant event region detection in wireless sensor networks. IEEE Trans. Comput. **53**, 241–250 (2004)
19. Kharisova, V.N., Perova, A.I., Boldina, V.A.: Global satellite navigation system GLONASS. IPGR (1998)
20. Ochodnicky, J.: Data filtering and data fusion in remote sensing systems. In: Amicis, R.D., Stojanovic, R., Conti, G. (eds.) GeoSpatial Visual Analytics. NAPSC, pp. 155–165. Springer, Dordrecht (2009). https://doi.org/10.1007/978-90-481-2899-0_12
21. Andreev, A.V., Skorinov, D.A.: Algorithms of data fusion in biometric systems and application of neural network technologies in them. Intelligence (4), 253–263 (2006)
22. Wu, S., Crestani, F., Bi, Y.: Evaluating score normalization methods in data fusion. In: Ng, H.T., Leong, M.-K., Kan, M.-Y., Ji, D. (eds.) AIRS 2006. LNCS, vol. 4182, pp. 642–648. Springer, Heidelberg (2006). https://doi.org/10.1007/11880592_57
23. Abdulhafiz, W.A., Khamis, A.: Handling data uncertainty and inconsistency using multisensor data fusion. Adv. Artif. Intell. (1), 1–11 (2013)
24. Pitsikalis, V., Katsamanis, A., Papandreou, G., Maragos, P.: Adaptive multimodal fusion by uncertainty compensation. In: Ninth International Conference on Spoken Language Processing, Pittsburgh (2006)
25. Fei, X., Hao, Z., Longhu, X., Daogang, P.: D-S evidence theory in the application of turbine fault diagnosis. In: International Conference on Machine Learning and Computing, vol. 13, pp. 178–182 (2011)
26. Khoshelhama, K., Nardinocchia, C., Nedkova, S.: A comparison of Bayesian and evidence-based fusion methods for automated building detection in aerial data. In: The International Archives of the Photogrammetry, Remote Sensing and Spatial Information Sciences, vol. 7, p. 1184 (2008)
27. Gautam, K.K., Bhuria, V.: Fuzzy logic application in power system fault diagnosis. Ind. J. Comput. Sci. Eng. **2**(4), 554–558 (2011)
28. Fan, D., Wang, S.: Application of fuzzy data fusion in storage environment security monitoring. Am. J. Appl. Math. **4**(5), 197–203 (2016)
29. Takagi, T., Sugeno, M.: Fuzzy identification of systems and its application to modeling and control. IEEE Trans. Syst. Man Cybern. (1985)

Neuro-Fuzzy System Based on Fuzzy Truth Value

Vasily Grigorievich Sinuk[(⊠)]
and Sergey Vladimirovich Kulabukhov[Ⓘ]

Belgorod State Technological University named after V.G. Shukhov, 46
Kostyukova Street, 308012 Belgorod, Russian Federation
vgsinuk@mail.ru

Abstract. The paper introduces neuro-fuzzy systems with inference based on fuzzy truth value and learning by means of evolution strategy (μ, λ). Paper also represents results of benchmarking, which is approximation of a functional dependency using described neuro-fuzzy system and its quality assessment.

Keywords: Fuzzy systems · Neuro-fuzzy systems · Evolution strategies

1 Introduction

Theoretical basis of fuzzy inference was introduced by Zadeh in [1]. After that other solutions of this problem were proposed. Approaches introduced by Mamdani [2], Larsen [3], Takagi and Sugeno [4], Tsukamoto [5] are most popular in practice applications. These solutions are generally used due to simplicity and efficiency of their implementation. However, these methods do not follow at all Zadeh's theory due to significant simplifications they make. The reason of these simplification can be noticed for systems having multiple fuzzy inputs.

In modern fuzzy modeling packages [6] the inference can only be performed for crisp input values. Nevertheless, in many applied problems input data contain either non-numeric (linguistic) assessments [7, 8], or the received input signals contain noise [9, 10]. In both cases the input data are formalized by membership functions, i.e. represent fuzzy inputs. Inference with fuzzy inputs and polynomial computational complexity was considered in [9, 10]. However, such solution is only possible for Mamdani-type fuzzy systems, when minimum or production operation is used as t-norm.

At the same point, the computational complexity of inference according to [1] for individual rule in case of fuzzy input values exponentially depends on the number of premises in the antecedent of the rule. For example, solving problems of gene classification by means of data mining produces inference rules with hundreds of premises [11]. This fact makes Zadeh's inference method inapplicable in its original form for such cases, because the inference cannot be performed in a reasonable amount of time.

An important advantage of the fuzzy inference approach to that proposed in this article is the inference within a single space of truthfulness for all premises. This is achieved by means of transforming relation between fact and premise into a so-called

S. O. Kuznetsov et al. (Eds.): RCAI 2018, CCIS 934, pp. 91–101, 2018.
https://doi.org/10.1007/978-3-030-00617-4_9

fuzzy truth value. Bringing all the relationship between different facts and premises into a single fuzzy space of truthfulness simplifies the computation of compound truthfulness function, reducing its complexity to linear dependency on the number of inputs. Therefore, this approach is devoid of the problems of multidimensional analysis and is more applicable to solving data mining problems.

2   Fuzzy Inference Method Based on Fuzzy Truth Value

2.1   Statement of the Problem of Fuzzy Inference

The problem which is solved using fuzzy inference system is formulated as follows. Consider a system having n inputs $x = [x_1, ..., x_n]$ and a single output y. The relation of inputs and output is defined using N fuzzy rules represented as

$$R_k : \text{If } x_1 \text{ is } A_{1k} \text{ and } \dots \text{ and } x_n \text{ is } A_{nk} \text{ then } y \text{ is } B_k, \quad k = \overline{1,N}, \tag{1}$$

where $x \in X = X_1 \times X_2 \times \dots \times X_n$, $y \in Y$ and $A_k = A_{1k} \times A_{2k} \times \dots \times A_{nk} \subseteq X$, $B_k \subseteq Y$ are fuzzy sets.

The feature of implicative systems, according to classification that was proposed in [1] is that rules (1) are formalized using fuzzy $(n + 1)$-ary relation $R_k \subseteq X_1 \times \dots \times X_n \times Y$ in following way:

$$R_k = A_{1k} \times \dots \times A_{nk} \times Y \rightarrow X_1 \times \dots \times X_n \times B_k, \quad k = \overline{1,N}, \tag{2}$$

where «$\rightarrow$» denotes fuzzy implication or causal relationship between antecedent «x_1 is A_{1k} and $\dots$ and x_n is A_{nk}» and consequent «y is B_k». The problem is to define fuzzy inference result $B'_k \subseteq Y$ for a system represented in the form (1) with fuzzy sets $A' = A'_1 \times \dots \times A'_n \subseteq X$ as input values. The computational complexity of (2) is $O(|X|^n|Y|)$.

2.2   Inference Method Based on Fuzzy Truth Value

Using truth modification rule [12]:

$$\mu_{A'}(x) = \tau_{A/A'}(\mu_A(x)),$$

where $\tau_{A/A'}(\cdot)$ denotes fuzzy truth value of fuzzy set A with respect to A', meaning the compatibility $CP(A, A')$ of term A with respect to A' [13, 14]:

$$\tau_{A/A'}(t) = \mu_{CP(A,A')}(t) = \sup_{\substack{\mu_A(x) = t \\ x \in X}} \{\mu_{A'}(x)\}, \quad t \in [0,1]. \tag{3}$$

Defining new variable t as $t = \mu_A(x)$ one obtains:

$$\mu_{A'}(x) = \tau_{A/A'}(\mu_A(x)) = \tau_{A/A'}(t).$$

Hence fuzzy modus ponens rule for single-input systems takes the form of

$$\mu_{B'_k}(y) = \sup_{t \in [0,1]} \left\{ \tau_{A/A'}(t) \overset{T}{*} I(t, \mu_{B_k}(y)) \right\}, \quad k = \overline{1, N}. \tag{4}$$

In order to generalize (4) for systems with multiple independent inputs, in [11, 15] the following was proven: fuzzy truth value of the rule (1) antecedent A_k with respect to A' is defined as

$$\tau_{A_k/A'}(t) = \underset{i=\overline{1,n}}{\mathbf{T}}\, \tau_{A_{ki}/A'_i}(t), \quad t \in [0, 1], \tag{5}$$

where $\mathbf{T}$ is an n-ary t-norm extended by means of the generalization principle.

Expression (5) has computational complexity of $O(n|t|^2)$. Using (5), the inference of the output value B' based on fuzzy truth value can be represented as

$$\mu_{B'_k}(y) = \sup_{t \in [0,1]} \left\{ \tau_{A_k/A'}(t) \overset{T}{*} I(t, \mu_{B_k}(y)) \right\}, \quad k = \overline{1, N}. \tag{6}$$

Expression (6) shows that the use of fuzzy truth value enables us to avoid exponential computational complexity when using logic-type inference method based on (2), which is an algorithm of complete enumeration. Besides, the proof of (5) in [11, 15] does not imply any simplifications, except for input independence.

2.3 Output Value Inference for a Rule Base

Consider output value inference for N rules (1), using center of gravity defuzzification method [9, 10]

$$\bar{y} = \sum_{k=\overline{1,N}} \bar{y}_k \cdot \mu_{B'}(\bar{y}_k) \Big/ \sum_{k=\overline{1,N}} \mu_{B'}(\bar{y}_k), \tag{7}$$

where $\bar{y}$ denotes crisp output value of a system containing N rules (1); $\bar{y}_k$, $k = \overline{1, N}$ stand for centers (mode) of membership functions $\mu_{Bk}(y)$, i.e. in which the value is

$$\mu_{B_k}(\bar{y}_k) = \max_y \{ \mu_{B_k}(y) \} = 1.$$

With logic-type approach, the fuzzy set B' is derived using intersection operator [9] as follows:

$$B' = \bigcap_{j=\overline{1,N}} B'_j.$$

Hence membership function of B' is derived using t-norm:

$$\mu_{B'}(y) = \underset{j=\overline{1,N}}{\mathrm{T}}\{\mu_{B'_j}(y)\}. \tag{8}$$

Using (4), (6) we get

$$\bar{y} = \frac{\sum\limits_{k=\overline{1,N}} \bar{y}_k \cdot \underset{j=\overline{1,N}}{\mathrm{T}} \left\{ \sup\limits_{t\in[0,1]} \left[\tau_{A_j/A'}(t) \overset{\mathrm{T}}{*} I\left(t, \mu_{B_j}(\bar{y}_k)\right) \right] \right\}}{\sum\limits_{k=\overline{1,N}} \underset{j=\overline{1,N}}{\mathrm{T}} \left\{ \sup\limits_{t\in[0,1]} \left[\tau_{A_j/A'}(t) \overset{\mathrm{T}}{*} I\left(t, \mu_{B_j}(\bar{y}_k)\right) \right] \right\}}.$$

3 Neuro-Fuzzy System Training Using Evolution Strategy (μ, λ)

3.1 Statement of Neuro-Fuzzy System Learning Problem

Some fuzzy systems applications may include data represented in form of a training set, i.e., a set of M pairs $\left\{ <\mu_{A'}^{(t)}(x), d^{(t)}> \right\}_{t=\overline{1,M}}$, each containing a desired output value $d^{(t)}$ when input values are $\mu_{A'}^{(t)}(x)^{(t)} = <\mu_{A'_1}^{(t)}(x_1), \mu_{A'_2}^{(t)}(x_2), \ldots, \mu_{A'_n}^{(t)}(x_n)$, along with a fuzzy rule base built from expert knowledge. $\mu_{A'_j}^{(t)}(x_j)$ denotes membership function of a fuzzy set (fact) at j-th input. A way to combining these two types of knowledge is adjustment of parameters present in rule base. This procedure can be formulated as finding minimum of an objective function that is a measure of deviation of system's actual output values for specific values of parameters from desired ones for all input vectors in the training set. Arguments of this function are all parameters that undergo learning (adjustment). Below such systems will be referred to as neuro-fuzzy systems.

In the article the use of the following objective function is assumed:

$$R = \frac{1}{l} \sqrt{\frac{1}{M} \sum_{t=1}^{M} (\bar{y}^{(t)} - d^{(t)})^2}, \tag{9}$$

where $\bar{y}^{(t)}$ is system's inference result for given parameter values and input values $<\mu_{A'_1}^{(t)}(x_1), \mu_{A'_2}^{(t)}(x_2), \ldots \mu_{A'_n}^{(t)}(x_n)>$; $l = \sup(Y) - \inf(Y)$ denotes the width of output variable y value range Y. This function represents mean square error of learning for the entire training set, which is often used in machine learning and statistical analysis, divided by the length of output variable's domain of definition. The division does not affect learning process but makes it easier to compare learning of different neuro-fuzzy systems, since the value no more depends on abovementioned width.

Let us consider possible parameters of neuro-fuzzy system that can be involved into learning procedure. First, (7) implicitly assumes the weights of all rules equal. As an example, they can be taken into account by multiplying the expression under summation sign in both numerator and denominator by the weight w_k of the rule. Second, parameters of membership functions can also be involved into learning. This requires to determine the type of membership functions and to include them (all or some) into the set of learned parameters. Let us consider the case when all terms are defined by Gaussian membership functions with mean value m and standard deviation σ. For clarity, these parameters will be present as arguments of membership functions. The subscript of parameters will denote the term, to membership function of which this parameter belongs. Hence, substituting expression (7) into (8) with abovementioned parameters:

$$R(\vec{w}, \vec{m}, \vec{\sigma}) = \frac{1}{l} \sqrt{\frac{1}{M} \sum_{t=1}^{M} \left(\frac{\sum\limits_{k=\overline{1,N}} w_k m_{B_k} \mu_{B'}(\bar{y}_k, \vec{m}, \vec{\sigma})}{\sum\limits_{k=\overline{1,N}} w_k \mu_{B'}(\bar{y}_k, \vec{m}, \vec{\sigma})} - d^{(t)} \right)^2}, \qquad (10)$$

where $\mu_{B'}(\bar{y}_k, \vec{m}, \vec{\sigma})$, according to (8), (3), (5) and (6), equals to

$$\mathop{\text{T}}_{j=\overline{1,N}} \left\{ \sup_{t \in [0,1]} \left[\left(\mathop{\text{T}}_{i=\overline{1,n}} \sup_{\substack{\mu(x, m_{A_{ji}}, \sigma_{A_{ji}}) = t \\ x \in X}} \left\{ \mu_{A_i'}(x) \right\} \right)^{\text{T}} * I\left(t, \mu\left(m_{B_k}, m_{B_j}, \sigma_{B_j}\right)\right) \right] \right\}, \qquad (11)$$

where $\mu(x, m, \sigma) = \exp(-(x - m)^2/\sigma^2)$ is value of Gaussian membership function with mean value m and standard deviation σ in x, $\vec{w} = (w_1, w_2, \ldots, w_N)$ are weights of the rules, $\vec{m} = (m_{A11}, \ldots, m_{A1n}, \ldots, m_{AN1}, \ldots, m_{ANn}, m_{B1}, \ldots, m_{BN})$ are mean values of membership functions, and $\vec{\sigma} = (\sigma_{A11}, \ldots, \sigma_{A1n}, \ldots, \sigma_{AN1}, \ldots, \sigma_{ANn}, \sigma_{B1}, \ldots, \sigma_{BN})$ are standard deviations of membership functions.

As it was mentioned above, the problem of learning is to find the conditional minimum of R $(\vec{w}, \vec{m}, \vec{\sigma})$ function within the set of valid values. This article considers learning by means of evolution strategy (μ, λ) [9]. The choice in favor of this method is caused by the fact that in general case the shape of R $(\vec{w}, \vec{m}, \vec{\sigma})$ is not known, along with its derivatives. Computation of results of individual stages of proposed inference method cannot be represented analytically, especially if the input values are fuzzy. Hence the analysis of R $(\vec{w}, \vec{m}, \vec{\sigma})$ in general case is difficult. Evolutionary algorithms do not require any information about the objective function other than possibility to compute its value in arbitrary point. Within this family of algorithms, the evolution strategies were chosen, since they operate real numbers and not their binary encoding. Strategy (μ, λ) has the preference due to its resistance to convergence in local minima, which is the best within the family of evolution strategies. The structure of an individual in this problem has the form of real number vector $\vec{u} = \langle \vec{w}, \vec{m}, \vec{\sigma} \rangle$ that contains all parameters.

3.2 Constraints of Learning

Evolutionary algorithms belong to random methods of optimum search. They require a possibility to compute value of the objective function in any arbitrary point $\vec{u}$. However, objective function may have its domain of definition D, a set of valid parameter values combinations. For example, a rule weight is often limited to [0; 1]. This problem had been considered in [9], and the solution introduced was to remove $\vec{u}$ from population if $\vec{u} \notin D$. However, the more parameters undergo adjustment, the smaller is the probability that a random solution $\vec{u}$ is valid. This can result in rejection of entire population. Another approach is considered in this article.

Common constraints have form of p_k R a (e.g. rule weight must be non-negative) or p_i R p_j (e.g. relations between coordinates of triangular membership function vertices), where R denotes relation operator; p_i, p_j and p_k are adjusted parameters and a is a constant real number. Such constraints are linear with respect to all parameters, each of them represents a half-space of valid parameter values. Intersection of all half-spaces is D (by definition) and is convex. Such constraint systems are considered in linear programming. This means that a linear combination of two valid solutions (a result of crossover in evolution strategy) is also valid.

Mutation operator «moves» a solution $\vec{u}$ in a random direction $\delta\vec{u}$ of search space. It is probable that $\vec{u} + \delta\vec{u} \notin D$, while $\vec{u} \in D$. There must be $\alpha \in [0; 1]$, for which $\vec{u} + \alpha\delta\vec{u} \in D$. The value of α can be found analytically by replacing $\vec{u}$ with $\vec{u} + \alpha\delta\vec{u}$ in constraint system. Another method is to iteratively check if $\vec{u} + \alpha\delta\vec{u} \in D$, starting with some initial value of α, and if it does not, divide α by some number (e.g. 2) and check the condition again. Procedure of checking if $\vec{u} \in D$ usually has much lower computational complexity than computation of the objective function value, therefore this method cannot significantly increase overall computational complexity. The simplest alternative way is to repeat mutation until the condition $\vec{u} + \delta\vec{u} \in D$ is satisfied.

Generation of initial population must also take constraint system into account. Initial solutions may be derived from constraint system, such solutions may belong to boundary of D. If a valid solution $\vec{u}_0 \in D$ is known, then the remaining part of initial population may be generated by means of mutating $\vec{u}_0$.

Adjusted rule base parameters may have unexpected values, which lead to loss of original semantics of the rule base. These situations may be caused by bad training sets, generalization effect [9] or other reasons. Some of these situations can be prevented by extending constraint system with additional constraints. For example, a linear order of terms (centers of their membership functions) can be altered after adjustment. The initial order can be sustained by adding constraints to coordinates of membership functions centers.

4   Benchmarking

The proposed approach has been tested according to [7], i.e. approximation of a known function by a fuzzy model with its subsequent learning on a training set obtained from this function. For this purpose the function $f(x_1, x_2) = (x_1 - 7)^2 \cdot \sin(0.61x_2 - 5.4)$ had been picked, which was considered within the domain $[0;10] \times [0;10]$. The function is shown in Fig. 1.

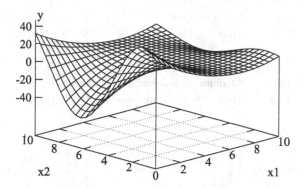

Fig. 1. Plot of the approximated function

Fuzzy rule base depends on the shape of membership function and passes learning procedure, where both crisp and fuzzy training sets are considered. The crisp training set is generated by random selection of a point $\left(x_1^{(i)}, x_2^{(i)}\right)$ in the considered domain of definition and computation of the expected output value $d^{(i)}$ in this point as $f\left(x_1^{(i)}, x_2^{(i)}\right)$. Testing set is generated in the same way. Fuzzy training set is generated by the replacement of one of coordinates of each point by the term of the corresponding linguistic variable that has the greatest membership function value in this point among all terms.

Neuro-fuzzy system that was used in the benchmarking is defined by expressions (10) and (11). It takes minimum operation t-norm and Aliev's implication, defined as:

$$I\left(t, \mu_{B_k}(y)\right) = \begin{cases} 1, & \text{if } t \le \mu_{B_k}(y) \\ \mu_{B_k}(y)/\left(t+1-\mu_{B_k}(y)\right), & \text{if } t > \mu_{B_k}(y) \end{cases}$$

Inputs x_1 and x_2 are assigned linguistic variables described by linguistic terms "Low", "Medium" and "High". The output y is also assigned a linguistic variable, containing "Below medium" and "Above medium" terms as well. Table 1 describes fuzzy rule base, including initial rule weights (w_0), rule weights after learning on crisp (w_c) and fuzzy (w_f) training set.

All membership functions are Gaussian. Their graphs before and after learning are shown in Fig. 2.

The learning was performed by means of evolution strategy (μ, λ). For this particular case, $\mu = 40$ and $\lambda = 160$. Initial weights were assumed equal to 1. The means of membership functions were evenly distributed within value ranges, and standard deviations were assumed equal to 3/20 of corresponding range length.

Table 1. Rules and their weights

x_1	x_2	y	w_0	w_c	w_f
Low	Low	High	1	1	1
Low	Medium	Low	1	1	0.829
Low	High	High	1	0.143	0.257
Medium	–	Medium	1	1	0.727
High	Low	Above medium	1	0.403	0.521
High	Medium	Below medium	1	0.685	0.414
High	High	Above medium	1	0.078	0.395

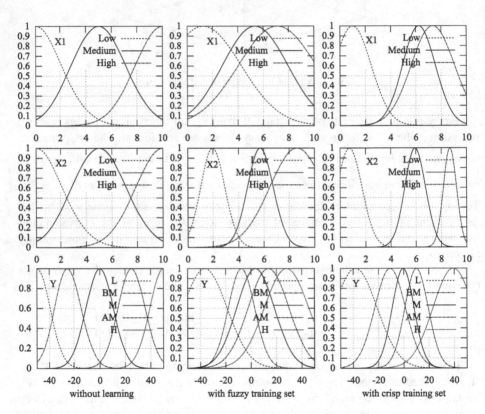

Fig. 2. Membership functions before and after learning

The above results were obtained for training set of size 80. Testing set contained 1000 points. Dependence of minimal achieved value of function (9) on training set size is shown in Fig. 3.

Learning was performed throughout 300 generations. In Fig. 4 learning curves are represented. Objective function value of best solution and average value for the entire population are drawn in linear scale of left axis, average value of standard deviation of genes drawn in logarithmic scale of right axis. Figure 5 illustrates the dispersion of

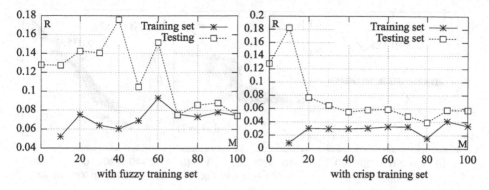

Fig. 3. Dependence of minimal error on the size of training set

expected and achieved results for input values from testing set. Every point corresponds to an element of testing set; its abscissa denotes the expected result, derived from the approximated function, and the ordinate corresponds to the output of neuro-fuzzy system for same input values.

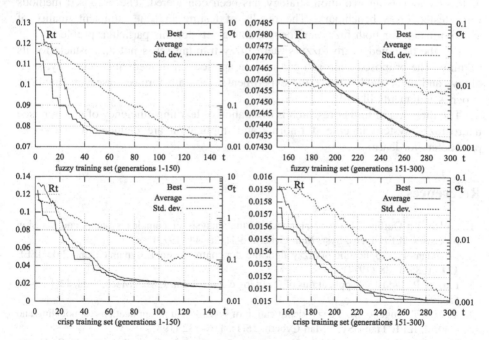

Fig. 4. Learning curves

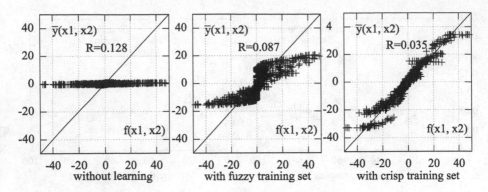

Fig. 5. Dispersion of the expected and achieved results for testing set

5 Conclusion

The fuzzy inference method proposed in this article can be implemented with polynomial computational complexity for implicative systems with fuzzy inputs. Learning of this system using evolution strategy has been considered. The proposed methods were tested on a benchmark. The results of testing indicate sufficient quality of approximation for both fuzzy and crisp training sets for this particular problem.

For the designed neuro-fuzzy system fuzzy training set is not an obstacle for its training. The proposed approach can find an application in medicine, economy, sociology, and other fields, where the experimental data are formed based on linguistic expert's judgments like «If …, then …».

The direction of further research is evaluation of learning efficiency of the proposed neuro-fuzzy system in case if t-norms and implication are included in the set of parameters for learning.

References

1. Zadeh, L.: Outline of a new approach to the analysis of complex systems and decision processes. IEEE Trans. Syst. Man Cybern. **3**(1), 28–44 (1973)
2. Mamdani, E.: Application of fuzzy algorithm for control a simple dynamic plant. Proc. IEEE **121**, 1585–1588 (1974)
3. Larsen, P.: Industrial applications of fuzzy logic control. Int. J. Man Mach. Stud. **12**(1), 3–10 (1980)
4. Takagi, T., Sugeno, M.: Fuzzy identification of systems and its applications to modeling and control. IEEE Trans. Syst. Man Cybern. **15**(1), 116–132 (1985)
5. Tsukamoto, Y.: An approach to fuzzy reasoning method. In: Gupta, M., Ragade, R., Yager, R. (eds.) Advances in Fuzzy Sets Theory and Applications. North-Holland, Amsterdam (1979)
6. Leonenkov, A.: Fuzzy modeling in MATLAB and FuzzyTech environment. BHV – Petersburg, Saint Petersburg (2003). (in Russian)

7. Rothstein, A., Shtovba, S.: Identification of nonlinear dependence by fuzzy training set. Cybern. Syst. Anal. **2**, 17–24 (2006). (in Russian)
8. Borisov, V., Kruglov, V., Fedulov, A.: Fuzzy models and networks. Hot Line – Telecom, Moscow (2007). (in Russian)
9. Rutkowsky, L.: Methods and techniques of computational intelligence. Hot Line – Telecom, Moscow (2010). (in Russian)
10. Rutkowska, D., Pilinsky, M., Rutkowsky, L.: Neural networks, genetic algorithms and fuzzy systems. Hot Line – Telecom, Moscow (2004). (in Russian)
11. Sinuk, V.G., Polyakov, V.M., Kutsenko, D.A.: New fuzzy truth value based inference methods for non-singleton MISO rule-based systems. In: Abraham, A., Kovalev, S., Tarassov, V., Snášel, V. (eds.) Proceedings of the First International Scientific Conference "Intelligent Information Technologies for Industry" (IITI 2016). AISC, vol. 450, pp. 395–405. Springer, Cham (2016). https://doi.org/10.1007/978-3-319-33609-1_36
12. Borisov, A., Alekseev, A., Krumberg, O., et al.: Decision making models based on linguistic variable. Zinatne, Riga (1982). (in Russian)
13. Zadeh, L.: PRUF—a meaning representation language for natural language. Int. J. Man Mach. Stud. **10**, 395–460 (1978)
14. Dobuis, D., Prade, H.: Possibility theory. Applications to the representation of knowledge in Informatics. Radio and Communication, Moscow (1990). (in Russian)
15. Kutsenko, D., Sinuk, V.: Inference method for systems with multiple fuzzy inputs. Bull. Russ. Acad. Sci. Control Theory Syst. **3**, 48–56 (2015). (in Russian), https://doi.org/10.7868/s0002338815030129

Architecture of a Qualitative Cognitive Agent

A. Kulinich[1,2](✉) [iD]

[1] V. A. Trapeznikov Institute of Control Sciences, RAS,
Moscow, Russian Federation
kulinich@ipu.ru
[2] National Research Centre "Kurchatov Institute", Moscow, Russian Federation

Abstract. An architecture of a qualitative cognitive agent is proposed, which includes: a knowledge representation model, a behavioral planning, and an agent cooperation subsystem. The agent knowledge model is represented in the form of a qualitative conceptual framework structuring the agents functioning environment on the subspaces defining the classes of possible states. In the agent behavior planning subsystem, the agent's choice of action is based on solving the inverse problem in the logical-linguistic equations of agent dynamics. The subsystem of agent cooperation is based on the criterion of mutual usefulness of agents. The proposed architecture of a qualitative cognitive agent allows implementing and exploring various principles of command behavior of set agent with a BDI-architecture with less labor, while maintaining the basic properties of intelligent agents.

Keywords: Cognitive architecture · BDI architecture · Conceptual framework
Behavior planning · Cooperative agents

1 Introduction

Trends in the development of group robotics now involve the creation of many simple robots with reactive architecture. The behavior of a large number of reactive agents is organized on the basis of their interaction, built, for example, on biological or social principles [1]. It is believed that a flock or swarm of agents due to self-organization will have emergent properties that allow such a team of agents to solve complex tasks.

As a mathematical model of the behavior of a flock or a swarm of robots, the behavior pattern of a flock of birds proposed by Reynolds [2] is used. The tasks that can be solved by swarms of robot - specific ones - are movement in the order, protection of the territory, monitoring of the state of objects, etc., do not presuppose the actions of robots beyond the actions created by the developer.

To expand the capabilities of robots groups it is possible to use intelligent agents with cognitive capabilities that are as close as possible to human cognitive capabilities. The cognitive architecture of such agents is built on the basis of psychologists' research of the human intellect, its cognitive functions: perception; representations in the memory of the surrounding world; beliefs; goal-setting; preferences, etc.

The architecture that allows implementing set of properties of intelligent agents is a mental BDI (Belief-Desire-Intention) agent architecture, where three components are

S. O. Kuznetsov et al. (Eds.): RCAI 2018, CCIS 934, pp. 102–111, 2018.
https://doi.org/10.1007/978-3-030-00617-4_10

distinguished: Beliefs, which characterizes knowledge of the subject domain; Desires, reflecting agents' goals and Intentions are possible actions of agents to meet their goals [3].

Despite that the creation of the basic elements of intelligent robots, i.e., subsystems of knowledge representation, language of communication, and knowledge sharing, etc., have been brought to the level of standards by the efforts of the FIPA (Foundation for Intelligent Physical Agents) Association, the complexity of creating teams of autonomous agents based on various principles of command behavior remains high.

In this work, the architecture of a qualitative cognitive agent is proposed that allows implementing and exploring various principles of command behavior of many agents with a BDI-architecture with less labor, preserving the basic properties of intelligent agents.

Simplification of the cognitive architecture is achieved through the use as a model of knowledge representation of the agent about the functioning environment, the qualitative ontology in the form of the so-called conceptual framework of the functioning environment.

2   Architecture of a Qualitative Cognitive Agent

The main elements of the architecture of intelligent agents are defined in the architecture of InteRRap, proposed in [4]. This is a hierarchical knowledge base and agent control component.

The hierarchical knowledge base includes three levels: the lower level is a model of the robot's functioning environment; the middle level is knowledge of the rules for planning the individual activities of robots; the upper level of the hierarchy of knowledge is knowledge about cooperative actions of robots, allowing to coordinate local plans of robots, and to develop a plan for solving a common problem.

The component of robot control is a multilevel control system, which functioning is determined by the content of knowledge presented at different levels of the hierarchical knowledge base. The lower control level, which realizes the elementary actions of the robot for various states of the functioning environment is reactive. The middle level is responsible for planning independent actions of the robot to achieve the goal and determines the sequence of elementary actions implemented at the lower level. The upper level is responsible for organizing the cooperative behavior of robots sets, whose resources are joined to achieve a common goal. The organization of cooperative behavior is plans, scenarios, individual actions implemented at the middle and lower levels of the robot's control system.

Further, when constructing the architecture of a qualitative cognitive agent, we will take into account the InteRRap architecture. Consider the organization of the main components of a quality agent: a knowledge representation model, behavior planning, cooperative agent interaction, and the control system for implementing the agent's individual and cooperative behavior.

3 Knowledge Representation Subsystem of Qualitative Cognitive Agent

In the knowledge representation subsystem of a qualitative cognitive agent, we will determine the quantitative and qualitative models of the dynamic system "robot group-environment".

3.1 Model of the Dynamic System "Robots Group - Environment"

In [5], a "robots group - environment" dynamic system was defined. Such a system is a set of agents R_i with a set of properties $F = \{f_i\}$ and a set of their possible values $Z = \{Z_i\}$ defined for each agent. The agent operation environment is defined by the direct product of the sets of values of all agent properties, $SF = \underset{ij}{\times} Z_{ij}$, and the state of the agents' functioning environment at time t is the vector of the values of the properties of all agents: $Z(t) = (z_{1e}, ...z_{1q} ..., z_{ne}, ...z_{nq})$. It is believed that the i-th agent has resources Z_{ij}^R, with which he can change his own j-th properties and properties of other agents.

Each agent selects $U_{iq} \in \underset{j}{\times} Z_{ij}^R$ strategy to achieve its goal $G_i = (z_{i1}, ...,z_{jh})$ or general purpose $G = (z_{1b} ...,z_{1h}; ...; z_{nb} ...,z_{nr})$, $z_{ij} \in Z_i$. Dynamics of the system "robots group-environment" is determined by the mapping

$$W : (Z(0) \underset{i}{\oplus} U_{ij}) \rightarrow G, \tag{1}$$

where $\underset{i}{\oplus}$ is the agent strategy aggregation rule.

According to mapping (1), agents select and coordinate their strategies to achieve a common goal G. As stated in [5], it is impossible to solve this problem in such a general formulation.

3.2 A Qualitative Model of the Dynamic System "Robots Group - Environment"

One of the possible ways to solve this problem is considered in this paper. It is based on the interpretation of the state space (SF) as a semantic space in which the states Z (t) (agent property values vectors) of the functioning environment are determined by the names and vectors of the characteristic values that determine their content (sense). That is, the states of the system "robots group - environment" are represented in symbolic form as signs-symbols. Recall that in the definition of G. Frege [6], the sign is called the triple: the name, sense and denotation of the sign. A name is a symbol (image, icon) denoting an object of the real world, sense defines the properties of this object, and the denotation is this object itself. In semantic spaces a real object (de-notation) has a name (symbol) and is represented in the characteristic space as a point, the coordinates of which determine the meaning of the signs.

In this paper, in the space of states of the dynamic system "robots group-environment", in essence, in the semantic space, nested subspaces determining the classes of possible states of the "robots group-environment" system [7, 8] are defined. These subspaces are structured as a conceptual framework in which all classes of possible states have the names d^H and form a partially ordered set of subspaces $SS(d^H)$ of the space of states of the dynamic system "robots group - environment" FS, i.e. SS $(d^H) \subseteq FS$. A formally qualitative conceptual framework is a partially ordered set of names of classes of states:

$$KK^W = (\{d^H\}, \wedge, \vee),$$

where $\{d^H\}$ is the set of names of the classes of states d^H that uniquely determine the subspaces of the functioning environment $d^H \Leftrightarrow SS(d^H)$, $SS(d^H) \subseteq FS$, and the volume of the class of states $V(d^H) = R_i | R_i = (z_{1e}, ... z_{1q}) \in SS(d^H)$.

Thus, in the state space of a dynamic system, we have defined a semiotic operating environment in which all the classes of its possible states are represented by signs-symbol. It should be noted that signs and sign systems were studied by well-known mathematicians, logicians and philosophers [9–11], etc. The science of signs and sign systems, called semiotics, has application in humanities in the field of control, decision-making and system design. In the field of control systems, the semiotic approach developed within the context of Pospelov's situational approach [12]. In the works [13] in terms of applied semiotics based on the models of the sign (the triangle Frege and the square of Pospelov), approaches to decision-making with the use of linguistic information were developed. In these works, the semiotic system (sign system) is defined as a formal system adapted to changing conditions [13].

In the field of semiotic decision support systems, models of the elements of the semiotic system architecture are developed. Many works are devoted to the study of the adaptive capabilities of applied semiotic systems. In [14], the architecture of a control system based on the principles of the semiotic approach is proposed. In [15], a fuzzy semiotic model consisting of fuzzy regulators is proposed and allows adjustment of T-norms and models of a particular user. In [16], the principles of constructing a semiotic decision support system are considered, in which the adaptation for solving the problem is carried out on the basis of taking into account the experience of decision-making in the non-Markov model with training. In [17], the author's main idea is to combine theoretical semiotics and evolutionary modeling to modeling the formation and evolution of signs (semiosis) in multi-agent systems.

Semiotic approach is used in the design of information and control systems. In [18], a computer semiotics is proposed in which the formalized person knowledge is represented in the form of a symbolic description of the computer system and its elements. In work [19] the information system is considered in three aspects: syntactic, semantic and pragmatic. Different syntactic, semantic and pragmatic structures allow us to classify different types of information systems. In [20] an approach is developed, called algebraic semiotics. The author introduces the notion of semiotic morphism similar to the morphism in category theory, to explore the possibility of representing interfaces of systems in different but equivalent sign systems. All these approaches and models of semiotic systems are difficult to adapt to the field of modeling the behavior of intelligent agents or robots. Therefore, mathematical models of knowledge representation in the form of a signs system are currently being developed.

Mathematical models of the sign picture of the world are presented in [21–23]. In these works, theoretical models of the sign picture of the world are created, in which its structural elements, relations between them are distinguished, models for the formation and self-organization of the sign (semiotic) picture of the world are constructed. Such models can form a theoretical basis of semiotic models of knowledge representation of intelligent agents and robots. It should be noted that an important feature of the sign representation of world pictures in the agent or robot conceptual system is the greater flexibility of representations, the multiplicity of possible interpretations of situations represented in a sign form, and as a result, the complexity of creating real agents (robots) on the basis of such models and determining their unambiguous behavior.

The main feature of the semiotic model of representation of agents' knowledge in this work is that their knowledge is represented in the form of a rough qualitative picture of the world. These qualitative picture results from the qualitative structuring of the space of states of the dynamic system "robot group-environment" into classes of states denoted by artificial names.

A qualitative representation of the environment of functioning in the model of an intelligent agent with BDI architecture allows as focusing not on building an agent knowledge model (ontology) that requires much effort, but on studying the mechanisms of formation and functioning of teams of intelligent agents with a simple qualitative model of knowledge. In this case, the BDI-architecture of the agent itself is preserved, but it is by definition qualitative.

3.3   BDI-Architecture of Qualitative Cognitive Agent

In [8], in terms of the conceptual framework (KK^W), elements of the BDI agent architecture are defined.

The knowledge (belief) of an agent's BDI is determined by a tuple:

$$\langle BEL_i, W_i^{BEL} \rangle,$$

where BEL_i are beliefs of the agents represented by a partially ordered set of environment class names of the FS functioning environment, i.e.

$$BEL_i = (\{d^H\}, \leq)\}, d^H \Leftrightarrow SS_i(d^H) \subseteq KK^W, BEL_i \subseteq KK^W;$$

W_i^{BEL} is knowledge of the agent about the laws of the functioning environment represented by a mapping

$$W_i^{BEL} : \underset{j}{\times} Z_{ji} \rightarrow \underset{j}{\times} Z_{ji},$$

where $\underset{i}{\times} Z_i$ are vectors of values of agent properties, $\underset{i}{\times} Z_{ji} \in SS_{ji}(d^H), d^H \in BEL_j; W_i^{BEL}$ is a set of production rules (If, Then), reflecting the laws of the functioning environment.

Desire goals are the vector $G = (z_{1b} ..., z_{1h}; ...; z_{nk} ..., z_{nr})$, $z_{ij} \in Z_i$, whose elements determine the desired values of the properties of each agent in the functioning environment (FS) are represented in terms of classes of the conceptual framework:

$$DES_i = (d_1^{Gi}; d_2^{Gi}; ...; d_n^{Gi}),$$

where $d_j^{Gi} \Leftrightarrow SS_j(d^H) \subseteq KK^W$ are the names of classes of goal states, $G_i \in SS_j(d^H)$ is the agent's goal, $\forall i,j$, i, $j = 1,...,N$.

The actions of the agents in the FS environment are represented by the set of vectors $Ui(t) = (u_{i1}, ..., u_{in}) \in \underset{j}{\times} Z_{ij}^R$, and in terms of the conceptual framework they are represented as a set of names of the classes of states of the functioning environment:

$$INT_i = \{d_j^{Ui}\},$$

where d_j^{Ui} is the name of the action state class that determines its content $SS(d^{Uj})|U_i \in \underset{j}{\times} Z_{ij}^R \in SS(d^{Uj})$.

A feature of the representation of the environment of functioning in the form of a conceptual framework is that the names of the classes of states can be artificial. If agents have agreed on the names of classes of all classes of states, then such an agent functioning environment can already be considered as a qualitative semiotic environment of functioning with artificial names.

3.4 The Relation of Quantitative and Qualitative Models of the Dynamic Systems "Robots Group - Environment"

The conceptual framework KK^W can be represented as a homomorphic mapping of the functioning environment of FS, i.e. $\Psi: FS \rightarrow KK^W$. Thus, any point of the functioning environment $Z(t) \in FS$ is uniquely mapped to a class of states with the name d^H, and the inverse mapping of the class of states d^H is represented by the set of points of the subspace $SS(d^H) \subseteq FS$. The same mappings are also valid for the beliefs $\langle BEL_i, W_i^{BEL} \rangle$, goals DES_i and the actions INT_i of the BDI-architecture of agents.

The main idea of a qualitative representation of an intelligent agent is that the description of the functioning environment using names of state classes allows us to offer simple logical conditions for the formation and operation of the agents' team. If these conditions are met, then for their implementation a mathematical dynamic system model of the type "robots group-environment" is used.

Thus, in the architecture of a qualitative cognitive agent within the qualitative model of the dynamic system "robots group - environment", the tasks of team interaction of agents are solved, and the realization of this interaction is carried out in terms of a quantitative model of the dynamic system. Next, consider the construction of the subsystems of planning and cooperative behavior of qualitative cognitive agents.

4 Quantitative Agent Behavior Planning Subsystem

The choice of the model for planning the agent's behavior depends on his model of knowledge representation. Currently known models of behavior planning such as situational calculus [24]; STRIPS [25] and others are based on logical models of knowledge representation and on methods of mathematical logic.

The general formulation of the task of planning the agent's behavior is as follows: the initial and target states are determined. The agent, performing the possible actions, must get into the target state. The basis of the behavioral planning systems in the logical models of knowledge representation resides in theorem proving, allowing to find the necessary actions for meeting the goal in the state space.

In the dynamic system "robot group - environment" with the model of representation of knowledge in the form of a qualitative conceptual framework, a solution to the problems of planning the agent's behavior can be represented as a solution of the inverse problem for the equations of dynamics of this system.

Thus, the change in the state of the environment by the i-th agent, taking into account his knowledge W_i^{BEL}, can be expressed by a system of logical-linguistic equations [8]:

$$W_i^{BEL} : \ (Z^*(t), INT_i^*)) \rightarrow Z^*(t+1), \ \forall i,$$

where W_i^{BEL} is the knowledge of the i-th agent (the "If, Then" rule system), $Z^*(t)$, is the initial state vector of the environment in terms of the names of the state classes, INT_i^* is the control (agent action), $Z^*(t + 1)$ is the state of the environment in terms of the names of the state classes after the agent performs the actions.

Formally, the search for actions for achieving a given target vector by each agent is reduced to solving an inverse problem [8]:

$$INT_i^* = DES_i \stackrel{\circ}{=} W_j^{BEL},$$

where INT_i^* is the set of agent actions that allow to achieve the goal DES_i, $\stackrel{\circ}{=}$ is the procedure of reverse inference. The actions of INT_i^* are the set of names of the state classes (d_j^{Ui}), to which the agent must go, in order to achieve the goal DES_i.

In the control component in the qualitative agent architecture, the name of the action class d_j^{Ui} is represented as the subspace $SS(d_j^{Ui})$ and for the real action $U_i(t)$ the agent solves the inverse problem in which goal $G = (z_{1b}, ..., z_{1h}; ...; z_{nk}, ..., z_{nr})$, $z_{ij} \in Z_i$, any point of $SS(d_j^{Ui})$ is chosen, i.e., $G \in SS(d_j^{Ui})$. It is believed that the intelligent agent, falling into the subspace $SS(d_j^{Ui})$, independently selects the desired action of $U_i(t)$ that solves the inverse problem.

In the most general form, if we abstract from the agent's current state $(Z(0))$, the control action U can be found as a solution of the inverse problem:

$$W^{-1} : \ G \rightarrow U \tag{2}$$

Despite the roughness of the proposed method of planning behavior, it allows you to significantly reduce the complexity of creating an intelligent agent.

It should be noted that the model of agent behavior planning in solving the command behavior problem becomes an element of the agents' command behavior system based on the agent communication system.

5 Subsystem of Cooperative Interaction of a Quantitative Agent

There are two possible approaches to planning the cooperative behavior of agents to meet a common goal. The first involves building a common plan before the beginning of solving the problem of achieving a common goal. Collaboration-based algorithms based on the construction of a general plan [26] do not work well in dynamic agent environments, when unpredictable changes in its state can lead to a restructuring of the overall plan.

The second approach is situational cooperation between agents, when the decision on joint actions is taken by agents on the basis of an analysis of the state of the functioning environment and the availability of conditions for cooperation among the agents. Selecting an agent action based on solving the inverse problem (2) allows you to choose the best action.

As a condition for agents cooperating, the criterion of mutual utility of agents, proposed in [27], is considered. By this criterion, an agent is chosen as the partner, the interaction with which is most useful to both agents.

In [8], the conditions for cooperation of agents were formulated:

1. Agent j is considered attractive for cooperation for agent i if the element of its goal $d_i^H \notin BEL_i$ exists in the belief system (knowledge) of agent j, i.e., $d_i^H \in BEL_j$.
2. Agent j is considered attractive for cooperation for agent i if in its belief system there are elements $d_i^H \in BEL_i$, which are also elements of agent's belief system j, i.e. $d_i^H \in BEL_j$.
3. Agent j can change the properties of agent i if agent i is included in the volume of one of the possible state classes of agent j, i.e. $a_i \in V(d_j^H)|d_j^H \in BEL_j$.

The first condition determines that the potential partner should have the resources to provide assistance. Second, potential partners should have the intersection of their conceptual systems knowledge (beliefs). Then they will be able to speak one language they understand. The third condition, if the first two conditions are met, then the agent who asked for help should himself create conditions for joint action to be possible.

The command work of the agents in this case consists in exchanging information in terms of the names of states classes about their beliefs and goals, and checking the conditions for mutual usefulness, and, therefore, the possibility of cooperation. If such an opportunity is established, then the cooperation is carried out at the planning level of individual actions (Eq. 2) to create conditions for cooperation.

6 Conclusion

The architecture of a qualitative cognitive agent was proposed that includes: a knowledge model in the form of a conceptual framework of the functioning environment, an agent model with a BDI architecture, subsystems of behavior planning and agent cooperation, and a control system of agent behavior based on methods for solving inverse problems in the dynamic model "robots group - environment". The proposed qualitative architecture was studied on the simulation model of the mechanisms of formation and functioning of qualitative cognitive agents.

Acknowledgement. This reported study was partly funded by Russian Science Foundation (project No. 16-11-00018) and RFBR (project No. 16-29-04412) grants.

References

1. Karpov, V.E.: Modeli sotsial'nogo povedeniya v gruppovoy robototekhnike [Models of social behavior in group robotics]. Upravleniye bol'shimi sistemami [Large Scale Syst. Control] **59**, 165–232 (2016). Vypusk 59. M: IPU RAN, 2016. S. 165–232
2. Reynolds, C.: Flocks, herds, and schools: a distributed behavioral model. Comput. Graph. **21** (4), 25–34 (1987)
3. Rao, A.S., Georgeff, M.P.: BDI agents: from theory to practice. In: Lesser, V. (eds.) Proceedings of First International Conference on Multi-agent Systems, pp. 312–319. AAAI Press/The MIT Press (1995)
4. Müller, J.P., Pischel, M.: The Agent Architecture InteRRaP: Concept and Application, p. 99 (1993). https://books.google.ru/books?id=gTNFOAAACAAJ. Accessed 13 July 2017
5. Kalyayev, I.A., Gayduk, A.R., Kapustyan, S.G.: Modeli i algoritmy kollektivnogo upravleniya v gruppakh robotov [Models and algorithms of collective control in groups of robots]. Fizmatlit, Moscow (2009)
6. Biryukov, B.V.: Teoriya smysla Gotloba Frege [Theory of the meaning of Gottlob Frege]. V kn.: Primeneniye logiki v nauke i tekhnike. [In: Application of Logic in Science and Technology]. Izd-vo AN SSSR, Moscow (1960)
7. Kulinich, A.A.: Model' komandnogo povedeniya agentov v kachestvennoy semioticheskoy srede. Chast' 1. Kachestvennaya sreda funktsionirovaniya. Osnovnyye opredeleniya i postanovka zadachi. [Model of command behavior of agents in a qualitative semiotic environment. Part 1. Qualitative environment of functioning. Basic definitions and statement of the problem.]. Iskusstvennyy Intellekt i Prinyatiye Resheniy [Artif. Intell. Decis. Mak.] (3), 38–48 (2017)
8. Kulinich, A.A.: Model' komandnogo povedeniya agentov v kachestvennoy semioticheskoy srede. Chast' 2. Modeli i algoritmy formirovaniya i funktsionirovaniya komand agentov [Model of command behavior of agents in a qualitative semiotic environment. Part 2. Models and algorithms for the formation and operation of teams of agents]. Iskusstvennyy Intellekt i Prinyatiye Resheniy [Artif. Intell. Decis. Mak.] (3), 29–40 (2018)
9. Frege, G.: On sense and nominatum. Readings in philosophical analysis (1949)
10. Morris, C.W.: Ders. Foundations of the Theory of Signs. Writings on the General Theory of Signs (1971)
11. Peirce, C.S.: The Collected Papers of Charles Sanders Peirce. Search (1931)

12. Pospelov, D.A.: Situatsionnoye upravleniye: teoriya i praktika [Situational control: theory and practice]. Nauka, Moskva (1986)
13. Pospelov, D.A., Osipov, G.S.: Prikladnaya semiotika [Applied semiotics]. Novosti iskusstvennogo intellekta [News Artif. Intell.] **1** (1999)
14. Erlikh, A.I.: Prikladnaya semiotika i upravleniye slozhnymi ob"yektami [Applied semiotics and control of complex objects]. Programmnyye produkty i sistemy[Softw. Prod. Syst.] **3**, 2–6 (1997)
15. Averkin, A.N., Golovina, E.Yu.: Nechetkaya semioticheskaya sistema upravleniya [Fuzzy semiotic control system]. In: Intellektual'noye upravleniye: novyye intellektual'nyye tekhnologii v zadachakh upravleniya (ICIT 1999) [Intelligent Control: New Intelligent Technologies in Control Tasks], pp. 141–145. Nauka, Fizmatlit, Moskva (1999)
16. Yeremeyev, A.P., Tikhonov, D.A., Shutova, P.V.: Podderzhka prinyatiya resheniy v usloviyakh neopredelennosti na osnove nemarkovskoy modeli [Support of decision-making under conditions of uncertainty based on the non-Markov model]. Izv. RAN Teoriya sistem i upravleniya [Theor. Syst. Control] **5**, 75–88 (2003)
17. Tarasov, V.B.: Ot mnogoagentnykh sistem k intellektual'nym organizatsiyam [From multi-agent systems to intelligent organizations]. Editorial URSS, Moscow (2002)
18. Andersen, P.B.: A Theory of Computer Semiotics: Semiotic Approaches to Construction and Assessment of Computer Systems. Cambridge University Press, Cambridge (2006)
19. Barron, T.M., Chiang, R.H.L., Storey, V.C.: A semiotics framework for information systems classification and development. Decis. Support Syst. **25**(1), 1–17 (1999)
20. Goguen, J.: An introduction to algebraic semiotics, with application to user interface design. In: Nehaniv, C.L. (ed.) CMAA 1998. LNCS (LNAI), vol. 1562, pp. 242–291. Springer, Heidelberg (1999). https://doi.org/10.1007/3-540-48834-0_15
21. Osipov G.S.: Sign-based representation and word model of actor. In: Yager, R. (eds.) IEEE 8th International Conference on Intelligent Systems (IS), pp. 22–26. IEEE (2016)
22. Osipov, G.S., Panov, A.I.: Otnosheniya i operatsii v znakovoy kartine mira sub"yekta povedeniya [Relationships and operations in a landmark picture of the subject's world of behavior]. Iskusstvennyy intellekt i prinyatiye resheniy [Artif. Intell. Decis. Mak.] **4**, 5–22 (2017)
23. Osipov, G.S., Panov, A.I., Chudova, N.V.: Upravleniye povedeniyem kak funktsiya soznaniya. I. Kartina mira i tselepolaganiye [Managing behavior as a function of consciousness. I. The picture of the world and the goal setting]. In: Izvestiya Rossiyskoy Akademii Nauk. Teoriya i sistemy upravleniya [Proceedings of the Russian Academy of Sciences. Theory and Control Systems], vol. 4, pp. 49–62 (2014)
24. McCarthy J.: Situations, Actions and Causal Laws. Stanford Artificial Intelligence Project: Memo 2. Comtex Scientific (1963)
25. Fikes, R.E., Nilsson, N.J.: STRIPS: a new approach to the application of theorem proving to problem solving. In: IJCAI 1971, pp. 608–620 (1971)
26. Grosz, B., Kraus, S.: Collaborative plans for complex group actions. Artif. Intell. **86**, 269–358 (1996)
27. Kulinich, A.A.: A model of agents (robots) command behavior: the cognitive approach. Autom. Remote Control **77**(3), 510–522 (2016)

Recognition of Multiword Expressions Using Word Embeddings

Natalia Loukachevitch[1,2](✉) and Ekaterina Parkhomenko[2]

[1] Tatarstan Academy of Sciences, Kazan, Russia
louk_nat@mail.ru
[2] Lomonosov Moscow State University, Moscow, Russia
parkat13@yandex.ru

Abstract. In this paper we consider the task of extracting multiword expressions (MWE) for Russian thesaurus RuThes, which contains various types of phrases, including non-compositional phrases, multiword terms and their variants, light verb constructions, and others. We study several embedding-based features for phrases and their components and estimate their contribution to finding multiword expressions of different types comparing them with traditional association and context measures. We found that one of the distributional features has relatively high results of MWE extraction even when used alone. Different forms of its combination with other features (phrase frequency, association measures) improve both initial orderings. Besides, we demonstrate significant potential of an existing thesaurus for recognition of new multiword expressions for adding to the thesaurus.

Keywords: Thesaurus · Multiword expression · Embedding

1 Introduction

Automatic recognition of multiword expressions (MWE) having lexical, syntactic or semantic irregularity is important for many tasks of natural language processing, including syntactic and semantic analysis, machine translation, information retrieval, and many others. Various types of measures for MWE extraction have been proposed. These measures include word-association measures comparing frequencies of phrases and their component words, context-based features comparing frequencies of phrases and encompassing groups [17]. For multiword terms, such measures as frequencies in documents and a collection, contrast measures are additionally used [1].

But currently there are new possibilities of applying distributional and embedding-based approaches to MWE recognition. Distributional methods allow representing lexical units or MWE as vectors according to the contexts where the units are mentioned [2]. Embedding methods use neural network approaches to improve vector representation of lexical units [16]. Therefore, it is possible to use embedding characteristics of phrases trying to recognize their irregularity,

© Springer Nature Switzerland AG 2018
S. O. Kuznetsov et al. (Eds.): RCAI 2018, CCIS 934, pp. 112–124, 2018.
https://doi.org/10.1007/978-3-030-00617-4_11

which makes it important to fix them in computational vocabularies or the-saurus. The distributional features were mainly evaluated on specific types of multiword expressions as non-compositional noun compounds [5] or verb-direct object groups [9,12], but they were not studied on a large thesaurus containing various types of multiword expressions.

In this paper we consider several measures for recognition of multiword expressions based on distributional similarity of phrases and component words. We compare distributional measures with association measures and context mea-sures and estimate the contribution of distributional features in combinations with other measures. As a gold standard, we use RuThes thesaurus of the Rus-sian language [14], which comprises a variety of multiword expressions. In the current study, only two-word phrases are considered.

The structure of this paper is as follows. Section 2 considers related work. Section 3 briefly describes the thesaurus RuThes and principles for including multiword expressions. Section 4 introduces distributional features for extracting multiword expressions. Section 5 presents the results of extracting multiword expressions using the introduced features and their combinations, Sect. 6 studies an additional feature of the similarity of a MWE candidate and thesaurus units.

2   Related Work

Distributional (embedding) features have been studied for recognizing several types of multiword expressions. Fazly et al. [9] studied verb-noun idiomatic constructions combining two types of features: lexical fixedness and syntactic fixedness. The lexical fixedness feature compares pointwise mutual information (PMI) for an initial phrase and PMI of its variants obtained with substitution of component words to distributionally similar words according to the Lin the-saurus [13].

In [20], the authors study the prediction of non-compositionality of multiword expressions comparing traditional distributional (count-based) approaches and word embeddings (prediction approaches). They test the proposed approaches using three specially prepared datasets of noun compounds and verb-particle constructions in two languages. All expressions are labeled on specialized scales from compositionality to non-compositionality. The authors [20] compared dis-tributional vectors of a phrase and its components and found that the use of word2vec embeddings outperforms traditional distributional similarity with a substantial margin.

The authors of [5] study prediction of non-compositionality of noun com-pounds on four datasets for English and French. They compare results obtained with distributional semantic models and embedding models (word2vec, glove). They use these models to calculate vectors for compounds and their component words and order the compounds according to less similarity of compound vector and the sum of component vectors. They experimented with different parameters and found that the obtained results have high correlations with human judge-ments. All the mentioned works tested the impact of embedding-based measures of specially prepared data sets.

In [18], the authors study various approaches to recognition of Polish MWE using 46 thousand phrases introduced in Polish wordnet (plWordNet) as an etalon set. They utilized known word association measures described in [17] and proposed their own association measures. They also tested a measure estimating fixedness of word-order of a candidate phrase. To combine the proposed measures, weighted rankings were summed up. The weights were tuned at a separate corpus, the tuned linear combination of rankings was transferred to another test corpus and was still better than any single measure.

Authors of [15] study several types of measures and their combinations for term recognition using real thesauri as gold standards: EUROVOC for English and Banking thesaurus for Russian. They use 88 different features for extracting two-word terms. The types of the features include: frequency-based features; contrast features comparing frequencies in target and reference corpora; word-association measures; context-based features; features based on statistical topic modeling, and others. They studied the contribution of each group of features in the best integrated model and found that association measures do not have positive impact on term extraction for two-word terms.

Both works with real thesauri [15,18] did not experiment with distributional or embedding-based features.

3   Multiword Expressions in RuThes

The thesaurus of Russian language RuThes [14] is a linguistic ontology for natural language processing, i.e. an ontology, where the majority of concepts are introduced on the basis of actual language expressions. As a resource for natural language processing, RuThes is similar to WordNet thesaurus [10], but has some distinctions. One of significant distinctions important for the current study is that RuThes includes terms of so-called sociopolitical domain.

The sociopolitical domain is a broad domain describing everyday life of modern society and uniting many professionals domains, such as politics, law, economy, international relations, finances, military affairs, arts and others. Terms of this domain are usually known not only to professionals, but also to ordinary people [14], they are often met in news reports and newspaper articles, and therefore the thesaurus representation of such terms is important for effective processing of news flows. Currently, RuThes contains almost 170 thousand Russian words and expressions.

As a lexical and terminological resource for automatic document processing, RuThes contains a variety of multiword expressions needed for better text analysis:

- traditional non-compositional expressions (idioms),
- constructions   with   light   verbs   and   their   nominalizations: помочь – оказать по-мощь – оказание помощи (*to help – to provide help – provision of help*),
- terms of the sociopolitical domain and their variants. According to terminological studies [6], domain-specific terms can have large number of variants in texts;

these variants are useful to be included into the thesaurus to provide better term recognition: экономика – экономическая сфера – экономическая область (*economy – economic sphere – sphere of economy*),

- multiword expressions having thesaurus relations that do not follow from the component structure of the expression, for example, *traffic lights* [19] is a *road facility*, *food courts* consist of *restaurants* [8],
- geographical and some other names.

Recognition of such diverse multiword expressions requires application of non-compatible principles. Non-compositional expressions often do not have synonyms or variants (lexical fixedness according to [9]), but domain-specific terms often have variations useful to describe in the thesaurus.

The development of RuThes, introduction of words and expressions into the thesaurus, are based on expert and statistical analysis of the current Russian news flow (news reports, newspaper articles, analytical papers). Therefore, we suppose that RuThes provides a good coverage for MWEs extracted from news collections and gives us possibility to evaluate different measures used for automatic recognition of MWE from texts.

4   Distributional Features

We consider three distributional features computed using word2vec method [16].

The First Feature (DFsum) is based on the assumption that non-compositional phrases can be distinguished with comparison of the phrase distributional vector and distributional vectors of its components: it was supposed that the similarity is less for non-compositional phrases [5,12]. For the phrases under consideration, we calculated cosine similarity between the phrase vector $v(w_1 w_2)$ and the sum of normalized vectors of phrase components $v(w_1 + w_2)$ according to formula from [5].

$$v(w_1 + w_2) = (\frac{v(w_1)}{|v(w_1)|} + \frac{v(w_2)}{|v(w_2)|})$$

$$DFsum = cos(v(w_1 w_2), v(w_1 + w_2))$$

In the DFsum ordering, the phrases are ordered according to increasing value of DFsum, because non-compositional phrases should have minimal values of DFsum.

The Second Feature (DFcomp) calculates the cosine similarity of the component word vectors to each other.

$$DFcomp = cos(v(w_1), v(w_2))$$

This means the similarity of contexts of component words. Examples of the thesaurus entries with high DFcomp include: симфонический оркестр (*symphony orchestra*), зерноуборочный комбайн (*combine harvester*), отрасль промышлен-ности (*branch of industry*), тройская унция (*troy ounce*), etc.

This measure is a form of numerically weighted association between words. The phrases are ordered according to the decreasing values of DFcomp.

The Third Feature (DFsing) is calculated as the similarity between the phrase and the most similar single word; the word should be different from the phrase components words.

$$DFsing = max_i(cos(v(w_1w_2), v(w_i)))$$

where w_i is a word of the text collection, distinct from w_1 and w_2.

The phrases were ordered according to decreasing value of DFsing. It was found that most words in the top of the list (the most similar to phrases) are abbreviations (Table 1). It can be seen that some phrases have quite high similarity values with their abbreviated forms (more than 0.9).

Table 1. DFsing measure orders phrases according to the maximum similarity to a single word

MWE	Single Word	translation	Score	Type
детский сад	детсад	kindergarten	0.961	abbr.
Европейский союз	евросоюз	European Union	0.942	abbr.
атомная электростанция	АЭС	nuclear station	0.933	abbr.
атомная станция	АЭС	nuclear station	0.925	abbr.
генеральная прокуратура	генпрокуратура	prosecution office	0.923	abbr.
районный суд	райсуд	district court	0.923	abbr.
государственный бюджет	госбюджет	state budget	0.917	abbr.
следственный изолятор	сизо	detention center	0.917	abbr.
Государственная дума	Госдума	State duma	0.914	abbr.
федеральный закон	ФЗ	federal law	0.911	abbr.

5 Experiments

We used a Russian news collection (0.45 B tokens) and generated phrase and word embeddings with word2vec tool. In the current experiments, we used default parameters of the word2vec package, but after the analysis of the results we do not think that the conclusions can be significantly changed.

We extracted two-word noun phrases: adjective+noun and noun+noun in Genitive with frequencies equal or more than 200 to have enough statistical data. From the obtained list, we removed all phrases containing known personal names. We obtained 37,768 phrases. Among them 9,838 are thesaurus phrases, and the remaining phrases are not included into the thesaurus. For each measure, we created a ranked list according to this measure. At the top of the list, there should be multiword expressions, at the end of the list there should be free, compositional, or non-terminological phrases.

5.1 Extraction of MWE with Association Measures

We generated ranked lists for the following known association measures: pointwise mutual information (PMI, formula 1), its variants (cubic MI, normalized PMI, augmented MI, true MI), Log-likelihood ratio, t-score, chi-square, Dice and modified Dice measures [15,17]. Table 2 presents the top of the ordered phrase candidates according to the PMI measure. It can be seen that many specific names not included in the thesaurus are in the top of the list.

Some used association measures presuppose the importance of the phrase frequency in a text collection for MWE recognition and enhance its contribution to the basic measure. For example, Cubic MI (2) includes the cubed phrase frequency if compared to PMI (1), and True MI (3) utilizes phrase frequency without logarithm.

$$PMI(w_1, w_2) = log(\frac{freq(w_1, w_2) \cdot N}{freq(w_1) \cdot freq(w_2)}) \tag{1}$$

$$CubicMI(w_1, w_2) = log(\frac{freq(w_1, w_2)^3 \cdot N}{freq(w_1) \cdot freq(w_2)}) \tag{2}$$

$$TrueMI(w_1, w_2) = freq(w_1, w_2) \cdot log(\frac{freq(w_1, w_2) \cdot N}{freq(w_1) \cdot freq(w_2)}) \tag{3}$$

Table 2. 10 top phrases extracted according to PMI

Phrase	Translation	In Thes?	Type
всенощное бдение	all-night vigil	Y	domain term
ларго винч	largo winch	N	movie title
централ партнершип	central partnership	N	movie company
едиот ахронот	yedioth ahronoth	N	newspaper title
нон грата	non grata	N	term fragment
бесславный ублюдок	inglourious basterd	N	movie title
торо россо	toro rosso	N	racing team
мумий тролль	mumiy troll	N	russian rock group
имеретинская низменность	imereti lowlands	Y	geoname
клещевой энцефалит	tick-borne encephalitis	Y	domain term

Modified Dice (5) measure also enhances the contribution of the phrase frequency in Dice measure (4).

$$Dice(w_1, w_2) = \frac{2 \cdot freq(w_1, w_2)}{freq(w_1) + freq(w_2)} \qquad (4)$$

$$ModifiedDice(w_1, w_2) = log(freq(w_1, w_2)) \cdot \frac{2 \cdot freq(w_1, w_2)}{freq(w_1) + freq(w_2)} \qquad (5)$$

Also we calculated c-value measure, which is used for extraction of domain-specific terms [11]. To evaluate the list rankings, we utilized uninterpolated average precision measure (AvP), which attains the maximal value (1) when all multiword expressions are located in the beginning of a list without any interruptions [17]. AvP at level of n first candidates is calculated as follows:

$$AvP@n = \frac{1}{m} \cdot \sum_{k=1}^{n} (r_k \cdot (\frac{1}{k} \cdot \sum_{1 \le i \le k} r_i)) \qquad (6)$$

$r_k = 1$ – if k-th candidate is a phrase from the etalon set, and $r_k = 0$ in another case, m is the maximal possible number of the etalon phrases between candidates. The average precision is the mean of the precision scores obtained after each etalon phrase appears in the list.

The Table 3 shows the AvP values at the level of the 100 first thesaurus phrases (AvP (100)), 1000 first thesaurus phrases (AvP (1000)) and for the full

Table 3. Average precision measure at 100, 1000 thesaurus phrases and for the full list

Measure	AvP (100)	AvP (1000)	AvP (Full)
Frequency	0.73	0.70	0.43
PMI and modifications			
PMI	0.52	0.54	0.44
CubicMI	**0.91**	**0.80**	**0.52**
NPMI	0.64	0.65	0.47
TrueMI	0.77	0.77	0.50
AugmentedMI	0.55	0.59	0.45
Other association measures			
LLR	0.78	0.78	0.51
T-score	0.73	0.71	0.46
Chi-Square	0.68	0.69	0.50
DC	0.68	0.67	0.48
ModifiedDC	0.81	0.71	0.49
C-value	0.73	0.70	0.43
Distributional features			
DFsum	0.20	0.19	0.24
DFcomp	0.47	0.42	0.35
DFsing	0.85	0.69	0.42

Table 4. 10 top phrases extracted according to CubicMI

Phrase	Translation	In Thes?	Type
правоохранительный орган	law-enforcement body	Y	domain term
уголовное дело	criminal case	Y	domain term
точка зрения	point of view	Y	idiom
лишение свободы	deprivation of liberty	Y	domain term
заработная плата	employee wages	Y	domain term
ближний восток	middle east	Y	geograph. name
взрывное устройство	explosive device	Y	domain term
учебное заведение	educational institution	Y	domain term
стихийное бедствие	natural disaster	Y	domain term
детский сад	kindergarden	Y	domain term

list for all mentioned measures and features. It can be seen that the results of PMI is lowest in comparison to all association measures. This is due to extraction of some specific names or repeated mistakes in texts (i.e. words without spaces between them). Even the high frequency threshold preserves this known problem of PMI-based MWE extraction (Table 2).

Normalized PMI extracts MWE much better as it was indicated in [4]. Modified Dice measure gives better results in comparison with initial Dice measure. The best results among all measures belong to the Cubic MI measure proposed in [7]. Table 4 shows the top of the CubicMI list of candidate phrases. All ten top phrases are thesaurus phrases.

It is important to note that there are some evident non-compositional phrases that are located in the end of the list according to any statistical measures. This is due to the fact that both words are very frequent in the collection under consideration but the phrase is not very frequent. Examples of such phrases include: игра слов (*word play*), человекгода (*person of the year*), план счетов (*chart of accounts*), государственнаямашина (*state machinery*), and others.

5.2   Extraction of MWE with Distributional Features and Combinations

In the previous subsection, it was found that some evident MWEs are not revealed by association measures. For all lists of association measures, these phrases were located in the last thousand of the lists. According to the distributional feature DFsum, these phrases significantly shifted to the top of the list. Their positions became as follows: игра слов (561), человек года (1346), план счетов (1545), государственнаямашина (992). Thus, it could seem that this feature can generate a qualitative ranked list of phrases.

But the overall quality of ordering for the DFsum distributional feature is rather small (Table 3). Thus, when we work with candidates extracted from a raw corpus these measure is not so useful as for specially prepared lists of compositional and non-compositional phrases as it was described in [5,12].

Table 5. Combining distributional features with the phrase frequency

Measure	AvP (100)	AvP (1000)	AvP (Full)
DF multiplied by Frequency			
DFsum	0.70	0.69	0.43
DFcomp	0.75	0.72	0.46
DFsing	0.76	0.73	0.46
DF multiplied by log (Frequency)			
DFsum	0.57	0.56	0.36
DFcomp	0.70	0.57	0.39
DFsing	**0.93**	**0.83**	**0.50**

On the other hand, we can see that another distributional feature – maximal similarity with a single word (DFsing) showed a quite impressive result, which is the second one at the first 100 thesaurus phrases (Table 3). As it was indicated earlier, the first 100 thesaurus phrases are most similar to their abbreviated forms, which means that for important concepts, reduced forms of their expressions are often introduced and used.

As it was shown for association measures, additional accounting of phrase frequency can improve a basic feature. Table 5 presents the results of multiplying initial distributional features to phrase frequency or its logarithm. It can be seen that DFsing more improved when multiplied by log (Freq). DFcomp multiplied by the phrase frequency became better then initial frequency ordering (Table 3).

Then we try to combine the best distributional DFsing measure with association measures in two ways: (1) multiplying values of initial measures, (2) summing up ranks of phrases in initial rankings (Table 6). In both cases,

Table 6. Combining DFsing with traditional association measures

Measure	Multiplying values			Summing up rankings		
	AvP (100)	AvP (1000)	AvP (Full)	AvP (100)	AvP (1000)	AvP (Full)
DFsing*CubicPMI	**0.95**	**0.84**	**0.54**	0.94	0.83	0.53
DFsing*PMI	0.62	0.62	0.47	0.62	0.64	0.48
DFsing*NPMI	0.78	0.73	0.50	0.79	0.73	0.50
DFsing*augMI	0.64	0.65	0.48	0.69	0.67	0.48
DFsing*TrueMI	0.80	0.80	0.52	**0.96**	**0.86**	0.53
DFsing*Chi-Square	0.73	0.71	0.50	0.85	0.77	0.51
DFsing*LLR	0.82	0.81	0.53	**0.96**	**0.86**	0.53
DFsing*DC	0.74	0.70	0.49	0.84	0.77	0.50
DFsing*ModifiedDC	0.85	0.73	0.50	0.88	0.79	0.51
DFsing*T-score	0.80	0.78	0.49	0.95	0.84	0.51
DFsing*C-value	0.76	0.74	0.46	0.95	0.83	0.49

Table 7. 10 top phrases extracted according to combined measure CubicMI*DFsing

Phrase	Translation	In Thes?	Type
уголовный розыск	criminal investigation	Y	domain term
конституционный суд	constitutional court	Y	domain term
нефтяная компания	oil company	Y	domain term
электронная карта	digital map	Y	domain term
физическая культура	physical culture	Y	domain term
противовоздушная оборона	anti-air defense	Y	domain term
добыча нефти	oil mining	Y	domain term
санитарный врач	health officer	Y	domain term
сельское поселение	rural settlement	Y	domain term
колесный диск	wheel disk	Y	domain term

AvP of initial association measures significantly improved. For Cubic MI, the best result was based on values multiplying. For LLR and True MI, the best results were attained with summing up ranks. In any case, it seems that the distributional similarity of a phrase with a single word (different from its component) bears important information about MWEs.

Table 7 shows 10 top phrases according to the combination of CubicMI and DFsing. It can be seen that all proposed candidates are correct and the top is quite different from the single CubicMI top phrases (Table 4).

6   Thesaurus-Based Similarity Measure

In many works the process of extracting multiword expressions is considered from scratch as if no resources are available. But in fact, for many languages, thesauri exist. For example, large WordNet-like thesauri have been created at least for 26 natural languages, and small wordnets exist for 57 languages [3].

Therefore, for languages with existing thesauri, it is important to extract new multiword expressions for augmenting these thesauri. Thus, we can introduce additional feature DFthes, which calculates the maximal cosine similarity of a phrase with the existing text entries of the basic thesaurus (RuThes in our case) and orders phrase candidates according to decreasing value of similarity with the thesaurus entries (single words or phrases).

$$DFthes = max_i(cos(v(w_1w_2), v(te_i)),$$

where te_i is a thesaurus entry (word or phrase). Table 8 shows that DFthes as a single feature gives the best result if compared to all other single measures in our experiments. Its multiplication with log of the phrase frequency improves the results.

Combination of DFthes with such measures as TrueMI, CubicMI, and LLR improves the extraction of the thesaurus phrases in the middle and in the end of the candidate lists (Table 9).

Table 8. Results of DFthes measure and its combinations with frequencies

Measure	AvP (100)	AvP (1000)	AvP (Full)
DFThes	**0.95**	0.82	0.55
DFthes in combination with frequency			
DFThes*freq	0.77	0.76	0.47
DFThes*log(freq)	**0.95**	**0.88**	**0.57**

Table 9. Combining DFthes with other measures

Measure	Multiplying values			Summing up rankings		
	AvP (100)	AvP (1000)	AvP (Full)	AvP (100)	AvP (1000)	AvP (Full)
DFthes*PMI	0.81	0.72	0.52	0.81	0.79	0.57
DFthes*AugmentedMI	0.80	0.74	0.52	0.85	0.81	0.57
DFthes*NormalizedMI	0.93	0.83	0.55	0.91	0.85	0.59
DFthes*TrueMI	0.82	081	0.53	**0.97**	**0.91**	0.60
DFthes*CubicPMI	**0.96**	**0.91**	**0.61**	**0.96**	**0.91**	**0.61**
DFthes*DC	0.82	0.73	0.50	0.93	0.87	0.59
DFthes*ModifiedDC	0.91	0.76	0.51	0.94	0.89	0.59
DFthes*T-Sxore	0.80	0.80	0.52	0.96	0.90	0.58
DFthes*Chi-Square	0.78	0.73	0.51	0.94	0.88	0.60
DFthes*LLR	0.83	0.82	0.54	**0.97**	**0.91**	**0.61**
DFthes*C-value	0.77	0.76	0.47	**0.96**	0.90	0.57

7   Conclusion

In this paper we have considered the task of extracting multiword expressions for Russian thesaurus RuThes, which contains various types of phrases, including non-compositional phrases, multiword terms and their variants, light verb construction, and others. We study several embedding-based features for phrases and their components and estimate their contribution to finding multiword expressions of different types comparing them with traditional association and context measures.

We have found that one of the most discussed distributional features (DFsum) provides low quality of a ranked MWE list when extracted from a raw corpus. However, another distributional feature (DFsing) has relatively high results of MWE extracting even when used alone. Different forms of its combination with other features (phrase frequency, association measures) achieve even higher results.

We have also found that the similarity of phrase candidates with the entries of the existing thesaurus improves the recognition of multiword expressions as a single measure, as well as in combination with other measures.

Acknowledgments. This work was partially supported by Russian Science Foundation, grant N16-18-02074.

References

1. Astrakhantsev, N.: ATR4S: toolkit with state-of-the-art automatic terms recognition methods in scala. Lang. Resour. Eval. **52**(3), 853–872 (2018)
2. Biemann, C., Giesbrecht, E.: Distributional semantics and compositionality 2011: shared task description and results. In: Proceedings of the Workshop on Distributional Semantics and Compositionality, pp. 21–28. Association for Computational Linguistics (2011)
3. Bond, F., Foster, R.: Linking and extending an open multilingual wordnet. In: Proceedings of the 51st Annual Meeting of the Association for Computational Linguistics: Long Papers, vol. 1, pp. 1352–1362 (2013)
4. Bouma, G.: Normalized (pointwise) mutual information in collocation extraction. In: Proceedings of GSCL, pp. 31–40 (2009)
5. Cordeiro, S., Ramisch, C., Idiart, M., Villavicencio, A.: Predicting the compositionality of nominal compounds: giving word embeddings a hard time. In: Proceedings of ACL-2016g Papers. vol. 1, pp. 1986–1997 (2016)
6. Daille, B.: Term Variation in Specialised Corpora: Characterisation, Automatic Discovery and Applications, vol. 19. John Benjamins Publishing Company, Amsterdam (2017)
7. Daille, B.: Combined approach for terminology extraction: lexical statistics and linguistic filtering. Ph.D. thesis, University Paris 7 (1994)
8. Farahmand, M., Smith, A., Nivre, J.: A multiword expression data set: annotating non-compositionality and conventionalization for English noun compounds. In: Proceedings of the 11th Workshop on Multiword Expressions, pp. 29–33 (2015)
9. Fazly, A., Cook, P., Stevenson, S.: Unsupervised type and token identification of idiomatic expressions. Comput. Linguist. **35**(1), 61–103 (2009)
10. Fellbaum, C.: WordNet. Wiley Online Library (1998)
11. Frantzi, K., Ananiadou, S., Mima, H.: Automatic recognition of multi-word terms: the C-value/NC-value method. Int. J. Digit. Libr. **3**(2), 115–130 (2000)
12. Gharbieh, W., Bhavsar, V.C., Cook, P.: A word embedding approach to identifying verb-noun idiomatic combinations, pp. 112–118 (2016)
13. Lin, D.: Automatic retrieval and clustering of similar words. In: Proceedings of ACL-1998, pp. 768–774. Association for Computational Linguistics (1998)
14. Loukachevitch, N., Dobrov, B.: RuThes linguistic ontology vs. Russian wordnets. In: Proceedings of the Seventh Global Wordnet Conference, pp. 154–162 (2014)
15. Loukachevitch, N., Nokel, M.: An experimental study of term extraction for real information-retrieval thesauri. In: Proceedings of TIA-2013, pp. 69–76 (2013)
16. Mikolov, T., Chen, K., Corrado, G., Dean, J.: Efficient estimation of word representations in vector space. arXiv preprint arXiv:1301.3781 (2013)
17. Pecina, P.: Lexical association measures and collocation extraction. Lang. Resour. Eval. **44**(1–2), 137–158 (2010)
18. Piasecki, M., Wendelberger, M., Maziarz, M.: Extraction of the multi-word lexical units in the perspective of the wordnet expansion. In: RANLP-2015, pp. 512–520 (2015)

19. Sag, I.A., Baldwin, T., Bond, F., Copestake, A., Flickinger, D.: Multiword expressions: a pain in the neck for NLP. In: Gelbukh, A. (ed.) CICLing 2002. LNCS, vol. 2276, pp. 1–15. Springer, Heidelberg (2002). https://doi.org/10.1007/3-540-45715-1_1
20. Salehi, B., Cook, P., Baldwin, T.: A word embedding approach to predicting the compositionality of multiword expressions. In: Proceedings of NAACL-2015, pp. 977–983 (2015)

Hierarchical Aggregation of Object Attributes in Multiple Criteria Decision Making

Alexey B. Petrovsky[1,2,3,4(✉)]

[1] Federal Research Center "Informatics and Control", Russian Academy of
Sciences, Prospect 60 Letiya Octyabrya, 9, Moscow 117312, Russia
pab@isa.ru
[2] Belgorod State National Research University, Belgorod, Russia
[3] V.G. Shukhov Belgorod State Technological University, Belgorod, Russia
[4] Volgograd State Technical University, Volgograd, Russia

Abstract. The paper presents a multi-method technology PAKS-M for a choice
of objects with many numerical and/or verbal attributes. The technology provides an aggregation of object attributes by a reduction of the attribute space
dimension; a construction of hierarchical systems of composite criteria and an
integral index of quality; ordering and/or classification of objects using several
decision making methods. This technology was applied for solving hard problems of multiple criteria decision making, in particular, the group multiple
criteria selection of perspective computing complex.

Keywords: Multiple criteria decision making · Multi-attribute objects
Attribute space · Reduction of dimension · Hierarchical aggregation of attributes
Composite criteria · Integral index

1 Introduction

One of the hard problems of multiple criteria decision making is a choice among a
small number of objects described by a large set of numerical and/or verbal attributes,
characterizing the object properties. Such a task is, e.g., a selection of the place of
airport or power plant, the route of gas or oil pipeline, the prospective scheme of city or
railway development, the configuration of a complicate technical system, and so on. In
these cases, the objects are usually incomparable by their attributes. And for an expert
or decision maker (DM), it is very difficult to select the best object, rank, or to classify
objects. The famous methods of multiple criteria decision making [1–6, 9, 11–14]
require significant labor costs and time to receive and process big volumes of data
about objects, knowledge of experts and/or the preferences of DMs. So, these methods
are less suitable and sometimes useless for solving choice tasks in the attribute space of
large dimension.

The paper describes a new multi-method technology PAKS-M (Progressive
Aggregation of the Classified Situations by many Methods) for a solution of manifold
problems of individual and collective decision making, ordering or classifying multi-
attribute objects based on knowledge of experts and/or preferences of DMs [10]. This
technology provides a hierarchical granulation of information [15] by a reduction of

S. O. Kuznetsov et al. (Eds.): RCAI 2018, CCIS 934, pp. 125–137, 2018.
https://doi.org/10.1007/978-3-030-00617-4_12

dimension of the initial attribute space and a consecutive aggregation of a large set of numerical, symbolical, or verbal characteristics of objects into a single integral index of quality or a small number of composite criteria with verbal scales. The grades of criteria scales are constructed using different methods of verbal decision analysis [5, 9]. The final quality criteria or index present the initial characteristics of objects in a compact form, which experts/DMs apply for solving the considered task of multi-attribute choice by many different techniques of decision making and comparing the obtained results to select the most preferable and valid one.

The multi-method technology PAKS-M allows users to compare the hierarchical lists of criteria, select the preferable criteria, solve the considered task by several methods, analyze the obtained results, and assess the quality of the made decision. This technology has been applied for the multi-criteria evaluation and selection of a perspective computing complex of the personal level [10]. Currently, such complexes are increasingly used to solve various scientific and practical problems and are alternatives to expensive supercomputers and high-performance clusters.

2   Main Stages of Technology for Multi-attribute Choice

The problem of individual or group multi-attribute choice is formulated generally as follows. A collection of objects (alternatives, options) $A_1, \ldots, A_p$, which are evaluated by one or several experts/DMs with many criteria $Q_1, \ldots, Q_t$, is given. Each criterion Q_s has a scale $Z_s = \{z_s^1, \ldots, z_s^{d_s}\}$, $s = 1, \ldots, t$, the discrete numerical or verbal grades of which are ordered in some cases. Based on knowledge of experts and/or preferences of DMs, it is required: (1) to select the best objects; (2) to order all objects; (3) to classify all objects by several categories.

Let us consider the main stages of solving a multi-attribute choice task using the PAKS-M technology. First, an expert and/or a DM determines the set $K_1, \ldots, K_m$ of initial characteristics of objects and their scales $X_1, \ldots, X_m$, $m \geq 2$, which reflect the basic properties of the considered objects. The scale $X_i = \{x_i^1, \ldots, x_i^{g_i}\}$, $i = 1, \ldots, m$ of each initial indicator has numerical (point, interval) or verbal evaluation grades.

Further, user reduces the dimension of the attribute space, which is the Cartesian product $X_1 \times \ldots \times X_m$ of the indicator scale grades, by constructing a hierarchical system of criteria. An expert/DM, based on knowledge, experience and intuition, forms the aggregation-tree, establishing the structure, number and content of criteria. At each level of hierarchy, including the highest level, the actor defines which of indicators are considered as the independent final criteria and which are combined into particular intermediate composite criteria. The various combinations of initial attributes $K_1, \ldots, K_m$ are aggregated into smaller sets of new attributes (intermediate or final criteria) $L_1, \ldots, L_n$, $n < m$. The attributes are aggregated consecutively step by step. The groups of criteria of the previous level of hierarchy are combined serially in new criteria of the next level of hierarchy.

The formation of the rating scale $Y_j = \{y_j^1, \ldots, y_j^{h_j}\}$, $j = 1, \ldots, n$ of a composite criterion L_j at every hierarchical level is considered as the task of ordinal classification [5, 9]. Each classification block of i-th level includes any set of attributes and a single composite criterion. Tuples of initial attribute estimates are the classified objects. Grades of scale of the composite criterion are the decision classes. In a classification block of the next $(i + 1)$-th level, composite criteria of the i-th level are considered as new attributes. Tuples of their scale grades represent new classified objects in the reduced attribute space, whereas scale grades of new composite criterion will be now decision classes of the $(i + 1)$-th level. As a rule, the aggregation of criteria is a multi-stage procedure. The upper level of the aggregation tree can consist of several final criteria or be a single integral index with verbal grades of scales, which characterize a quality of the compared objects and have the concrete semantic content for the expert/DM.

Rating scales of composite criteria can be constructed with different methods [9]. The simplest method is the tuple stratification, when the multi-attribute space is cut with parallel hyper-planes. Each layer (stratum) consists of combinations of the homogeneous (for example, with the fixed sum of grade numbers) initial estimates and represents any generalized grade on the scale of composite criterion. Methods of the verbal decision analysis are more complicated [5]. The ZAPROS (the abbreviation of Russian words: CLosed PRocedures nearby Reference Situation) method allows to construct a joint ordinal scale of composite criterion from initial estimates. The ORCLASS (ORdinal CLASSification) method builds a complete and consistent classification of all tuples of initial estimates where classes of initial estimates form grades of an ordinal scale of composite criterion. A number of layers or classes (scale grades) is determined by the expert/DM.

Consider a small illustrative example. Suppose that a scale of a composite criterion D is formed from the grades of three initial attributes A, B, C. Let, for instance, the scales of the attributes A, B, C and criterion D have three following verbal grades: $A = \{a^0, a^1, a^2\}$, $B = \{b^0, b^1, b^2\}$, $C = \{c^0, c^1, c^2\}$, $D = \{d^0, d^1, d^2\}$, where x^0 is the 'high' grade, x^1 is the 'medium' grade, x^2 is the 'low' grade.

By building a scale of the criterion D with the tuple stratification technique, a user can combine grades of the initial attributes A, B, C as follows: the high initial grades form the high generalized grade, medium initial grades form the medium generalized grade, and the low initial grades form the low generalized grade (Fig. 1).

Layers of tuples						
		$a^0b^0c^1$	$a^0b^0c^2$	...	$a^1b^1c^2$	$a^1b^2c^2$
	$a^0b^0c^0$	$a^0b^1c^0$	$a^0b^2c^0$	$a^1b^1c^1$	$a^1b^2c^1$	$a^2b^1c^2$ $a^2b^2c^2$
		$a^1b^0c^0$	$a^2b^0c^0$	...	$a^2b^1c^1$	$a^2b^2c^1$
	d^0		d^1		d^2	
			Composite criterion D			

Fig. 1. Scale of the composite criterion obtained by the tuple stratification technique.

By building a scale of the criterion D with the ZAPROS method, a user can mark the generalized grades on the joint ordinal scale constructed from all tuples of the initial grades (Fig. 2).

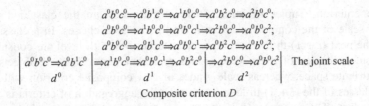

$$a^0b^0c^0\Rightarrow a^0b^1c^0\Rightarrow a^1b^0c^0\Rightarrow a^0b^2c^0\Rightarrow a^2b^0c^0;$$
$$a^0b^0c^0\Rightarrow a^1b^0c^0\Rightarrow a^0b^0c^1\Rightarrow a^2b^0c^0\Rightarrow a^0b^0c^2;$$
$$a^0b^0c^0\Rightarrow a^0b^1c^0\Rightarrow a^0b^0c^1\Rightarrow a^0b^2c^0\Rightarrow a^0b^0c^2;$$

$a^0b^0c^0\Rightarrow a^0b^1c^0$	$\Rightarrow a^1b^0c^0\Rightarrow a^0b^0c^1\Rightarrow a^0b^2c^0$	$\Rightarrow a^2b^0c^0\Rightarrow a^0b^0c^2$	The joint scale
d^0	d^1	d^2	

Composite criterion D

Fig. 2. Scale of the composite criterion obtained by the ZAPROS method.

By building a scale of the criterion D with the ZAPROS method, a user can determine the generalized grades as the upper and lower boundaries the formed classes of all corteges of initial grades (Fig. 3).

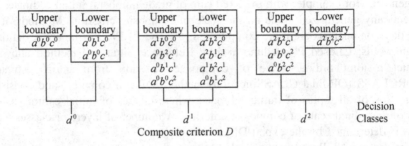

Upper boundary	Lower boundary	Upper boundary	Lower boundary	Upper boundary	Lower boundary	
$a^0b^0c^0$	$a^0b^1c^0$	$a^1b^0c^0$	$a^2b^2c^0$	$a^2b^2c^1$	$a^2b^2c^2$	
	$a^0b^0c^1$	$a^0b^2c^0$	$a^2b^1c^1$	$a^1b^0c^2$		
		$a^0b^1c^1$	$a^1b^2c^1$	$a^0b^2c^2$		
		$a^0b^0c^2$	$a^0b^1c^2$			
d^0		d^1		d^2		Decision Classes

Composite criterion D

Fig. 3. Scale of the composite criterion obtained by the ORCLASS method.

For building rating scales of composite criteria, a user can apply different techniques simultaneously. For example, generate scales of some composite criteria with the tuple stratification technique, and scales of other criteria with the multicriteria ordinal classification.

The multi-stage construction of hierarchical systems of criteria and formation of verbal rating scales of criteria are subjective non-formalized procedures. In the PAKS-M technology, several hierarchical systems of composite criteria with different rating scales, that variously aggregate initial characteristics of objects, are built. In order to reduce the influence of the methods' features and simplify the construction of criteria scales, it is suggested to form the scales with a small number of verbal grades applying various methods and/or their combinations at different stages of the aggregation procedure.

While using several aggregation trees and several different ways to construct the scales of composite and final criteria, the original problem of choice is transformed into a problem of group multi-attribute choice that is solved at the conclusive stage. Every object is presented now in several versions (copies) with different sets of estimates, which corresponds to each hierarchical system of criteria. We consider such presentation of objects as different judgments or viewpoints of several experts/DMs, and therefore solve a problem of group choice.

In order to increase the validity of the final decision, we use several methods of group multi-attribute choice. The expert/DM analyzes the results obtained with different techniques, and makes the conclusive choice: selection of the best object(s), ranking objects, division of objects into categories (classes).

The flowchart for solving a multiple criteria choice task with the PAKS-M technology consists of the following steps (Fig. 4).

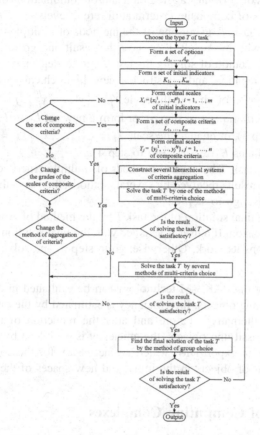

Fig. 4. The flowchart of the multi-method technology PAKS-M.

Step 1. Choose the type T of a task: T_1 – select the best option; T_2 – order options; T_3 – divide options into ordered groups.

Step 2. Form a set $A_1, \ldots, A_p$, $p \geq 2$, of options of the task T solution.

Step 3. Form a set $K_1, \ldots, K_m$, $m \geq 2$, of the initial attributes (indicators).

Step 4. Form the rating scales $X_i = \{x_i^1, \ldots, x_i^{g_i}\}$, $i = 1, \ldots, m$, of initial attributes depending on the type T of task.

Step 5. Form a set $L_1, \ldots, L_n$, $n < m$, of the composite criteria, which aggregate the initial characteristics $K_1, \ldots, K_m$, and define new properties of options chosen by a user.

Step 6. Form the rating scales $Y_j = \{y_j^1, \ldots, y_j^{h_j}\}$, $j = 1, \ldots, n$ of composite criteria. Each scale grade of a composite criterion is a combination of the grades of initial attributes. The following methods of attributes aggregation are possible: W_1 – the stratification of tuples; W_2 – the multi-criteria ordinal classification of tuples; and W_3 – ranking tuples.

Step 7. Construct some hierarchical systems for the aggregation of criteria using the different methods of attributes aggregation and/or combinations of methods in order to form the scales of composite criteria at different levels.

Step 8. Solve the task T, using one of the methods of multiple criteria choice. If a user is satisfied with the result, then save the result and go to step 9. Otherwise, either change the method of solving the task T (go to step 8), or change the method of attributes aggregation and construct a new hierarchical system of composite criteria (go to step 7), or change the scale $Y_j = \{y_j^1, \ldots, y_j^{h_j}\}$, $j = 1, \ldots, n$ of one or more composite criteria (go to step 6), or form a new set $L_1, \ldots, L_n$ of the composite criteria (go to step 5), or change the scale $X_i = \{x_i^1, \ldots, x_i^{g_i}\}$, $i = 1, \ldots, m$ of one or more initial attributes $K_1, \ldots, K_m$ (go to step 4).

Step 9. Solve the task T by several methods of multiple criteria choice. If a user is satisfied with the results received by several methods, then save the results and go to step 10. Otherwise, go to step 8.

Step 10. Find the final solution of the task T by the method of group choice, analyze and prove the decision. If a user is satisfied with the final solution of the task T, then the algorithm stops its work. Otherwise, go to step 3, and solve the task T again.

The efficiency of the PAKS-M technology can be evaluated in different ways. For example, in the ranking objects, the efficiency is estimated by the ratio of the numbers of not-comparable alternatives before and after the reduction of the attribute space dimension. In the classifying objects, the efficiency is evaluated by the comparison of numbers of calls to an expert/DM, which are necessary for the construction of consistent classifications of objects in the initial and new spaces of the attributes.

3   Evaluation of Computing Complexes

The comparison and choice of complicated technical systems, in particular, computing complexes of the personal level, is a poorly formalized and ill-structured task. These complexes are characterized by many initial quantitative and qualitative indicators and, as a rule, are not comparable by their parameters. This does not allow one to use many of the known methods of decision making to choose the best option(s). 30 indicators, chosen as the initial characteristics of computing complexes, are as follows [10].

TI. The technical characteristics of computing module (frequency of processor core; bits of processor core; quantity of streams; number of processor cores; volume of memory supported by the processor; number of processors in the module; volume of the memory in the module; presence of accelerator for universal calculations; the memory of the module; presence of optical data store).

CI. The computational characteristics of complex (number of modules in the complex; rate of exchange between modules; presence of built-in input–output facilities; presence of not-interruptible power supply; software characteristics; possibility of modernizing the hardware and software).

SI. The structural characteristics of complex (sizes of the complex (height, depth, and width); mass of the complex; protection against noise).

OI. The operational characteristics of complex (energy consumption; noise level; thermal emission; service conditions (temperature, humidity); time between failures).

CP. The complex productivity.

CM. The cost of complex manufacture.

A rating scale with three verbal grades was formed for each initial indicator. For example, the 'Complex Productivity' can be graded as CP^0 – high (>2000 Gflops); CP^1 – medium (2000–500 Gflops); and CP^2 – low (>500 Gflops), and the 'Cost of Complex Manufacture' as CM^0 – high; CM^1 – medium; and CM^2 – low (the concrete values are omitted).

A construction of aggregation-trees of initial characteristics and formation of rating scales of criteria are the most difficult and important stages in the PAKS-M technology. These not formalized procedures are based on knowledge and preferences of an expert/DM. When objects have many characteristics (tens in our case), it is rather difficult to say in advance which characteristics will be important for choosing the preferable object, how to aggregate characteristics into criteria. It is also difficult to determine which criteria would be final and which intermediate, how to combine characteristics and criteria in aggregation-trees. All of these aspects become clear only when the considered task will be solved.

For a comparison of computing complexes and selection the most preferable one, we constructed six hierarchical systems of criteria by various ways of aggregation of initial characteristics. In our case, the expert/DM established the "Complex Productivity" (CP) and "Cost of Complex Manufacture" (CM) as independent final criteria. The "Computational Indicators" (CI), "Structural Indicators" (SI), and "Operational Indicators" (OI) of complex were aggregated into the correspondent composite criteria. All of "Technical Indicators" (TI) of a computing module were included in the computational characteristics of complex as a single special attribute. The aggregation-trees of the computational, structural and operational characteristics of a complex are represented in Figs. 5, 6, and 7.

In various hierarchical systems of criteria, the criteria CP, CM, CI, SI, and OI are considered as the final criteria, and as intermediate criteria, combined differently into a single integral index of the complex perspectivity.

The first aggregation scheme includes five final criteria: CP, CM, CI, SI, and OI. We constructed scales for composite criteria CI, SI, OI using the method of tuple stratification where three variants of each scale were formed with different numbers and ranges of grades' variables. So, in this and others hierarchical systems of criteria, every computing complex is presented in three different versions (copies) with various criteria estimates, which we consider as the separate evaluations given by three independent experts.

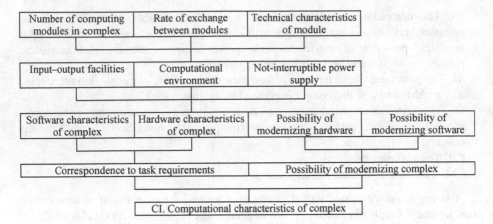

Fig. 5. Aggregation-tree of the computational characteristics of a complex.

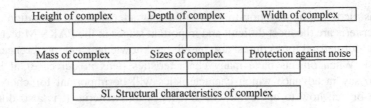

Fig. 6. Aggregation-tree of the structural characteristics of a complex.

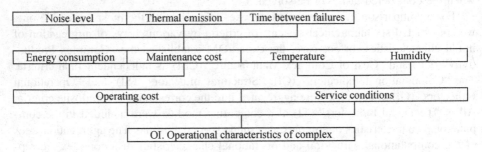

Fig. 7. Aggregation-tree of the operational characteristics of a complex.

The second aggregation scheme includes three final criteria: CP, CM, and GC. The composite criterion "Generalized Characteristics of Complex" GC = (CI, SI, OI) unites the computational, structural and operational characteristics of a complex and has the following verbal rating scale:

GC^0 – high (CI^0, SI^0, OI^0), (CI^0, SI^0, OI^1), (CI^0, SI^1, OI^0), (CI^1, SI^0, OI^0), (CI^0, SI^1, OI^1), (CI^1, SI^0, OI^1), (CI^1, SI^1, OI^0), (CI^0, SI^0, OI^2), (CI^0, SI^2, OI^0), (CI^2, SI^0, OI^0);

GC^1 – medium (CI^1, SI^1, OI^1), (CI^2, SI^0, OI^1), (CI^2, SI^1, OI^0), (CI^0, SI^2, OI^1), (CI^1, SI^2, OI^0), (CI^0, SI^1, OI^2), (CI^1, SI^0, OI^2), (CI^2, SI^1, OI^1), (CI^1, SI^2, OI^1), (CI^1, SI^1, OI^2), (CI^0, SI^2, OI^2), (CI^2, SI^2, OI^0), (CI^2, SI^0, OI^2);

GC^2 – low (CI^2, SI^2, OI^1), (CI^2, SI^1, OI^2), (CI^1, SI^2, OI^2), (CI^2, SI^2, OI^2).

We formed an integral index "Complex Category" (CC) by the various aggregations of the criteria CP, CM, CI, SI, and OI. This index characterizes the complex preference for a user and has three verbal evaluation grades: CC^0 – a perspective complex, CC^1 – a modern complex, and CC^2 – an outdated complex.

In the third aggregation scheme (Fig. 8a) we combined the criteria CP and CI into the composite criterion "Computing Potential of Complex" PC = (CP, CI). We combined the criteria SI and OI into the composite criterion "Maintenance cost of complex" MC = (SI, OI). Further, we combined the criteria PC, MC, and CM into the integral index "Complex Category" CC = (PC, MC, CM).

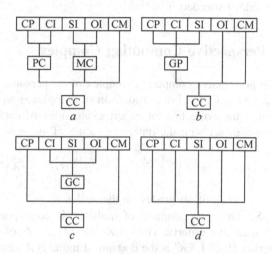

Fig. 8. Aggregation of criteria into the integral index of complex quality: (a) the third scheme, (b) the fourth scheme, (c) the fifth scheme, and (d) the sixth scheme.

In the fourth aggregation scheme (Fig. 8b), we combined the criteria CP, CI, and SI into the composite criterion "Generalized Potential of Complex" GP = (CP, CI, SI). Further, we combined the criteria GP, OI, and CM into the integral index "Complex Category" CC = (GP, OI, CM).

In the fifth aggregation scheme (Fig. 8c), we combined the criteria CI, SI, and OI into the composite criterion "Generalized Characteristics of Complex" GC = (CI, SI, OI). Then we combined the criteria CP, CM, and GC into the integral index "Complex Category" CC = (CP, CM, GC), with the following verbal rating scale:

CC^0 is a perspective complex (CP^0, CM^0, GC^0), (CP^0, CM^0, GC^1), (CP^0, CM^1, GC^0), (CP^1, CM^0, GC^0), (CP^0, CM^1, GC^1), (CP^1, CM^0, GC^1), (CP^1, CM^1, GC^0), (CP^0, CM^0, GC^2), (CP^0, CM^2, GC^0), (CP^2, CM^0, GC^0);

CC^1 is a modern complex (CP^1, CM^1, GC^1), (CP^0, CM^1, GC^2), (CP^1, CM^0, GC^2), (CP^0, CM^2, GC^1), (CP^1, CM^2, GC^0), (CP^2, CM^0, GC^1), (CP^2, CM^1, GC^0);

CC^2 is an outdated complex (CP^1, CM^1, GC^2), (CP^1, CM^2, GC^1), (CP^2, CM^1, GC^1), (CP^0, CM^2, GC^2), (CP^2, CM^0, GC^2), (CP^2, CM^2, GC^0), (CP^1, CM^2, GC^2).

In the sixth aggregation scheme (Fig. 8d), we combined five final criteria into the integral index "Complex Category" $CC = (CP, CI, SI, OI, CM)$. The ordinal scale of CC index is not given here due to its large size. We performed the grades of this scale as follows. The complex was stated as perspective (CC^0) if the sum of the criteria grades did not exceed 4. The complex was stated as modern (CC^1) if the sum of the criteria grades was equal to 5 or 6. The complex was stated as outdated (CC^2) if the sum of the criteria grades exceeded 6.

4 Selection of Perspective Computing Complex

We selected the most perspective computing complex of the personal level (CPL) from three options: CPL_1, CPL_2, CPL_3. For comparison of complexes and selection of the best one, we presented the collection of expert evaluations of each complex CPL_i, $i = 1, 2, 3$ as a multiset or set with the repeating values of attributes [7]:

$$A_i = \{k_{A_i}(z_1^1) \circ z_1^1, \ldots, k_{A_i}(z_1^{f_1}) \circ z_1^{f_1}; \ldots; k_{A_i}(z_s^1) \circ z_s^1, \ldots, k_{A_i}(z_s^{f_s}) \circ z_s^{f_s}\}.$$

The set $Z = Z_1 \cup \ldots \cup Z_5$ of all grades on the scales $Z_s = \{z_i^1, \ldots, z_i^{f_s}\}$ of the criteria CP, CM, CI, SI, OI is the domain of multiset A_i corresponding to the first aggregation scheme with five criteria. The set $Z = Z_1 \cup \ldots \cup Z_3$ of all grades on the scales Z_s of the criteria CP, CM, GC is the domain of multiset A_i corresponding to the second aggregation scheme with three criteria. The number $k_{A_i}(z_s^{e_s})$ indicates how many times the grade $z_s^{e_s} \in Z_s$, $s = 1, \ldots, 5$ or $s = 1, 2, 3$, $e_s = 0, 1, 2$ occurs in the description of a complex CPL_i. The sign $\circ$ denotes the multiplicity of the grade $z_s^{e_s}$

In order to select the most preferable option of a computing complex, in our case we used three methods of group multi-attribute choice: the weighted sum of ranks, the lexicographic ordering by grades of the evaluations, and the ARAMIS method. We built the final generalized ordering of complexes by Borda procedure [2, 8, 9].

The method of the weighed sum of ranks is based on expert evaluations of multi-attribute objects and combines different grades of rank and weights corresponding to these ranks. In this paper, the high grade has the weight 3, the medium grade has the weight 2, and the low grade has the weight 1. The objects $A_1, \ldots, A_p$ are ordered by value function Σ that is a sum of products of the grade number and the rank weight. The most preferred object has the maximum weighed sum.

The method of lexicographic ordering multi-attribute objects is based on the sequential comparison of objects by the total number of the corresponding grades of estimates. First, the objects $A_1,...,A_p$ are arranged by the number of high grades or the first places (1p), then by the number of medium grades or the second places (2p), and further by the number of low grades or the third places (3p). The most preferred object has the greatest number of high grades.

The ARAMIS method (Aggregation and Ranking Alternatives nearby the Multi-attribute Ideal Situations) is based on the comparison of multi-attribute objects $A_1,...,A_p$, which are considered as points of the Petrovsky multiset metric space [7–9]. The method allows us to arrange objects that are evaluated by several experts by many quantitative and/or qualitative criteria without pre-constructing individual rankings. The best A_+ and worst A_- objects (may be hypothetical) have, respectively, the highest and lowest scores according to judgments of all experts by all criteria. The objects are ordered by the value of the index $l_+(A_q) = d_+/(d_+ + d_-)$ of relative proximity to the best object, where $d_+ = d(A_+, A_q)$ is the distance from the object A_q to the best object A_+ and $d_- = d(A_-, A_q)$ is the distance to the worst object A_- in the metric space. The most preferred object has the minimal value $l_+(A_q)$ of relative proximity.

The multiset representations of complexes and the results of comparing complexes by many different methods are shown in Tables 1 and 2.

Table 1. The multiset representation of complexes evaluated by five criteria, and the results of comparing the complexes (the first scheme of aggregation).

Complex	CP^0	CP^1	CP^2	CM^0	CM^1	CM^2	CI^0	CI^1	CI^2	SI^0	SI^1	SI^2	OI^0	OI^1	OI^2	Σ	1p	2p	3p	l_+
CPL_1	0	3	0	0	3	0	0	0	3	3	0	0	3	0	0	33	6	6	3	0.43
CPL_2	0	0	3	3	0	0	1	2	0	0	0	3	0	0	3	28	4	5	6	0.55
CPL_3	3	0	0	0	0	3	3	0	0	0	0	3	0	0	3	27	6	0	9	0.60

Table 2. The multiset representation of complexes evaluated by three criteria, and the results of comparing the complexes (the second scheme of aggregation).

Complex	CP^0	CP^1	CP^2	CM^0	CM^1	CM^2	GC^0	GC^1	GC^2	Σ	1p	2p	3p	l_+
CPL_1	0	3	0	0	3	0	3	0	0	21	3	6	0	0.40
CPL_2	0	0	3	3	0	0	0	1	2	16	3	1	5	0.60
CPL_3	3	0	0	0	0	3	0	0	3	15	3	0	6	0.67

The results of comparing complexes according to the first aggregation scheme with five criteria are as follows (Table 1). By the weighed sums of ranks, the complexes 2 and 3 are approximately equivalent, and the complex 1 is preferable to both of them: $CPL_1 > CPL_2 \approx CPL_3$. By the lexicographic ordering, the complex 1 is preferable to the complex 3, and the complex 3 is preferable to the complex 2: $CPL_1 > CPL_3 > CPL_2$. By the ARAMIS method, the complex 1 is preferable to the complex 2, and the complex 2 is preferable to the complex 3: $CPL_1 > CPL_2 > CPL_3$. The generalized ordering complexes obtained by the Borda procedure looks as $CPL_1 > CPL_2 \approx CPL_3$.

The results of comparing the complexes according to the second aggregation scheme with three criteria are as follows (Table 2). By the weighed sums of ranks, the complexes 2 and 3 are approximately equivalent, and the complex 1 is preferable to both of them: $CPL_1 > CPL_2 \approx CPL_3$. By the lexicographic ordering, the complex 1 is preferable to the complex 2, and the complex 3 can be considered as approximately equivalent to the complex 2: $CPL_1 > CPL_2 \approx CPL_3$. By the ARAMIS method, the complex 1 is preferable to the complex 2, and the complex 2 is preferable to the complex 3: $CPL_1 > CPL_2 > CPL_3$. The generalized ordering complexes obtained by the Borda procedure looks as $CPL_1 > CPL_2 > CPL_3$ and differs from the ordering by the first scheme.

The results of comparing complexes by the integral index "Complex Category" are as follows. According to the third scheme of criteria aggregation, $CPL_1 \approx CPL_2 > CPL_3$. This result is not quite consistent with comparing the complexes using five or three criteria. According to the fourth scheme of the criteria aggregation, $CPL_1 > CPL_2 > CPL_3$. This result is generally coordinated with comparing the complexes by three criteria. According to the fifth scheme of criteria aggregation, $CPL_1 > CPL_2 \approx CPL_3$. This result is generally coordinated with comparing the complexes by five criteria. The sixth scheme of criteria aggregation shows that $CPL_1 > CPL_2 \approx CPL_3$. This result coincides with comparing the complexes by five criteria and the fifth scheme of criteria aggregation.

So, the results of selection of the most preferable complex by the first, fifth and sixth aggregation schemes are consistent with each other. These results have a clear explanation. Really, the complex 1 has two high estimates (zero grade on the scale) by the criteria SI and OI, two medium estimates (the first grade on the scale) by the criteria CP and CM, as well as one low estimate (the second grade on the scale) by the criterion CI. The high and medium estimates by majority of criteria stay the complex 1 at the first place in the preference compared with the complexes 2 and 3. The complexes 2 and 3 yield to the complex 1 by the final estimates and are approximately equivalent. In contrast, the results of selection of the most preferable complex by the third and fourth aggregation schemes differ from each other and from the first and second aggregation schemes. Thus, using various ways of criteria aggregation, we obtained the same result in most cases. Namely, the computing complex 1 is the most preferable.

5 Conclusions

The new multi-method PAKS-M technology for solving problems of multi-attribute choice of a large dimension was proposed. In this technology, in order to aggregate object attributes we use the procedures for the reduction of the attribute space dimension together with various methods of decision making and/or their combinations. An important feature of the PAKS-M technology is the opportunity to create several hierarchical systems with different degrees of criteria aggregation, in particular, with a single integral index of quality having a verbal scale, solve an initial problem of choice by many methods, and give a clear explanation of the decision made. The analysis of the results obtained for different aggregation schemes allows us to choose the most convenient systems of criteria and apply these criteria for increasing the validity of the choice of the preferred option.

The PAKS-M technology is certainly universal, because it allows us to operate both with symbolical (qualitative) and numerical (quantitative) data. An attractive feature of PAKS-M technology is also the ability to use various methods of individual and collective decision making and procedures of processing information. Practical applications demonstrated the efficiency of the developed tools for solving the hard problems of multiple criteria choice.

Acknowledgments. This work was supported by the Russian Foundation for Basic Research (projects 16-29-12864, 17-07-00512, 17-29-07021, 18-07-00132, 18-07-00280).

References

1. Doumpos, M., Zopounidis, C.: Multicriteria Decision Aid Classification Methods. Kluwer Academic Publishers, Dordrecht (2002)
2. Hwang, C.L., Lin, M.J.: Group Decision Making Under Multiple Criteria. Lecture Notes in Economics and Mathematical Systems, vol. 281. Springer, Berlin (1987). https://doi.org/10.1007/978-3-642-61580-1
3. Hwang, C.L., Yoon, K.: Multiple Attribute Decision Making – Methods and Applications: A State of the Art Survey. Lecture Notes in Economics and Mathematical Systems, vol. 186. Springer, New York (1981). https://doi.org/10.1007/978-3-642-48318-9
4. Keeney, R.L., Raiffa, H.: Decisions with Multiple Objectives: Preferences and Value Tradeoffs. Cambridge University Press, Cambridge (1993)
5. Larichev, O.I.: Verbal Decision Analysis. Nauka, Moscow (2006). in Russian
6. Nogin, V.D.: Decision-Making in Multicriteria Environment: Quantitative Approach. Fizmatlit, Moscow (2005). in Russian
7. Petrovsky, A.B.: Spaces of Sets and Multisets. Editorial URSS, Moscow (2003). in Russian
8. Petrovsky, A.B.: Group verbal decision analysis. In: Adam, F., Humphreys, P. (eds.) Encyclopedia of Decision Making and Decision Support Technologies, pp. 418–425. IGI Global, Hershey, New York (2008)
9. Petrovsky, A.B.: Decision Making Theory. Publishing Center Academiya, Moscow (2009). in Russian
10. Petrovsky, A.B., Lobanov, V.N.: Multi-criteria choice in the attribute space of large dimension: multi-method technology PAKS-M. Sci. Tech. Inf. Process. **42**(5), 76–86 (2015). Artificial Intelligence and Decision Making (Iskusstvennyiy intellekt i prinyatiye resheniy) 3, 92–104 (2014). in Russian
11. Roy, B., Bouyssou, D.: Aide Multicritère à la Décision: Méthodes et Cas. Economica, Paris (1993)
12. Saaty, T.: Multicriteria Decision Making: The Analytic Hierarchy Process. RWS Publications, Pittsburgh (1990)
13. Steuer, R.: Multiple Criteria Optimization: Theory, Computation, and Application. Wiley, New York (1986)
14. Vincke, P.: Multicriteria Decision Aid. Wiley, Chichester (1992)
15. Zadeh, L.A.: From computing with numbers to computing with words – from manipulation of measurements to manipulation of perceptions. IEEE Trans. Circ. Syst. **45**(1), 105–119 (1999)

Fuzziness in Information Extracted from Social Media Keywords

Shahnaz N. Shahbazova[1](✉) and Sabina Shahbazzade[2]

[1] Azerbaijan Technical University, Baku 1073, Azerbaijan
shahbazova@gmail.com, shahbazova@berkeley.edu
[2] The George Washington University, Washington DC 20052, USA
shahbazzade@gwu.edu

Abstract. Social media becomes a part of our lives. People use different form of it to express their opinions on variety of ideas, events and facts. Twitter, as an example of such media, is commonly used to post short messages – tweets – related to variety of subjects.

The paper proposes an application of fuzzy-based methodologies to process tweets, and to interpret information extracted from those tweets. We state that the obtained knowledge is fully explored and better comprehend when fuzziness is used. In particular, we analyze hashtags and keywords to extract useful knowledge. We look at the popularity of hashtags and changes of their popularity over time. Further, we process hashtags and keywords to build fuzzy signatures representing concepts associated with tweets.

1 Introduction

In this paper, we describe a simple methodology of analyzing a set of Twitter hashtags. The main focus of the method is investigation of temporal aspects of this data. We are interested in analysis of hashtags from the point of view of their dynamics. We identify groups of hashtags that exhibit similar temporal patterns, look at their linguistic descriptions, and recognize hashtags that are the most representative of these groups, as well as hashtags that do not fit the groups very well. The presented and used method is based on a fuzzy clustering process. Once the clusters are created we examine obtained clusters in detail and draw multiple conclusions regarding variations of hashtags over time. Further, we construct fuzzy signatures of political parties based on analysis of hashtags and noun-phrases extracted from a set of tweets associated with US elections of 2012. We use obtained signatures to analyze similarities between issues and opinions important for each party.

The paper is divided into the following sections. We start with a brief introduction to the concepts of tweets, fuzzy sets, and fuzzy clustering, Sect. 2. Section 3 provides a brief description of used data: hashtags collected from *Hashtagify.me*; and tweets associated with US elections 2012. Further, we focus on analysis of hashtags – we provide some examples of hashtag popularity; describe a data pre-processing leading to representation of popularity changes. Section 4 contains discussion and conclusion.

© Springer Nature Switzerland AG 2018
S. O. Kuznetsov et al. (Eds.): RCAI 2018, CCIS 934, pp. 138–144, 2018.
https://doi.org/10.1007/978-3-030-00617-4_13

2 Hashtags and Clustering

2.1 Tweet and Hashtags

Twitter – one of the most popular online message systems – allows its users to post short messages called tweets. According to *dictionary.com* [14], the definition of a tweet is:

> *"... 2. (Digital Technology) a very short message posted on the Twitter website: the message may include text, keywords, mentions of specific users, links to websites, and links to images or videos on a website."*

The users posting these messages include special words in the text. These words – hashtags – are easily recognizable and play the role of "connectors" between messages. An informal definition of hashtags – obtained from Wikipedia [15] – is as follows:

> *"A **hashtag** is a type of label or metadata tag used on social network and microblogging services which makes it easier for users to find messages with a specific theme or content. Users create and use hashtags by placing the hash character (or number sign) # in front of a word or unspaced phrase, either in the main text of a message or at the end. Searching for that hashtag will then present each message that has been tagged with it."*

As it can be induced, hashtags carry quite a weight regarding marking and identifying topics the users wants to talk about or draw attention to. The spontaneous way hashtags are created – there are no restrictions regarding what a hashtag can be – is their crucial feature. This allows for forming a true image of the users' interests, things important for them, and things that draw their attention. As the result, any type of analysis of hashtag data could lead to a better understanding of the users' attitudes, as well as detection of events, incidents, and calamities.

2.2 Fuzzy Sets

Fuzzy set theory [13] aims at handling imprecise and uncertain information in various domains. Let D represents a universe of discourse. A fuzzy set F with respect to D is defined by a membership function $\mu_F\colon D \to [0,1]$, assigning a membership degree $\mu(d)$ to each $d \in D$. This membership degree represents the level of belonging of d to F. A fuzzy set can be represented as pairs:

$$F = \left\{ \frac{\mu(d_1)}{d_1}, \frac{\mu(d_2)}{d_2}, \dots \right\}$$

For more information on fuzzy sets and systems, please consult [17, 18].

2.3 Fuzzy Clustering

One of the most popular methods of analysis of data focuses on identifying clusters of data-points that exhibit substantial levels of similarity. There are multiple methods of clustering data that differ in their ability to find data clusters, and their complexity [4, 6, 7, 12].

Among many clustering algorithms there are ones that utilize fuzzy methodology [1, 2, 5]. In such a case, clusters of data-points do not have sharp boarders. In general, data-points belong to clusters to a degree. In the fuzzy terminology, we talk about a degree of belonging (membership) of a data-point to a given cluster. As the result, there are points that fully belong to a given cluster – membership value of 1, as well as points that belong to a cluster to a degree – membership values between 0 and 1. Such an approach provides more realistic segregation of data – very rarely we deal with a situation that everything is clear, and data can be divided into sets of data-points that are "clean", i.e., contain points that simply belong or do not belong to clusters.

The method used here is based on a fuzzy clustering method called FANNY [9]. The optimization is performed via minimizing the following objective function

$$\sum_{v=1}^{k} \frac{\sum_{i=1}^{n} \sum_{j=1}^{n} \mu_{iv}^{r} \mu_{jv}^{r} d(i,j)}{2 \sum_{j=1}^{n} \mu_{jv}^{r}}$$

where n is a number of data-points, k is a number of clusters, μ is a membership value of a data-point to a cluster, $d(i,j)$ is a distance or difference between points i and j.

The selection of that approach has been dictated by the fact that we do not want to create fictitious centers of clusters, as it happens in widely popular fuzzy clustering method FCM [3]. Additionally, there is its new implementation in R programming language [15] that is used here.

2.4 Cluster Quality and Visualization

Clusters contain multiple data-points that are distributed in the space embraced by the clusters' boundaries. Some of these points are quite inside – have high values of membership, while some are close to the boundaries – have small values of membership while at the same time they have comparable values of membership to other clusters. An interesting measure indicating quality of a cluster, i.e., demonstrating that data-points that belong to this cluster are well fitted into it, is called silhouette width [11]. This measure is represented by the following ratio for a given element i from a cluster k:

$$s(i,k) = \frac{OUT(i) - IN(i,k)}{\max(OUT(i), IN(i,k))}$$

with

$$OUT(i) = \min_{j \neq k} \left(\frac{\sum_{m=1}^{N_j} d(i,m)}{N_j} \right), \qquad IN(i,k) = \frac{\sum_{m=1}^{N_k} d(i,m)}{N_k}$$

where $d(i,m)$ is a distance (or a difference) between data-points i and m, N_k is a size of cluster k, N_j is a size of any other cluster. The value of $s(i,k)$ allows us to identify the closest cluster to a point i outside the cluster k. Positive values of silhouette indicate good separation of clusters.

The process of visualization of multi-dimensional clusters is fairly difficult. A possible solution could be a projection of clusters into selected dimensions. But then, the issue is which dimensions to choose. In his paper, we use an approach introduced in [10]. The approach called CLUSPLOT is based on a reduction of the dimension of data by principal component analysis [8]. Clusters are plotted in coordinates representing the first two principal components, and are graphically represented as ellipses. To be precise, each cluster is drawn as a spanning ellipse, i.e., as a smallest ellipse that covers all its elements.

3   Collected Data

3.1   Hashtag Data

The process of data analysis is performed on real data representing popularity of hashtags. The data are obtained from the website *Hashtagify.me*, and contain information about 40 different hashtags. The popularity ratings have been obtained for the period of nine weeks. The sample of data for a few selected hashtags is shown in Table 1.

Table 1. Popularity of selected hashtags

Hashtag	Popularity (# weeks in the past)									
	-nine	-eight	-seven	-six	-five	-four	-three	-two	-one	zero
#KCA	100.0	100.0	100.0	100.0	100.0	93.0	84.3	85.2	81.1	75.9
#callmebaby	0.0	0.0	30.3	20.9	65.5	98.0	95.1	89.1	85.7	87.3
#SoFantastic	56.9	71.5	72.7	73.5	75.8	97.4	92.3	36.6	37.7	48.8
#Nepal	44.0	44.5	42.3	42.5	42.6	42.5	43.2	73.8	80.8	63.6
#NepalQuake	0.0	0.0	0.0	0.0	0.0	0.0	0.0	59.9	72.5	53.1
#iphone	80.4	80.8	81.9	81.2	79.8	81.9	83.2	84.6	84.1	82.8

The values presented in Table 1 show the popularity (as % relative to other hashtags) for every week. For example, *#KCA* was the most popular hashtag for the first five weeks. However after the fifth week, its popularity has started decreasing. The hashtag *#callmebaby* did not even exist for the first few weeks, than rapidly gained popularity, and after two weeks its popularity has been around 85%. Very similar behavior can be observed for the *#NepalQuake*. Its popularity in the last few weeks has been in the range from 53% to 72% [13, 16]. The hashtag *#iphone*, on the other hand, is characterized via a continuous – with some small fluctuations – level of popularity: 80 to 84%.

Table 2. Changes in popularity of selected hashtags

Hashtag	Popularity (# weeks in the past)				
	zero vs nine	zero vs seven	zero vs five	zero vs thee	zero
#KCA	−24.1	−24.1	−24.1	−8.4	75.9
#callmebaby	87.3	57.0	21.8	−7.8	87.3
#SoFantastic	−8.1	−23.9	−27.0	−43.5	48.8
#Nepal	19.6	21.3	21.0	20.4	63.6
#NepalQuake	53.1	53.1	53.1	53.1	53.1
#iphone	2.4	0.9	3.0	−0.4	82.8

In order to analyze behavioral patterns of hashtags a simple processing of data has also been performed. Here, we are interested in the percentage of changes of popularity of hashtags. The data are presented in Table 2. Here, the calculations have been done using a very simple formula [19, 20]:

$$change_{zero\,vs\,N} = popularity_{week:zero} - popularity_{week:N}$$

The calculated change represents a difference between the popularity value for the current week *week:zero* and the popularity value for a considered *week:N*. For example, the value $change_{zero\,vs\,nine} = -24.1$ means that the popularity of *#KCA* in week **zero** is *−24.1* lower than its popularity in week *-nine*.

3.2 Presidential Election 2012 Data

The created data set focuses on elections in United States. We have selected elections of 2012. The main reason for such a selection is an importance and scope of the 2012 elections. The elections were a very large event in the US history. They consist of the following elections: (1) the 57th presidential election; (2) Senate elections; and (3) House of Representative elections.

The first step in collecting tweets of the members of parties has been creation of a list of Twitter accounts of members of a parliament and most important members of parties. Twitter has a feature called *twitter list* where you can create a collection of Twitter accounts for people to follow. Almost all parties share lists of party members, parliament members or party related accounts. Such lists enable to promote the party's ideas, and make it easy to follow news related to the party. Also, some websites offer such lists for individuals to follow. In order to create our own lists for each party, we merge all accounts that appear in those party lists in one list. Such created list will be used to collect tweets.

The collection process has been done using the Twitter Search API. We have constructed a program – twitter search collector – that periodically collects and stores tweets using eight different API keys. The program requests only tweets that have an ID higher than the last tweets we collected with the previous/last usage of the collector.

The details regarding number of tweets associated with each party are presented in Table 3.

Table 3. Collected tweets statistics

Party	Number of tweets	Number of accounts
Republican	95193	560
Democratic	95731	361
Libertarian	13202	63
Green	8625	175
Justice	62612	43
Socialism and liberation	2128	16

4 Conclusion

The presented here analysis of temporal aspects of hashtags – their popularities over time and the changes of these popularities – is an attempt to look at dynamic nature of the user-generated data. The application of fuzzy clustering shown here provides a number of interesting benefits related to fact that categorization of hashtags is not crisp. The further investigation of fuzzy-based measures leads to interesting conclusions.

The construction of fuzzy signatures based on frequency of occurrence of hashtags is an interesting approach to express importance of opinions and issues represented via tweets' hashtags and noun-phrases. A simple process of constructing such signatures is presented here. Once the signatures are obtained, they are used to compare the importance of opinions/issues articulated by groups of individuals represented by the signatures. These processes have been applied to tweets representing US elections 2012.

References

1. Wu, K.L., Yang, M.S.: Alternative c-means clustering algorithms. Pattern Recogn. **35**, 2267–2278 (2002)
2. Zadeh, L.A.: Fuzzy sets. Inf. Control **8**, 338–353 (1965)
3. http://dictionary.reference.com. Accessed 8 May 2015
4. http://www.r-project.org. Accessed 8 May 2015
5. https://twitter.com/. Accessed 8 May 2015
6. http://www.wikipedia.org. Accessed 8 May 2015
7. Klir, G., Yuan, B.: Fuzzy Sets and Fuzzy Logic: Theory and Applications. Prentice Hall, Upper Saddle River (1995)
8. Pedrycz, W., Gomide, F.: Fuzzy Systems Engineering: Toward Human-Centric Computing, Wiley-IEEE Press, New York (2007)
9. Pal, A., Mondal, B., Bhattacharyya, N., Raha, S.: Similarity in fuzzy systems. J. Uncertainty Anal. Appl. **2**(1), 18 (2014)
10. Pappis, C.P., Karacapilidis, N.I.: A comparative assessment of measures of similarity of fuzzy values. Fuzzy Sets Syst. **56**(2), 171–174 (1993)
11. Gerstenkorn, T., Manko, J.: Correlation of intuitionistic fuzzy sets. Fuzzy Sets Syst. **44**(1), 39–43 (1991)

12. Dumitrescu, D.: A definition of an informational energy in fuzzy sets theory. Stud. Univ. Babes-Bolyai Math. **22**(2), 57–59 (1977)
13. Dumitrescu, D.: Fuzzy correlation. Studia Univ. Babes-Bolyai Math. **23**, 41–44 (1978)
14. Atanassov, K.T.: Intuitionistic fuzzy sets. Fuzzy Sets Syst. **20**(1), 87–96 (1986)
15. Shahbazova, S.N.: Development of the knowledge base learning system for distance education. Int. J. Intell. Syst. **27**(4), 343–354 (2012). Wiley Periodicals, Inc., Wiley - Blackwell
16. Shahbazova, S.N.: Application of fuzzy sets for control of student knowledge. Appl. Comput. Math. Int. J. **10**(1), 195–208 (2011). Special Issue on Fuzzy Set Theory and Applications. ISSN 1683-3511
17. Koshelova, O., Shahbazova, S.N.: Fuzzy multiple-choice quizzes and how to grade them. J. Uncertain Syst. **8**(3), 216–221 (2014). www.jus.org.uk
18. Abbasov, A.M., Shahbazova, S.N.: Informational modeling of the behavior of a teacher in the learning process based on fuzzy logic. Int. J. Intell. Syst. **31**(1), 3–18 (2015). Wiley Periodicals, Inc., Wiley - Blackwell
19. Shahbazova, S.N.: Modeling of creation of the complex on intelligent information systems learning and knowledge control (IISLKC). Int. J. Intell. Syst. **29**(4), 307–319 (2014). Wiley Periodicals, Inc., Wiley - Blackwell
20. Zadeh, L.A., Abbasov, A.M., Shahbazova, S.N.: Fuzzy-based techniques in human-like processing of social network data. Int. J. Uncertainty Fuzziness Knowl.-Based Syst. **23**(1), 1–14 (2015). Special issue on 50 years of Fuzzy Sets

Some Approaches to Implementation of Intelligent Planning and Control of the Prototyping of Integrated Expert Systems

Galina V. Rybina(✉), Yury M. Blokhin, and Levon S. Tarakchyan

National Research Nuclear University MEPhI
(Moscow Engineering Physics Institute),
Kashiskoe sh. 31, Moscow 115409, Russian Federation
galina@ailab.mephi.ru

Abstract. The work discusses results of development of basic components of the intelligent software environment of the AT-TECHNOLOGY workbench intended for automation and rendering intelligent the processes of building integrated expert systems (IES) based on a problem-oriented methodology. The emphasis is on aspects of the implementation of the intelligent planner in the context of the problem of intelligent planning of prototyping processes of IES and the proposed method for its solution.

Keywords: Integrated expert system · Problem-oriented methodology
Intelligent program environment · Automated planning
Standard design procedure · Reusable component
Model of processes of prototyping

1 Introduction

Currently, integrated expert system (IES), with a single large-scale development strategies and solutions of various formalized and informalized tasks are most popular in many industrial and socially important areas. An analysis of the experience of developing internationally recognized IES (e.g., [4,5,13]), as well as domestic IES, created on the basis of a task-oriented methodology for constructing an IES and supporting the toolkit AT-TECHNOLOGY [7,8] has shown that the complexity and complexity of the stages of the analysis of requirements and general and detailed design of the IES, and the specific influence of the specific domain and the human factor play a significant role.

The main problems that show the need to create highly effective tools for automation and rendering intelligent the processes of IES development are as follows:

- The impossibility of knowledge engineers (knowledge analysts) to fully determine the requirements for the systems being developed, the lack of reliable

© Springer Nature Switzerland AG 2018
S. O. Kuznetsov et al. (Eds.): RCAI 2018, CCIS 934, pp. 145–151, 2018.
https://doi.org/10.1007/978-3-030-00617-4_14

methods for assessing the quality of verification and validation of prototypes of the IES, and the inapplicability of traditional tracing technology to the knowledge base (KB) of the IES;
- A significant increase in the number of intermediate stages and iterations in the life cycle models of the prototype construction of the IES, depending on the architecture of the IES being designed, the types of knowledge sources and the percentage of unreliable and temporal information;
- Practical absence of tools providing not only the automated design of software and information support for applied IES at all stages of the life cycle, but also reducing the intellectual burden on knowledge engineers.

Typically, the vast majority of commercial tools designed for developing intelligent systems, for example G2 (Gensym Corp.), RTWorks (Talarian Corp.), SHINE (NASA/JPL), RTXPS (Environmental Software & Services GmbH), etc. do not "know" what the knowledge engineer desings and develops with their help, so the effectiveness of the application of these tools is completely determined by the art of the developers. In the context of solving the above-mentioned problems, certain results were obtained by approaches such as KBSA (Knowledge Based Systems Assistent) and KBSE (Knowledge Base Software Engineering) integrating, at best, the capabilities of expert systems development tools and CASE tools.

One of the most well-known tool of the new generation, in which the KBSE approach is implemented, which includes elements of "intellectualization" of the processes of developing intelligent systems, is the AT-TECHNOLOGY [7,8] workbench, which not only provides automated support for all phases of the life-cycle building on the basis of a problem-oriented methodology, but also allows us to carry out intelligent assisting (planning) the actions of knowledge engineers through the use of so-called technological knowledge about the standard design procedures (SDP) and reusable components (RUC), which are in aggregate the basic components of the model of the intelligent software environment, the concept of which is described in detail in [7] and other works.

New results in the development of the intelligent software environment of the AT-TECHNOLOGY workbench were obtained by expanding the functionality and increasing the role in the prototyping processes of the IES of the main operational component, an intelligent planner. The analysis of the current versions of the intelligent planner [7,8] showed that with the complication of the models of IES architectures, as well as the appearance of a large number of SDP and RUC in the technological KB, the time and laboriousness of the search for solutions has increased significantly and the negative effect of the non-optimal choice of solutions became more significant.

Therefore, there was a need to improve the methods and planning algorithms used by the intelligent planner. The results of the system analysis of modern methods of intelligent planning and the cycle of experimental studies [9,10] have shown the effectiveness of applying a fairly well-known approach to the implementation of the new version of the intelligent planner related to state space planning.

A detailed description of the basic components of the intelligent software environment of the AT-TECHNOLOGY workbench, intended for automation and intellectualization of the processes of building IES of various architectural typologies on the basis of a problem-oriented methodology, is given in [10,12], and the focus of this work is a discussion of the general formulation of the task of generating plans for the development of prototypes of IES, approaches to its solution and the features of the software implementation of an intelligent planner.

2   Model of Prototyping Processes of Applied IES

An important feature of the problem-oriented methodology and intelligent software environment of the AT-TECHNOLOGY workbench is rendering intelligent rather complex and time-consuming processes of prototyping in applied IES at all phases of the life cycle, starting from requirement analysis up to creating a series of IES prototypes. To reduce the intellectual burden on knowledge engineers, to minimize possible erroneous actions and time risks in the prototyping of IES, according to [7], it is envisaged to use a technological KB containing a significant number of SDP and RUC reflecting the expertise of knowledge engineers in the development of applied IES (static, dynamic, tutoring).

Accordingly, the formal statement of the problem of intelligent planning of the prototyping processes of IES is considered in the context of the model of prototyping processes of IES in the following form [11]: $M_{proto} = \langle T, S, Pr, Val, A_{IES}, PlanTask_{IES} \rangle$, where T is the set of problem domains for which applied IES are created; S is the set of prototyping strategies; Pr is the set of created prototypes of IES based on a problem-oriented methodology; Val is the function of expert validation of the prototype of the IES, determining the need and/or the possibility of creating subsequent IES prototypes for a particular problem domain; A_{IES} is the set of all possible actions of knowledge engineers in the prototyping process; $PlanTask_{IES}$ is the function of planning knowledge engineer's actions to obtain the current prototype of IES for a particular problem domain. A detailed description of all the components of M_{proto} is given in [11], as well as the specification of some basic components of the model of the intelligent software environment.

For the effective implementation of the $PlanTask_{IES}$ component of the M_{proto} model, an analysis was made of modern methods of intelligent state space planning applied to the investigation of the mechanisms of actions of knowledge engineers in the construction of architecture models (M_{IES}) of different IES [7] at the initial stages of the lifecycle (requirements analysis, general and detailed design). Experiments have shown that the best results are achieved if the search space is formed by modeling the actions of the knowledge engineer while constructing fragments of the M_{IES} model using the appropriate SDPs (for a formal description of this process, graph theory is used by reducing the problem to the problem of covering of the M_{IES} model, represented as a labeled graph, with SDP fragments).

In general, the concept of the M_{IES} model plays a key role in the generation of plans for prototyping IES. In [7], the methods of constructing M_{IES} based on the mapping of interactions of real systems by means of structural analysis expanded by the inclusion of the special element "non-formalized operation" (NF operation), which points to the need of involving experts and/or other sources of knowledge. Therefore, M_{IES} model of an IES prototype is constructed as a hierarchy of extended data flow diagrams (EDFD).

An important role in the implementation of the PlanTaskIES component is played by the plan for constructing a specific coverage (i.e., the sequence of applied SDP fragments), which can be uniquely transformed into a plan for constructing a prototype of IES. In this case, the plan for constructing the prototype of the IES can be represented like [11]: $Plan = \langle A_G, A_{atom}, R_{prec}, R_{detail}, PR \rangle$ where A_G is the set of global tasks decomposable into subtasks); A_{atom} is the set of planned (atomic) tasks, the execution of which is necessary for the development of the IES prototype; R_{prec} is the function that determines the predecessor relation between the planned tasks; R_{detail} is the relation showing the affiliation of the planned task to the global tasks; PR is the representation of the plan.

Thus, with the help of the relation R_{prec} and the sets A_G and A_{atom} two task networks are formed: enlarged and detailed. The enlarged network of tasks obtained with R_{prec} and A_G is called the global plan (the relation of the precedence between the elements of the A_G is obtained on the basis of the R_{detail} detail relation), and the detailed network of tasks obtained on the basis of R_{prec} and A_{atom} is called the detailed plan, with each planned task associated with specific function by a function of a specific operational RUC.

Based on the experience of developing applied IES, a certain limitation is imposed on the structure of SDP: at least one compulsory fragment and an arbitrary number of optional fragments associated with the mandatory fragment of at least one data flow. As a rule, the first element of the set of fragments includes an non-formalized operation (NF-operation), which plays the main role in the construction of the model M_{IES} of the current prototype of IES.

3   Formulation of the Problem of Generation of Plans for the Development of Prototypes of IES

Let us consider the general formulation of the problem of generation of plans for developing prototypes of IES where the initial data is: the model of the architecture of the prototype of the IES (M_{IES}), described using the hierarchy of EDFD; the technological KB, containing a set of SDP and RUC. In addition, a number of restrictions and working definitions is introduced [11]:

1. From the composition of the model M_{IES} only a set of elements and a set of data flows represented in the form of a marked oriented graph G_{RIL}, where the labels determine the relationship between the elements of the hierarchy of the EDFD and the vertices and arcs of the graph. This graph is called a generalized EDFD, and it can be uniquely obtained from the initial model of the architecture M_{IES}.

2. The fragment of the generalized EDFD is an arbitrary connected subgraph contained in G_{RIL}. An SDP instance is an aggregate of TPP and a fragment of a generalized EDFD satisfying the conditions of applicability (the C component of the TPP model) of the corresponding SDP. The cover ($Cover$) of a generalized RDPA is the set of instances of SDP with mutually disjoint fragments containing all the vertices of G_{RIL} (or cover the entire G_{RIL}).
3. A coarse coverage of a generalized EDFD is a EDFD coverage in which all instances of the SDP contain only mandatory fragments. The fine coverage is an extension of the coarse coverage by including optional fragments in the coverage.
4. The inclusion of each fragment of the SDP in the cover is compared with a certain value determined by an expert evaluation.

Thus, the problem of generating a plan for developing a prototype of IES, taking into account the initial data and imposed constraints, is conveniently presented in terms of states and transitions [1–3,6] in the form of a model [11]: $PlanTask_{IES} = <S_{IES}, A_{IES}, \gamma, Cost, s_0, G_{IES}, F_{COVER}>$, where S_{IES} is the set of states of the graph G_{RIL} that describe the current coverage $Cover$; A_{IES} is the set of possible actions over G_{RIL}, which are the addition of fragments of specific instances of SDP to the cover (the total set is formed in the aggregate WKB and G_{RIL}); γ is the transition function between states; Cost is a function that determines the cost of a sequence of transitions; s_0 is the initial state describing the empty coverage; G_{IES} is the function of determining whether the state belongs to the target state; F_{COVER} is the function of generating a development plan ($Plan$) from the coverage ($Cover$). The solution of $PlanTask_{IES}$ is the plan of actions of the knowledge engineer (model $Plan$).

To solve the problem, a method was developed (described in details in [11]) with four stages: obtaining a generalized EDFD (G_{RIL}) from the model M_{IES}; generating an exact $Cover$ cover using heuristic search; generation of the plan of actions of the knowledge engineer ($Plan$) on the basis of the obtained fine coverage ($Cover$); generating a plan view (PR) based on coverage ($Cover$). For every stage necessary algorithms – including domain-specific heuristic function for reducing search space – were developed and experimentally studied.

4   General Characteristics of Basic Components of the Intelligent Software Environment of AT-TECHNOLOGY Workbench

The features of the implementation of the basic components of the intelligent software environment of the AT-TECHNOLOGY complex and the technology of the development of the IES based on it (1) were considered in detail in a plenty of works. Hence, we give here just a brief description of the software tools of the intelligent software environment, including the kernel, the user interface subsystem and the extension library for interaction with operational RUCs. The kernel implements all the basic functionality of automated support for the development of prototype IES, project file management, extension management, etc.

The technological KB is conventionally divided into an extension library that stores operational knowledge in the form of plug-ins that implement the relevant operational RUCs and the declarative part. The subsystem of the user interface has a convenient graphical interface, on the basis of which the RUC interacts with the knowledge engineer using screen forms (Fig. 1).

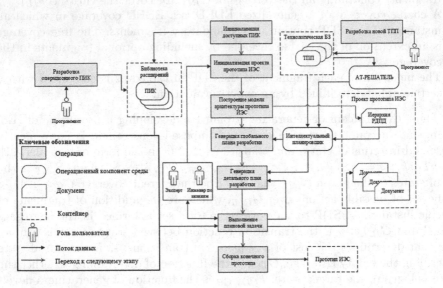

Fig. 1. The technology of the development of IES using intelligent software environment

The intelligent planner, being a part of the kernel, implements functionality related to planning the prototyping processes of IES: with the help of the preprocessor of the hierarchy of the EDFD, the pre-processing of the hierarchy of the EDFD is carried out by converting it into one generalized diagram of maximum detailing; the task of coverage of the detailed EDFD with the available SDP is performed with the help of a global plan generator that, on the basis of the technological KB and the constructed generalized EDFD, performs a fine coverage construction; generator of the detailed plan on the basis of the given coverage of the EDFD and the technological KB performs a detalization of each element of the coverage (thus forming a preliminary detailed plan); based on the analysis of available RUC, a detailed plan is formed in the plan interpretation component, where each task is related to a specific RUC and then with the help of the final plan building component, the required plan representation is generated.

By now, the technology of constructing prototypes of IES using a new version of an intelligent planner, three SDPs (including the SDP "Construction of dynamic IES"), as well as a set of operational and information RUC and other means of intelligent software environment of the AT-TECHNOLOGY workbench, has been experimentally studied.

5   Conclusion

Based on the data obtained as a result of joint testing of the intelligent planner and other components of the intelligent software environment of the AT-TECHNOLOGY workbench, it was shown that all the developed tools can be used quite effectively in the prototyping of applied IES. The software was integrated into the AT-TECHNOLOGY workbench, with the use of which several prototypes of a dynamic IES were developed, e.g., a prototype for managing medical forces and facilities for major road accidents and a prototype of dynamic IES for satellite network resources control.

Acknowledgements. The work was supported by the Russian Foundation for Basic Research (project №18-01-00457).

References

1. Geffner, H., Bonet, B.: A concise introduction to models and methods for automated planning. Online access: IEEE (Institute of Electrical and Electronics Engineers) IEEE Morgan & Claypool Synthesis eBooks Library, Morgan & Claypool (2013). https://books.google.ru/books?id=_KInlAEACAAJ
2. Ghallab, M., Nau, D., Traverso, P.: Automated Planning and Acting, 1st edn. Cambridge University Press, New York (2016)
3. Ghallab, M., Nau, D.S., Traverso, P.: Automated Planning - Theory and Practice. Elsevier, New York City (2004)
4. Giarratano, J.C., Riley, G.D.: Expert Systems: Principles and Programming, 4th edn. Course Technology, Boston (2004)
5. Meystel, A.M., Albus, J.S.: Intelligent Systems: Architecture, Design, and Control, 1st edn. Wiley, New York (2000)
6. Nau, D.S.: Current trends in automated planning. AI Mag. **28**(4), 43–58 (2007)
7. Rybina, G.V.: Theory and technology of construction of integrated expert systems. Monography. Nauchtehlitizdat, Moscow (2008)
8. Rybina, G.V.: Intelligent Systems: from A to Z. Monography Series in 3 Books. Knowledge-Based Systems. Integrated Expert Systems, vol. 1. Nauchtehlitizdat, Moscow (2014)
9. Rybina, G.V., Blokhin, Y.M.: Methods and means of intellectual planning: implementation of the management of process control in the construction of an integrated expert system. Sci. Tech. Inf. Process. **42**(6), 432–447 (2015)
10. Rybina, G.V., Blokhin, Y.M.: Use of intelligent planning for integrated expert systems development. In: IS 2016 - Proceedings of the 2016 IEEE 8th International Conference on Intelligent Systems, pp. 295–300 (2016)
11. Rybina, G.V., Blokhin, Y.M.: Methods and software of intelligent planning for integrated expert systems development. Artif. Intell. Decis. Mak. **1**, 12–28 (2018)
12. Rybina, G.V., Blokhin, Y.M.: Intelligent software environment for integrated expert systems development. In: Proceedings of the 2016 International Conference on Artificial Intelligence ICAI 2016. CSREA Press, USA (2016)
13. Schalkoff, R.: Intelligent Systems: Principles, Paradigms, and Pragmatics. Jones & Bartlett Learning (2011). https://books.google.ru/books?id=80FXUtF5kRoC

Ontology for Differential Diagnosis of Acute and Chronic Diseases

Valeriya Gribova, Dmitry Okun, Margaret Petryaeva, Elena Shalfeeva,
and Alexey Tarasov[✉]

Institute for Automation & Control Processes, FEBRUS, Vladivostok, Russia
atarasov@dvo.ru

Abstract. The paper presents an ontology of differential diagnosis of diseases with dynamics of symptoms. The ontology allows formalizing a disease as a continuously developing internal process. This can be useful for developers of medical diagnostic systems and for specialists in the area of theory and practical application of ontologies.

Keywords: Ontology · Differential diagnostics · Disease · Signs of patients
Decision-making support system

1 Introduction

The amount of knowledge in medicine is growing dramatically. For making diagnostic decisions a practical physician needs to take into account a number of factors: symptoms and syndromes of the disease, its nosological forms, etiologies, pathogenesis, clinical manifestations, taking into account individual features of patients. The differentiation of similar diseases often requires labor-intensive and time-intensive analysis or studies; precise diagnosis requires specification of the diagnosis according to the accepted classifier, there is always incompleteness of information about all the features of an individual patient. However, time for making decision by a doctor has not increased. As a result, the number of medical errors is increasing, which, according to the estimates of literary sources, reaches up to 30% in some countries [1]. The quality of medical care depends not only on the level of training (competence) of medical personnel, but also on systems that provide support for doctor decisions [2]. For the period from the first works on the creation of intelligent systems (IS) for medical diagnostics to the present time a huge spectrum of such systems has been implemented, but they are practically not introduced into the daily practice of a doctor.

A common cause of problems with implementation of IS is huge amount of work required for development of knowledge bases relevant to the real medical practice. It is not possible to assign this task to highly qualified experts without knowledge engineers as mediators.

More than that, the IS that are to make a diagnosis instead of a doctor (or to offer him a list of possible diagnoses ordered by their probability) are not required by experienced experts if these systems do not solve the problem no better than the experts themselves. Those IS that support decisions of the doctor in a difficult situation "at the

© Springer Nature Switzerland AG 2018
S. O. Kuznetsov et al. (Eds.): RCAI 2018, CCIS 934, pp. 152–163, 2018.
https://doi.org/10.1007/978-3-030-00617-4_15

level of council of physicians" are rare and the decision support systems usually do not take into account many of the signs and factors identified from the anamnesis and other sections of the case record with personal information.

The solution of these problems is the creation of specialized ontology-oriented shells for the development of differential diagnosis systems for diagnosing diseases at various stages of their development. In such shells, the model for representation of information (knowledge and data) should be expert-oriented, understandable by experts, and correspond to a wide range of diseases (rather than a specific medical profile).

The aim of the paper is to describe an ontology for differential diagnosis of diseases taking into account the dynamics of their development for the development of support systems for making diagnostic decisions in medicine.

2 Ontology Requirements

In medical intelligent systems as well as in other systems for supporting the solution of intellectual problems the main role is played by the knowledge base in which the problem-oriented and subject-oriented knowledge is concentrated. Since medical diagnosis is a rather complex subject area, there has been accumulated experience of representing medical diagnostic knowledge via multiple different ways: logical models, product models, frames, semantic and object-oriented models of domain terms, etc.

Both logical and production models are still applied for representing knowledge, in most cases representing knowledge for a single group of diseases; frames. Semantic networks are also a common representation model. The spectrum of types of semantic networks is very wide - from networks of arbitrary kind to networks that have a root node or networks similar to decision trees. In medicine using object-oriented models of terms is popular for the representation of hierarchies of terms, especially after the release of easy-to-use tool "Protégé" for defining classes of terms as classes with various properties that are descendants of a universal "Thing" class. However, DSS built on such models usually do not support the whole process of diagnostics and some stages associated with it, such as the issuing of answers to requests for information (e.g., "the vital sign with the largest value"), the assessment of some risks, etc. Semantic networks (of an arbitrary kind) are also used for visualization of content written in the OWL language, which is intended for storage or software processing. Flowcharts of diagnostic procedure are often used for visualization, while the knowledge about this process is implemented as procedural knowledge in the DSS. Tables can also be used as a complementary way of describing relationships between medical entities.

The complexity of medical diagnosis as a domain is due to the need of considering the development of internal processes in the body and their links (with external manifestations or among themselves) which are not always intuitively understandable, and to the fact that human body is often unpredictable. Since the limitations of some methods do not allow us to express all the knowledge of modern diagnostic process in a natural way, sometimes we turn to combining multiple ways of representing knowledge.

Most knowledge engineers explicitly record the result of conceptualization of the domain (referring to it as ontology - the ontology of the domain or the ontology of tasks).

The ontology of diagnostics includes the information description structure (meta-information) and the rules for its interpretation and application for diagnosis (ontological agreements). With the release of "Protégé" ontologies became often represented as hierarchies of classes of terms. Another perspective approach to knowledge representation is the ontology-based semantic approach developed by the team of Intelligent Systems Laboratory of IACP, which makes it possible to explicitly represent the ontology of knowledge in the form of a semantic network with a root node, cycles and loops, and under the control of this ontology to explicitly represent knowledge in the form of a semantic network with the root node.

The ontologies that most of the known medical diagnostic expert systems are based upon are essentially simplified in comparison with the real-world conceptualizations of the domain. Usually they do not consider the development of pathological processes in time and the interaction of different types of cause-effect relationships, as well as the possibility of combined and complicated pathologies.

One of the first ontologies of medical diagnostics close to the real notions of medicine and matching a wide range of diseases was the "Ontology of medical diagnostics of acute diseases" [3, 4]. It describes the clinical picture of diseases in the dynamics of the pathological process (over time), as well as impacts of therapeutic measures and other events on manifestations of diseases.

Using this ontology, the knowledge bases of diseases of some body systems have been developed: for respiratory organs (bronchial asthma, pneumonia), for digestive organs (peptic ulcer, acute appendicitis, acute and chronic pancreatitis, acute and chronic colitis), for eyesight (conjunctivitis, keratitis, glaucoma), etc. [5], respective software was also designed.

The experience of more than ten years of using the ontology for the formation of knowledge bases on the diagnosis of a number of diseases has made it possible to accumulate and reveal a number of limitations: impossibility of describing the clinical manifestations of disease for different groups of patients, impossibility of describing alternative diagnostics and taking into account in the final diagnosis the forms, variants of disease and its severity level, spectrum of modality values of features.

Modern systems should be able to diagnose and conduct differential diagnosis with other diseases at different periods of disease development, analyzing disease development in the period before visit to doctor and considering that patient can come to reception at different times from the onset of disease, as in the first hours, so and at a moment when semiology is quenching.

The ontology of medical diagnostics should have the following basic features:

1. The possibility of forming symptom-complex diseases taking into account the categories of users using reference ranges instead of certain "norms" for laboratory and instrumental indicators [6].
2. The possibility of forming alternative symptom complexes with different approaches to identifying reliable signs of the disease in order to choose the most sparing, fast or inexpensive diagnostic process.
3. Ability to clarify the diagnosis, taking into account the etiology, pathogenesis, variant of the flow, etc. for differential diagnosis of diseases and the selection of appropriate methods of treatment.

4. Uniform formalization of stages of chronic diseases and periods of development of acute diseases.
5. Expansion of a number of values of modality, previously represented by "obligation" and "possibility," the case of "specificity" (for a specific symptom).
6. Taking into account the values of features and features affected by events. The presence of such an element of cause-effect relationships allows us to account for external influences exerted on the patient's organism at different stages of the disease.
7. Accounting for different variants of the dynamics of the values of features. Such elements of knowledge allow to take into account (on the basis of medical experience) the diversity of the course of the same diseases in different patients [3, 7].

All these types of connections between the concepts of medical diagnostics make it possible to form a knowledge base sufficient not only to search for hypotheses about possible diagnoses, but also to reduce the set of hypotheses about the diagnosis. The latter is solved by searching for a request for such additional information about the patient's condition (using the results of observations of the situation and knowledge of internal processes and the influence of external events on them), the answer to which will refute previously accepted diagnoses (identifying among the known measurable signs of differentiators or "partial" [8]). It is possible to diagnose the combined and complicated pathologies.

3   Basic Features of the Ontology

The form of the representation of ontologies in general and medical ones in particular (hierarchical semantic networks) used by the team (authors) and the applied IACPaaS tools [9] allow to effectively modernize existing ontologies. New versions (modernized ontologies) are taken as a basis for designing new versions of shells for IS developers.

When assessing results of examinations of different groups of people it becomes obvious that "normal" values of an indicator for one group do not always turn out to be normal for another one. For example, during pregnancy many of the woman body's biochemical parameters change, so for this category specific reference ranges for the values of these indicators are defined. For children and adolescents high level of alkaline phosphatase is not only normal, but it is also desirable - because a child should grow healthy bones. However, the same level for an adult indicates diseases: osteoporosis, metastases of bone tumors etc. Description of reference values (taking into account sex, age, occupation, pregnancy, sports, etc.) for most laboratory and instrumental symptoms increases information significance of any symptom complex.

Modern systems should be able to determine diagnosis and perform differential diagnosis taking into account other diseases at different periods of the given disease, analyzing the progress of the disease before consulting the doctor. One should take into consideration that a patient may consult the doctor at different times from the start of the disease, as in the first hours, and at a time when the symptomatology is fading.

Registration of the values of features that were changed due to the influence of events. The presence of such an element of cause-effect relationships allows accounting for external influences on the patient's organism at different stages of the disease.

Each disease is represented by alternative symptom-complexes (complexes of complaints and objective studies, laboratory and instrumental studies), necessary conditions for this disease and can contain details of the corresponding diagnosis in form, variant, severity, stage, etc. Symptom complexes in different diseases can be of different amounts: depending on the course of the disease in different age groups of patients. A necessary condition for the disease is that event, without which the disease would not have happened. In complexes of complaints, objective studies, laboratory and instrumental studies, many features are presented, the changes in the values of which are symptoms of the disease. Possible causes of the disease are events or etiological factors that led or contributed to the development of the disease. They are described by modality and temporal features (interval before the onset of the disease, duration of the event, etc.). The detailed diagnosis is an additional set of symptoms (symptom complex), which allows to make an appropriate refinement to the main diagnosis, taking into account the etiology, pathogenesis, variant of the course, stage, etc. for more detailed (deep) diagnosis or differential diagnosis of the disease (see Fig. 1). In other words, the following relations (proposals) are introduced into the knowledge model:

Fig. 1. The screenshot of the fragment of the ontology of knowledge about diagnosis of diseases on the IACPaaS platform.

- variant of the symptom complex for some disease;
- variant of the process of changing the values of the feature, feature of some symptom complex;
- special conditions necessary for the onset of a disease;
- variant of reaction to the impact of the event;
- variant of reaction to the influence of a combination of factors and some others.

The dynamics of the values of the signs allows describing the diseases taking into account one of the main complexities of the diagnostic process in medicine - the need to determine a continuously developing process (disease). Each disease develops on its own time scale. From the point of view of the speed of the development of diseases, the sharpest - up to 4 days, acute - about 5–14 days, subacute – 15–40 days and chronic, lasting months and years are distinguished. In the development of the disease, it is almost always possible to distinguish the following stages: (1) the onset of the disease (sometimes called the latent period); (2) the stage of the disease itself; (3) the outcome of the disease. Diagnosis, as a rule, is carried out at the stage of "actual disease". At this stage, the following periods of development are distinguished: (1) the period of the increase in manifestations of the disease; (2) peak period (maximum severity of symptoms); (3) the period of extinction of manifestations of the disease (the gradual disappearance of clinical symptoms) (Fig. 2).

The symptom of the disease can be simple or compound, its values are presented by the period of the dynamics of the development of the trait or the disease as a whole; the necessary conditions for the consideration of the feature can be set. Each period of dynamics is characterized by an upper and lower limit of the duration of the period, a unit of measurement of boundaries. A simple symptom is modality (the entry of a symptom into a clinical picture); in each period the feature can have more than one variant of values. Each variant of the values specifies the set of possible values of the feature, the necessary conditions for the presence of the feature, and a description of the change in the value of this feature under the influence of certain events. A composite feature contains a description of the sets of its own time-varying features with the modality of occurrence of features in the clinical picture. A feature may have a number of intrinsic features and many values.

For the description of change of development of a disease at impact on an organism offers of a look "option of reaction of process of functioning on influence of an event".

The symptom is a compound sign, it contains a description of sets of features (Fig. 2) changing on the periods with a modality of entry of the feature of sign into a clinical picture). Each feature can have a set of own features and also a set of options of values of this feature. An example of a compound sign is the complaint of the patient: belly-ache. The sign has several features: character, localization, intensity, expressiveness, irradiation, frequency, strengthening reason. Each feature can have one or several various values. For example, the feature the nature of pain (sign belly-ache) has values: sharp, stupid, pricking, cutting, pulsing, pressing, pulling. The feature of localization has values: right hypochondrium, left hypochondrium, epigastriya, right iliac area, left iliac area, mezogastriya. Feature intensity: weak, moderate, strong, sharp, the sharpest.

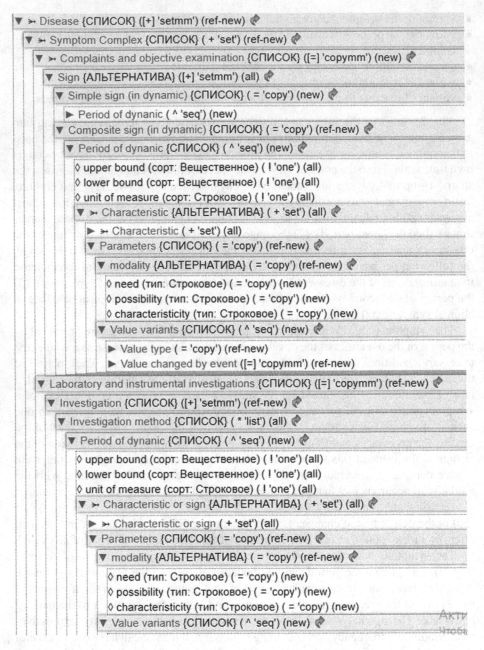

Fig. 2. The screenshot of the fragment of the structure of knowledge about symptom complexes (on the IACPaaS platform).

Example 1. The patient at the time of inspection has a complaint: an acute pain in the stomach. To the address to the doctor he took the anesthetizing drug (analginum), the acute pain in a stomach became dull.

Sign: "belly-ache";
Feature: "character";
Value of the feature: "sharp";
Influencing events: "intake of analginum";
Value the changed influence of an event (value of the feature): "dull".

Example 2. The patient with diphtheritic conjunctivitis has a hypostasis, hyperaemia, films on mucous a century. After removal of films from a conjunctiva bleeding has opened.

Sign: "conjunctiva century";
Feature: "changes on mucous a century";
Values of the feature: hypostasis, hyperaemia, diphtheritic films;
The influencing event: "removal of films";
Value the changed influence of an event (value of the feature): "bleeding".

Each variant of the feature values contains a lot of its possible values and a description of the change in the value of this feature under the influence of certain events. The term "value modified by the impact of the event" allows describing the change in the sign in the dynamics if, after the onset of the disease's development, before the patient has consulted the doctor, the patient himself took any measures, or the meanings of the symptoms (complaints, objective state) change under the influence of some events or manipulations undertaken by the doctor.

The ontology allows to describe diseases taking into account etiology, pathogenesis, option of a current, a stage, etc. by formation of additional complex of symptoms for more detailed (deep) diagnosis or differential diagnosis of a disease. In process of specification of the diagnosis for each unit of pathological process the additional or specific symptoms are described (Fig. 3).

Uniform representation and unambiguous interpretation of knowledge bases by participants of development and users (including from different institutions) requires the general set of all terms used in practice. Terminology needs to be standard and clear to medical experts, i.e. to result from ontological agreement in the field of medicine.

The general set of the terms used in practice contains names of observations, all their possible values and also their widespread synonyms used when filling stories of diseases (as "the mobility of lexical structure" and his continuous development is peculiar to medical terminology: the possibility of preservation of national and international terminological synonyms, doublets (short wind – dyspnoea) and partially coinciding synonyms, introduction to the name of a method or a symptom of a name of his opener is provided (a symptom of irritation of a peritoneum – Shchetkin Blyumberg's symptom, a sliding symptom - a symptom of "shirt" - Voskresensky's symptom).

In the "Base of medical terminology and observations" created by Laboratory of Intelligent Systems of IAPU FEB RAS such groups of observations as signs, events and factors (with the subgroups) are allocated. In particular, groups of signs are the

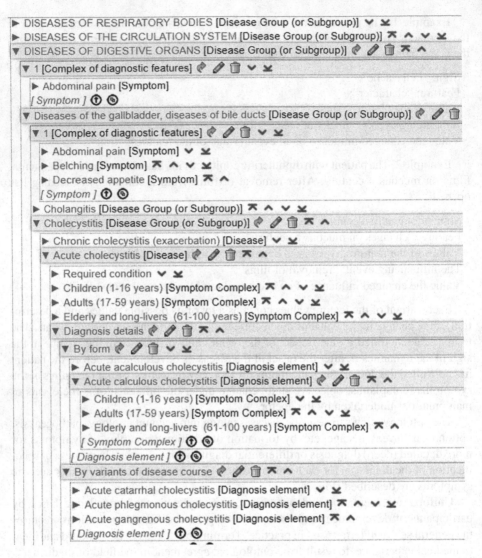

Fig. 3. The screenshot of the fragment of the knowledge about diagnosis of disease details (on the IACPaaS platform).

Complaints traditional for experts, Given an objective research, Data laboratory and Data of tool researches. In this base terms of the sections "pathogenic factors", "pharmacology", "measurement units" and others are grouped. Such "Base of observations" is the universal resource applied to formation diagnostic (and not only) knowledge bases of various areas (profiles) of medicine.

4   Application of the Formed Ontology

Based on the described ontology (using the "Database of Medical Terminology and Observations" IACPaaS), IACPaaS information resource "Knowledge Base on Diagnosis of Diseases" was formed, representing a variety of diseases (Figs. 3 and 4).

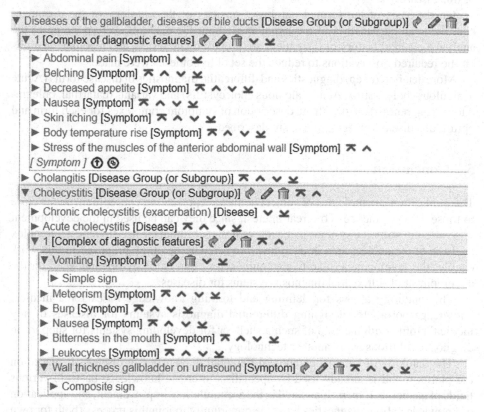

Fig. 4. The screenshot of the fragment of the knowledge about set of diagnostic features for group (on the IACPaaS platform).

The group of diseases "Diseases of the digestive system" consists of groups of diseases - "Diseases of the esophagus", "Diseases of the stomach and duodenum", "Bowel diseases", "Diseases of the liver", "Diseases of the gallbladder, biliary tract", etc. Each group includes diseases with a common set of diagnostic features, which contains those features that are feature of this group. So, the group "Diseases of the digestive organs" in the complex of diagnostic signs contains one sign "Pain in the stomach". For the group "Diseases of the gallbladder and bile ducts," the diagnostic complex includes the signs "Pain in the abdomen", "Nausea", "Skin itching", "Body temperature rise", "Stress of the muscles of the anterior abdominal wall". For the "Cholecystitis" group of diseases, more than 20 signs are already diagnostic: all the same signs plus "Meteorism", "Vomiting" "Burp", "Nausea", "Bitterness in the mouth",

"Leukocytes", "ESR", "Wall thickness gallbladder on ultrasound ", etc. The group of diseases "Cholecystitis" according to ICD 10 (International Classification of Diseases), includes the disease "Acute cholecystitis." The description of the disease includes a description of the symptom complexes for several age groups of patients (children 0–1 year, children 1–16 years, adults 17–59, etc.) and the necessary condition for the onset of this disease.

On the basis of the formed ontology, versions of the ontology of the document containing an explanation of the hypotheses about the diseases corresponding to the analyzed medical history of the patient are proposed and tested, and recommendations on the required observations to reduce the set of hypotheses were made.

More detailed (deep) diagnostics and differential diagnostics of disease forms (Acute acalculous cholecystitis, Acute calculous cholecystitis) and variants (catarrhal, phlegmonous, gangrenous), also with the description of symptom complexes for each form and option of a disease on age groups, are described.

5   Conclusion

The publication of ontologies of subject domains representing conceptualizations close to those used in practice is of great interest, since they can be used in the development of a variety of software systems in the relevant areas without the need to repeat a complex analysis of the subject area in each new project. The ontology proposed in this work enables the development of software components of an ontology-oriented shell for the development of differential diagnosis systems for diseases.

This ontology allows for defining and forming knowledge bases for a medical knowledge portal, for designing differential diagnosis systems for diseases of any medical profile with the help of such a shell, or for developing systems for supporting diagnostic solutions using another technology.

The ontology of knowledge about differential diagnostics of diseases is placed on the platform and is already used by specialists to create knowledge bases in various fields of medicine. Experts interested in the accumulation, improvement, and application of knowledge about diagnostics have the opportunity to join this process (both for own bases and for collectively used resources).

Acknowledgement. The work is carried out with the partial financial support of the Russian Foundation for Basic Research (projects nos. 18-07-01079, 17-07-00956).

References

1. Galanova, G.: Medical errors – a problem not only the doctor. Health Care Manag. **8**, 49–52 (2014). (in Russian)
2. Artificial intelligence for Medicine – the review. http://www.livemd.ru/tags/iskusstvennyj_intellekt. Accessed 20 Mar 2018. (in Russian)

3. Kleschev, A., Chernyakhovskaya, M., Moskalenko, P.: Model of ontology of subject domain "Medical diagnostics". Part 1. Informal description and definition of basic terms. NTI Ser. 2 J. **12**, 1–7 (2005). (in Russian)
4. Chernyakhovskaya, M., Moskalenko, P.: Formal description of a disease chronic pancreatitis. Inform. Control. Syst. **4**, 97–106 (2012). (in Russian)
5. Petryaeva, M., Okun, D.: Stomach ulcer in formal representation. In: XLVII of the International Scientific and Practical Conference "A Scientific Discussion: Medicine Questions", vol. 3, pp. 62–66. Internauk, Moscow (2016)
6. Kazakova, M., Lugovskaya, S., Dolgov, V.: Referensnye's values of indicators of the general blood test of the adult working population. Clinical laboratory diagnostics 6 (2012). https://cyberleninka.ru/article/n/referensnye-znacheniya-pokazateley-obschego-analiza-krovi-vzroslogo-rabotayuschego-naseleniya. Accessed 20 Mar 2018. (in Russian)
7. Zhmudyak, M., Povalikhin, A., Strebukov, A., Zhmudyak, A., Ustinov, G.: The automated system of medical diagnosis of diseases taking into account their dynamics. Polzunovsky Bull. **1**, 95–106 (2006). (in Russian)
8. Gribova, V., Kleschev, A., Shalfeeva, E.: Method of the solution of a problem of request of additional information. Des. Ontol. **3**(25), 310–322 (2017). [in Russian]
9. Gribova, V., Kleschev, A., Moskalenko, P., Timchenko, V., Fedorischev, L., Shalfeeva, E.: The IACPaaS cloud platform: features and perspectives. In: Second Russia and Pacific Conference on Computer Technology and Applications (RPC), vol. 3, pp. 80–84. IEEE (2017)

Using Convolutional Neural Networks for the Analysis of Nonstationary Signals on the Problem Diagnostics Vision Pathologies

Alexander Eremeev and Sergey Ivliev[✉]

NRU, Moscow Power Engineering Institute, Krasnokazarmennaya Street, 14,
Moscow 111250, Russia
eremeev@appmat.ru, siriusfrk@gmail.com

Abstract. The paper considers the use of convolutional neural networks for the analysis of non-stationary signals. Electroretinograms (ERG) are used to diagnose complex pathologies of vision. A method is proposed based on data clustering, that allows to extract knowledge from biophysical research data in situations where it is impossible to make an unambiguous diagnosis on the basis of available data, or there are several diseases simultaneously.

Keywords: Non-stationary signals · Convolutional neural networks
Clusterization · Medical diagnostics

1 Introduction

The task of processing non-stationary signals finds its place in many fields of science and technology. It is complicated by the fact that a set of methods that are successfully used when working with stationary signals (in particular, the allocation of fundamental harmonics by Fourier transform) is not applicable for non-stationary signals. The reason is that the signal spectrum can carry not any useful information. To solve this problem, a complex model is being built. To construct such a model, both the measurement group and the allocation of any numerical characteristics of the non-stationary signal can be used. Another solution is to develop existing methods for stationary signals for cases of non-stationary signals [10].

In this paper, the field of ophthalmology associated with the diagnosis of complex vision pathologies is considered, especially in the early stages of the disease. An important task in this area is to clarify the diagnosis. To clarify the diagnosis, methods based on the measurement of the biophysical potentials of the eyes - electroretinograms (ERGs) - in response to different light pulses are used [3]. When measuring the ERG, a non-stationary signal is obtained. The model with several types of waves that arise in the eye in response to a single light pulse (Fig. 1) are used to work with this signal.

The work was supported by RFBR (projects 17-07-00553, 18-51-00007, 18-01-00201).

Here, the component PI is composed of slowly varying values of different signs, PII and PIII reflect different ERG fronts.

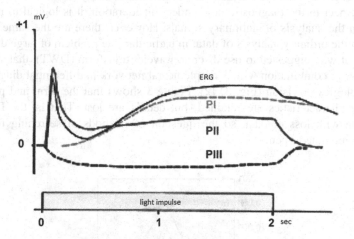

Fig. 1. Measurement of biological potentials in ERG and its division into components

This model showed its effectiveness for solving the problem of clarifying the diagnosis, but it was not effective for accurate diagnosis of diseases. To obtain new data, methods such as the ERG pattern (PERG) were used, which, however, only received a group of non-stationary signals. Of course, several methods have been proposed using several consecutive stimuli, but they allow us to obtain only a certain set of stationary signals.

As another method, the ERG research was proposed on the basis of simulations similar to those used in the analysis of complex dynamical systems [1]. The human eye, according to formal criteria, can be represented as a dynamic one with the decomposition of ERG into components (in Fig. 2). Blocks W1–W4 describe the transfer function of the eye reaction to the light pulse as a dynamic system.

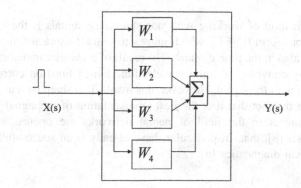

Fig. 2. The decomposition of ERG into components by analogy with the dynamical system

An important result of the work in this direction was the demonstration that the signal contains a large number of hidden components that can be used for further work with using artificial intelligence methods, in particular, fuzzy logic.

With respect to the diagnostic task under consideration, it is logical to use neural networks in the analysis of stationary signals. However, there are the same problems that arise in the ordinary analysis of data, in particular, the problem of large dimension. To solve it, it was suggested to use discrete wavelet transform (DWT), that showed its effectiveness in combination with a simple neural network in differential diagnostics of some pathologies of vision (Fig. 3) [2]. Figure 3 shows that the form and part of the quantitative characteristics are retained, but details are lost. That is, the DWT is a compression with loss of data, so the question of methods of searching for hidden regularities remained open.

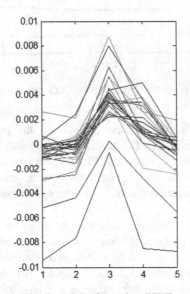

Fig. 3. ERG after using DWT

An effective method of working with non-stationary signals is the use of a continuous wavelet transform (CWT), which allows to extract characteristics not only in the frequency but also in the time domain. The result of a wavelet transform is a matrix that is obtained by convolution some wavelet function (a function corresponding to certain conditions) with the signal. However, this results in a sharp increase in the set of input data, despite the fact that it is the best representation of the signal [2, 4].

New developments in the field of neural networks are opened by promising methods of analysis [8], that, in particular, have already been successfully applied in problems of medical diagnostics [6].

2   Convolutional Neural Networks

In the past 5 years, convolutional neural networks (CNNs) have been widely used, based on splitting the matrix of input data into certain blocks and folding them with certain nuclei, that are matrices encoding a graphical representation of a feature. Such neural networks have shown themselves well in the problems of clustering complex images [7]. In Fig. 4 the generalized structure of CNN is shown [5]:

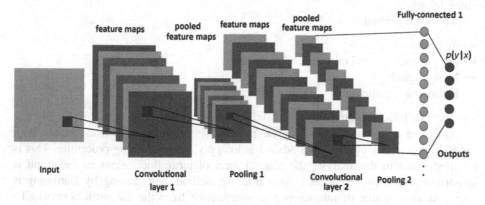

Fig. 4. Generalized structure of CNN

A convolution layer is processes an image by fragments, performing a convolution operation using the convolution kernel (matrix whose coefficients are selected in the learning process).

The voting layer calculate one value from several results of the previous layer with the goal of reducing the dimension, most often, simply using the maximum function, i.e. This layer, in fact, chooses the most presented features.

A fully connected network at the end serves to construct a mapping from the space of the selected features to the space of certain classes.

This network allows to classify the input data with a high accuracy. In addition, when CNN functions, a certain set of attributes is formed, that initially cannot be determined even by an expert, but manifested in the learning process.

This method, however, requires for its work a large sample of input data, that limits its using. In this case, the data should be assigned to a certain set of classes, that often requires painstaking work, and in some cases, it is not possible.

3   Dataset Analyzing

The sampling of the data for the analysis is the data obtained by means of the ERG apparatus. When the research is performed using the ERG apparatus, the electrodes are attached to the human eyes, that read the appearance of electrical impulses in the eye of a person when the light stimulus is applied. In this case, PERG studies are used that more accurately identify the presence of the disease, 1500 examples are used.

Each example consists of the results of reading the signal from an eye (using the right or left eye does not matter). The signal is digitized with a sampling frequency of 125 Hz. The amplitude of the signal is the potential difference that occurs when the light stimulus is applied.

Despite the fact that accurate researching are being carried out to define the exact diagnosis, it is still impossible to form a successful sample for direct using CNN potential.

The main problems are:

- inaccurate diagnoses;
- the presence of several concurrent diseases;
- Noisy signals;
- Errors during obtaining of data.

The first two problems can be solved by preprocessing.

Since the reception of a biological potential is often a problem associated with random muscle movements, it is advisable to apply any smoothing procedure. This is justified, since in the analysis the general form of large fluctuations in the graph is important, and not single peaks. As a filtering method, anti-aliasing by Hamming is used. Another source of interference is interference from the electrical network. To solve this problem, a Butter-band bandpass filter with a bandwidth of 0 to 45 Hz and from 55 Hz to the sampling frequency is used. This filter is selected because one of its properties is maximum smoothness at transmission frequencies [2].

Since errors in the actual data extraction are possible, the cases where the signal is radically visually different from ERG are excluded from the sample (unless it is indicated that the diagnosis corresponds to the retinal detachment).

Cases with erroneous diagnosis are a more significant problem for machine learning, since the classification problem requires a correspondence between the example and the class to which it belongs. However, there is an alternative way to solve this problem, which involves clustering rather than classification. Based on the results of the classification, we can post factum determine the list of possible diseases on the basis of the results obtained, since the list of diseases represents a finite set of pathologies. It is also important that later it can be easily used to consider cases of simultaneous presence of several diseases, then the example will be located on the border of clusters, or a group of such examples will fall into a separate small cluster. This idea opens the way to the possibility of using CNN in this task.

4   The Structure of the Clustering Algorithm

In terms of addressing the above tasks, an approach is proposed for researching the results of cluster-based ERGs in combination with using CNN. Such an integrated approach and the software system implemented on its basis allow successfully processing data with carrying out a wavelet transformation, i.e. with the maximum extraction of useful information.

For the combination of clustering and CNN, the modification of the last layer of CNN (a fully connected network) is suggested according to the k-means algorithm. The essence of the modification is that the elements of the vectors fed to the input are sorted in ascending order, that allows to take into account the situations connected with the fact that at different stages of the network operation different characteristics can change their position in the resulting vector [9].

It is worth noting that for the CWT (Fig. 5), Haar wavelet transformation is used, that allows to take into account the differences in the ERG graph, carrying important information [1, 2]. This allows you to solve problems with incorrect diagnosis and the presence of several diseases. Since the CWT performs different key factors at different scales and shifts, which can indicate the presence of any pathologies that are not visible in the one-dimensional signal. In particular, when using the haar wavelet, it is possible to determine the overall drop in the response of the eye to the light stimulus and the decrease in individual waves. This also allows you to transfer the analysis of results to the area of assessing the certainty of some pathologies based on the results of the algorithm.

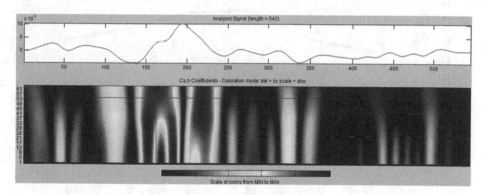

Fig. 5. CWT of ERG

The proposed algorithm includes the following steps:

1. For each result of the ERG researching, a CWT is conducted. The size of the matrix obtained as a result of the transformations is 250×250.
2. A random value for a characteristic vector is assigned to each received matrix.
3. The iteration of CNN training is performed (network sizes are 3 layers, at the first convolution 50×50, on the second – 25×25, on the third – 10×10). The error value of the k-means algorithm is used as a deviation value of the expected result from the obtained result.
4. An iteration of the k-averages algorithm is performed.
5. If the change in network operation error is greater than some constant (compared to the last iteration), then return to step 3.

The result of the algorithm is the clustering of results using a predetermined set of clusters. Another result of the algorithm's work is the selected features on the basis of which an additional analysis can be made.

5   Analyzing of Results

The work of the proposed method was tested on a sample of 1500 PERG studies, with 800 studies used for training, 200 for validation and 500 for determining the performance of the network.

With a different number of clusters used in the k-average algorithm, the best convergence of the results for N values in the range [3, 5] was found, which is due to the fact that most often presented ERGs were used to diagnose only the most common pathologies (Fig. 6).

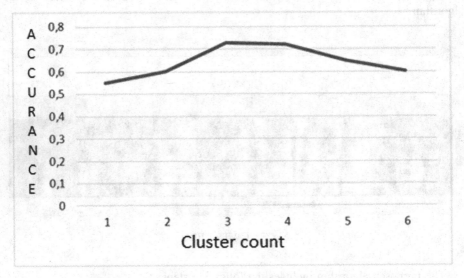

Fig. 6. Accuracy of classification by generated clusters depending on the number of clusters

In a subsequent review of the results obtained, it was found that for the case of N = 3, the graphs falling into separate clusters correspond to observations in such diseases as myopia, retinal detachment and glaucoma.

As an example, Figs. 7, 8 and 9 shows overlays of plots from individual clusters corresponding to myopia, retinal detachment and glaucoma.

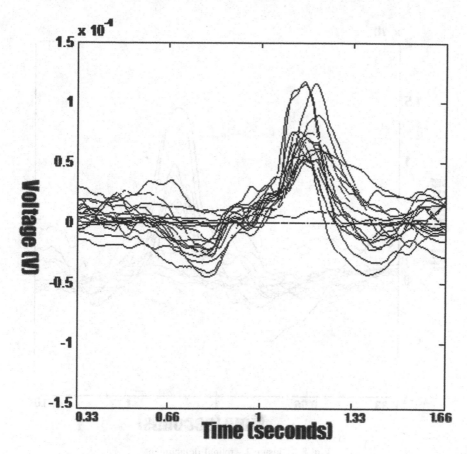

Fig. 7. Cluster 1 - Myopia

In Fig. 7, we can see the superposition of ERG records in one of the clusters, which can be associated with myopia. As can be seen, here the eye response to the light stimulus is reduced in the intervals of 0.6–1 s and 1.3–1.6 s, which corresponds to theoretical representations of changes in the response to light in the presence of myopia.

In Fig. 8, we can see the superposition of ERG records that fall into a cluster associated with retinal detachment. As can be seen, here there is a general reduction in the response of the eye to a light stimulus, which corresponds to the theoretical notions of changes in the response to a light pulse during detachment of the retina.

In Fig. 9, we can see the superposition of ERG records in one of the clusters, which can be associated with glaucoma. As you can see, here the eye response to the light stimulus in the interval and 1.3–1.6 s decreases, which corresponds to the theoretical notions of changes in the response to the light pulse in the presence of glaucoma.

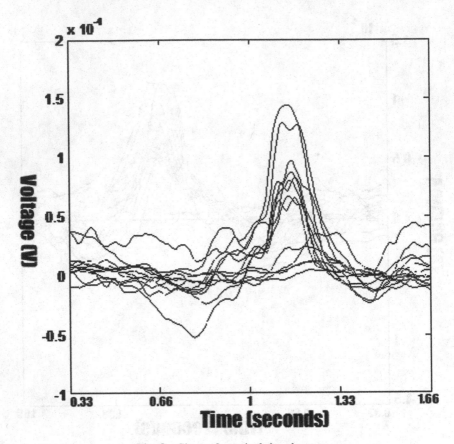

Fig. 8. Cluster 2 - retinal detachment.

As can be seen, it was possible to single out three clusters, that show well the changes in ERG in cases of certain diseases. It should be noted that in order to disseminate the results of researching conducted to an problem area of vision pathologies, in general, close contact with expert ophthalmologists is required in order to develop the relevant database.

At N = 4 (Fig. 10), another cluster appeared, in which complex cases occurred, for example, when the patient was diagnosed with both myopia and glaucoma, but also those cases when the data were clearly obtained with violation of the procedure.

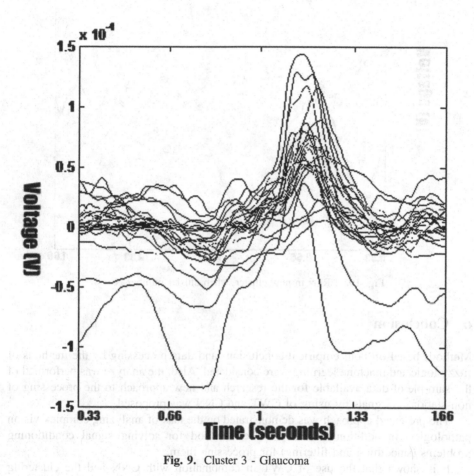

Fig. 9. Cluster 3 - Glaucoma

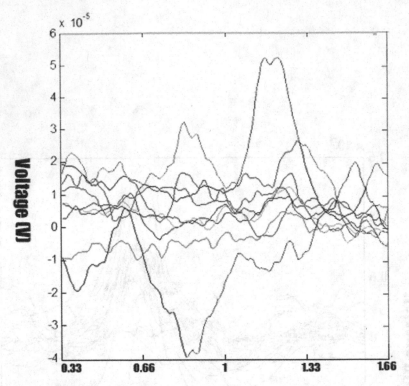

Fig. 10. ERGs in new cluster, when cluster number N = 4

6 Conclusion

Methods based on both empirical conclusions and data processing by the methods of fuzzy logic and machine learning were considered. Also, the analyze was performed of the sam-ple of data available for the research and new approach to the processing of non-stationary signals by using of CWT and CNN was proposed.

The presented approach was demonstrated on the task of analyzing complex vision pathologies. In addition to the proposed method for solving signal conditioning problems (smoothing and filtering) for processing them.

It is shown that the use of CWT in combination with CNN and the clustering algorithm allows for accurate cluster allocation even in the event of problems with the training sample, which is important for the use of CNN, since the sampling size is important for this type of neural networks.

It should be noted for further work that the method, in addition to clustering, can be used to determine those signs of non-stationary signal that are not noticeable in the "manual" analysis, but play a significant role in establishing the correct diagnosis.

The proposed approach, in addition to clustering, can be used to determine those signs of non-stationary signal that are not noticeable in "manual" analysis, but play a significant role in the formulation of the correct diagnosis. This is an important achievement, because, as it was said at the beginning, many customary models are built empirically and, despite the fact that they are time-tested, the question of the sources of regularities remains open.

An important factor is the further increase in the accuracy of classification, as well as the study of the possibility of applying the method developed for other types of biophysical data. As in [2] it might be consider the use of methods devel-oped to solve a similar problem using the example of heart pathologies.

References

1. Anisimov, D.N., Vershinin, D.V., Kolosov, O.S., Zueva, M.V., Tsapenko, I.V.: Diagnostics of the current state of dynamic objects and systems of complex structure using fuzzy logic methods using imitation models. Artificial Intelligence and Decision Making, no. 3, pp. 39–50 (2012). (In Russian)
2. Eremeev, A.P., Ivliev, S.A.: Analysis and diagnostics of complex pathologies of vision based on wavelet transformations and the neural network approach. In: Collection of Scientific Proceedings of the VIII International Scientific and Technical Conference, Integrated Models and Soft Computing in Artificial Intelligence, Kolomna, 18–20 May 2015, pp. 589–595. Fizmatlit, Moscow (2015). (In Russian)
3. Kurysheva, N.I.: Maslova EV electrophysiological studies in the diagnosis of glaucoma. Natl. J. Glaucoma **16**(1), 102–113 (2017). (In Russian)
4. Brynolfsson, J., Sandsten, M.: Classification of one-dimensional non-stationary signals using the Wigner-Ville distribution in convolutional neural networks. In: 25th European Signal Processing Conference, EUSIPCO 2017, pp. 326–330. Institute of Electrical and Electronics Engineers Inc. (2017). https://doi.org/10.23919/EUSIPCO.2017.8081222
5. Davies, E.R.: Computer Vision Principles, Algorithms, Applications, Learning, 5th edn, pp. 453–493. Academic Press, London (2018)
6. Litjens, G., Kooi, T., Bejnordi, B.E., Setio, A.A.A.: A survey on deep learning in medical image analysis. Med. Image Anal. **42**, 60–88 (2017)
7. Krizhevsky, A., Sutskever, I., Hinton, G.E.: ImageNet classification with deep convolutional neural networks. In: Advances in Neural Information Processing Systems, pp. 1097–1105 (2012)
8. Lee, H., Grosse, R., Ranganath, R., Ng, A.Y.: Convolutional deep belief networks for scalable unsupervised learning of hierarchical representations. In: Proceedings of the 26th Annual International Conference on Machine Learning (2009)
9. Bo, L., Ren, X., Fox, D.: Unsupervised feature learning for RGB-D based object recognition. In: Desai, J., Dudek, G., Khatib, O., Kumar, V. (eds.) Experimental Robotics. Springer Tracts in Advanced Robotics, vol. 88, pp. 387–402. Springer, Heidelberg (2013). https://doi.org/10.1007/978-3-319-00065-7_27
10. Ostrin, L.A., Choh, V.: The pattern ERG in chicks – Stimulus dependence and optic nerve section. Vis. Res. **128**, 45–52 (2016)

An Approach to Feature Space Construction from Clustering Feature Tree

Pavel Dudarin[✉], Mikhail Samokhvalov, and Nadezhda Yarushkina

Ulyanovsk State Technical University, Ulyanovksk 432027, Russian Federation
{p.dudarin,sam,jng}@ulstu.ru

Abstract. Generally, clustering feature tree consists of nodes given as vectors. In case of non-vector nodes a transformation into feature vectors is needed. Feature extraction algorithm determines the volume and quality of information enclosed in features and quality of clustering. Thus this kind of transformation is important part of clustering procedure. In this paper an approach to clustering feature space construction from clustering feature tree is proposed. Presented approach allows to save hierarchy information and reduce feature space dimension. An efficiency of proposed approach is shown in the experiment part with different clustering algorithms. Result analysis is provided at the end of the paper.

Keywords: Clustering · Clustering feature tree · Feature extraction

1 Introduction

Any machine learning algorithm consists of at least two steps. First one is constructing space of features and defining measure to calculate distance between feature vectors. The second part is a machine learning algorithm itself, for example, clustering, classification, etc. Feature extraction step is very important because the only information about objects available for the second step is the information contained in features. That is why machine learning algorithm results quality highly depends on features quality [13]. Generally, features are considered as non dependent and equally weighted. In case of non equally weighted features, there are two posibilites of work with weighted features: feature weights are included into feature vectors [1] or feature weights are included into distance function [17]. In some agglomerative and divisional clustering algorithms like BIRCH [29], Chameleon [15] and GATCH [16], special feature structures called *clustering feature trees* are used. In these algorithms clustering feature tree (CFT) is constructed as a result of clustering procedure which is performed for n-dimensional feature vectors. Such CFT, evidently, does not need to be transformed into feature vectors, as long as this transformation is trivial because all the tree nodes are already located in the same n-dimensional space. Hierarchical information are already automatically included into feature vectors.

S. O. Kuznetsov et al. (Eds.): RCAI 2018, CCIS 934, pp. 176–189, 2018.
https://doi.org/10.1007/978-3-030-00617-4_17

There is a special case of clustering feature tree with non vector nodes as an input for clustering procedure. Clustering objects are done as complex objects related to many nodes of CFT. The relation is defined by membership function. In this case a special transformation into feature vectors is necessary. In this paper an approach to feature space construction from clustering feature tree is proposed.

Nowadays datasets with a large and even huge amount of short text fragments become quite a common object. Different kinds of short messages, forum posts [11] and sms, paper or news headers are examples of this kind of objects. There is one more similar object which is a dataset of key process indicators (KPI) of Strategic Planning System of Russian Federation [10]. KPI is a shortly formulated target. Usually KPI consists of one sentence with 5–15 words. This system is quite new, KPI collection procedure stated in 2016, so the first steps of data mining are in progress [5].

There are some examples translated into English in the Table 1, full data could be found on the official government portal http://gasu.gov.ru.

Table 1. KPIs examples related to education

KPIs
Amount of educational organizations with electronic document management
Amount of new buildings for educational organizations
Average amount of personal computers per school
Edcucational organizations employees' salary growth rate
Math exam average rate over school graduate students

Usually, text clustering performed by using general clustering algorithms like $k - means$[2] and more complicated like $HDBScan$ [9] and many others [16,27,28]. The only additional step is needed - convert sentences or entire documents into features appropriate to the chosen clustering algorithm. One possible solution consists in using $doc2vec$ model [20]. But this approach oriented to large documents with many sentences within. More widely used approach to do this consists in using $word2vec$ model [18]. This model transforms each word into vector. To get vector feature for sentence, vectors could be summarized and optionally normalized. This approach as a result of neuron network studying does not allow to take into account additional information about current task which could be provided by experts or by ontology for the specific field.

Previously authors suggested an approach to construction hierarchical classifier from fuzzy graph obtained from KPI's dataset [8]. This hierarchical classifier could be corrected by experts and then treated as clustering feature tree (CFT) for the further clustering procedure. This approach is close to the first phase of BIRCH algorithm [29], where CFT is constructed, but it was generalized for the case of non-vector features with similarity measure by using fuzzy graph clustering [7].

In the paper [6] an approach to KPI's dataset clustering was proposed. But this method has some drawbacks: too many clusters in the result (5–6 times more than actual quantity), significantly big noise cluster (20–30% of total data amount) and high dimensionality of feature space (equals to amount of nodes). In this paper a new way of feature construction is proposed which allows to avoid these drawbacks.

The rest of the paper is organized as follows: in Sect. 2 input data for the task are described, feature construction approach is presented in Sect. 3, in Sect. 4 general experiment conditions are done, in Sect. 5 experiment results are discussed, and Sect. 6 concludes the paper.

2   Clustering Feature Tree

As it were shown in paper [8] any set of words could be transformed into fuzzy graph [23]. Fuzzy graph by means of $\epsilon - clustering$ algorithms [21,25] could be transformed into hierarchical classifier. Leaf nodes are words, and non terminal nodes form hierarchy of clusters. Any hierarchical classifier could be treated as clustering feature tree (CFT) for the given data: $CFT = <CFN, CFC>$, where CFN (CF Nodes) is a set of nodes, and CFC (CF Children) is a set of tuples of children nodes related to nodes. Nodes without children nodes called leaf nodes or terminal nodes. For each non terminal node is defined ϵ_i-related level of fuzzy graph clustering process, which means similarity degree among children nodes.

Let denote a set of KPIs as SL. Each KPI consists of words and membership relation could be defined. For leaf nodes: $R : (s \in SL, n \in CFN_{leaf} \rightarrow [0,1])$. And for non terminal nodes $R : (s \in SL, n \in CFN_{parent}) = max\{ R(s,k) \,|\, k \in CFN_{leaf} \cap CFN_{siblings} \}$.

The aim is to perform clustering procedure for KPIs based on feature space obtained from CFT: $Cluster(SL) = \{\cup c_i \,|\, \forall s \in SL \,\exists! \, i \in [1,C], s \in c_i\}$, where C is an amount of classes.

In Fig. 1 a sample of CFT is shown. Distinct words from sample sentences are hierarchically grouped according to their semantic distance. The root element contains all the words of the sample.

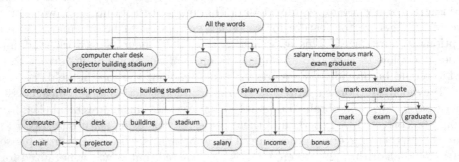

Fig. 1. Extract from hierarchical classifier for KPIs related to education

In general case CFT could be obtained form any weighted graph, topology, ontology or any other partially ordered set with defined similarity measure.

3   Feature Construction

Although there are many ways to construct feature vectors from CFT, only few of them are efficient. One possible form of features is a set of vectors where each vector coordinate equals to 0 or 1, value depends on whether or not the current sentence have at least one of the words in common with a set of words associated with the current CF Node. So the dimension of these vectors would be equal to the count of nodes in the CFT. But this transformation loses much information about hierarchy.

To preserve information about words relations the transformation discussed in [6] could be used. In this transformation parent nodes are included into the feature vectors as additional coordinates and a special weight function is used to smooth out the effect from the root and other nodes that are close to the root one. But this transformation also has some drawbacks. The first one is a necessity to calibrate weight function to each type of hierarchy. The second one is that all the siblings of each node are treated equally. For example, if words 'song' and 'singer' are children of the node 'song, singer', and words 'phone' and 'tv' are children of the node 'phone, tv, cable', parent nodes will have the same pattern in their vector form. Their vectors will be look like $[..., 1, 1, f(2), ..., 0, 0, 0, ...]$ and $[..., 0, 0, 0, ..., 1, 1, f(2), ...]$. This happens because of the function f only depend on amount of the children and all the leaves are treated as orthogonal vectors to each other. The third disadvantage is that the vectors' dimension is still high and equals to the nodes amount.

In this paper another transformation is proposed. Let assume that there are two sibling nodes in hierarchy attached to the same parent node, then instead of assigning them vectors $[0, ..., 0, 1, 0, 0, ..., 0]$ and $[0, ..., 0, 0, 1, 0, ..., 0]$, as it was in the first approach, the vectors $[0, ..., 0, x_1, x_2, 0, ..., 0]$ and $[0, ..., 0, x_2, x_1, 0, ..., 0]$ are used. Thus instead of distance $\sqrt{(1^2 - 0^2) + (0^2 - 1^2)} = \sqrt{2}$ more appropriate distance will have place $\sqrt{(x_1^2 - x_2^2) + (x_2^2 - x_1^2)} = d(x_1, x_2)$, where $d(x_1, x_2) \in [0, 1]$. This distance reflects the semantic distance between the words associated with the current leaf nodes. Proposed transformation is illustrated in Fig. 2 where initially orthogonal vectors are moved closer to each other.

Evidently, this transformation changes vector norm in order to simplify computer calculations. Distance between any two words with the same parent equals to $(1 - \epsilon) * \sqrt{2}$ and does not depend on amount of children. Geometrically correct transformation, which preserves vector norm, assume treating ϵ as $\cos \alpha$, where α is an n-dimensional angle between any two vectors. In this way coordinate re-calculations become much more complex for understanding and computer calculation, although experiments shows that there is no significant difference between clustering results.

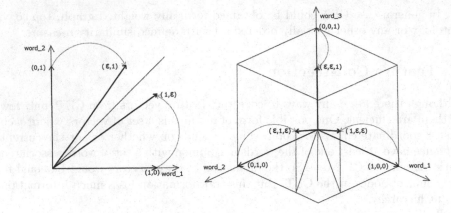

Fig. 2. Features constraction for 2- and 3- children nodes cases

The algorithm of the transformation is following:

1. Start with the root CF Node. Let it has k_1 children in CFC tuple, so for them vectors $([1, \epsilon_1, ..., \epsilon_1], [\epsilon_1, 1, \epsilon_1, ..., \epsilon_1], ..., [\epsilon_1, ..., \epsilon_1, 1])$ are defined, where each vector has k_1 coordinates and ϵ_i - parameter defined for each CF Node and characterize similarity degree between children nodes.
2. Get the first CF Node with vector $[1, \epsilon_1, ..., \epsilon_1]$. Let it has k_2 children nodes and similarity degree among them is ϵ_2. Thus the vector for the current node transformed to the set of vectors:

$$([\ \underbrace{1, \epsilon_2, ..., \epsilon_2}_{k_2\ coordinates}\ ,\ \underbrace{\epsilon_1, ..., \epsilon_1}_{k_1-1\ coordinates}\ ], \tag{1}$$

$$[\epsilon_2, 1, \epsilon_2, ..., \epsilon_2, \epsilon_1, ..., \epsilon_1], ..., [\epsilon_2, ..., \epsilon_2, 1, \epsilon_1, ..., \epsilon_1])$$

Vectors for the other $k_1 - 1$ CF Nodes are transformed one to one to the set:

$$([\ \underbrace{\epsilon_1, ..., \epsilon_1}_{k_2\ coordinates}\ ,\ \underbrace{1, \epsilon_1, ..., \epsilon_1}_{k_1-1\ coordinates}\ ],$$

$$[\ \underbrace{\epsilon_1, ..., \epsilon_1}_{k_2\ coordinates}\ , \epsilon_1, 1, \epsilon_1, ..., \epsilon_1], \tag{2}$$

$$..., [\ \underbrace{\epsilon_1, ..., \epsilon_1}_{k_2\ coordinates}\ , \epsilon_1, ..., \epsilon_1, 1]$$

3. Repeat step 2 for other CF Nodes of the level.
4. Recursively repeat steps 2 and 3 for all the levels of the CFT.

To give the intuition of this process, a short example could be discussed. Let us have a small sample hierarchy that will be vectored. This hierarchy is shown in the Fig. 3. Its vectored form could be found in Table 2, where the analytical

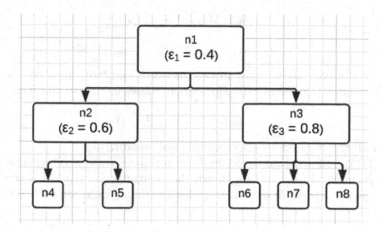

Fig. 3. Sample hierarchy to be vectored

Table 2. Vectors for sample hierarchy

Vectors/Nodes	n_4	n_5	n_6	n_7	n_8	Vectors/Nodes	n_4	n_5	n_6	n_7	n_8
$Vector\ n_4$	1	ϵ_2	ϵ_1	ϵ_1	ϵ_1	$Vector\ n_4$	1	0.6	0.4	0.4	0.4
$Vector\ n_5$	ϵ_2	1	ϵ_1	ϵ_1	ϵ_1	$Vector\ n_5$	0.6	1	0.4	0.4	0.4
$Vector\ n_6$	ϵ_1	ϵ_1	1	ϵ_3	ϵ_3	$Vector\ n_6$	0.4	0.4	1	0.8	0.8
$Vector\ n_7$	ϵ_1	ϵ_1	ϵ_3	1	ϵ_3	$Vector\ n_7$	0.4	0.4	0.8	1	0.8
$Vector\ n_8$	ϵ_1	ϵ_1	ϵ_3	ϵ_3	1	$Vector\ n_8$	0.4	0.4	0.8	0.8	1

Table 3. Euclidean and cosine distances for nodes of sample hierarchy

Nodes	n_4	n_5	n_6	n_7	n_8	Nodes	n_4	n_5	n_6	n_7	n_8
n_4	0.00	0.57	1.04	1.04	1.04	n_4	0.00	0.09	0.23	0.23	0.23
n_5	0.57	0.00	1.04	1.04	1.04	n_5	0.09	0.00	0.23	0.23	0.23
n_6	1.04	1.04	0.00	0.28	0.28	n_6	0.23	0.23	0.00	0.02	0.02
n_7	1.04	1.04	0.28	0.00	0.28	n_7	0.23	0.23	0.02	0.00	0.02
n_8	1.04	1.04	0.28	0.28	0.00	n_8	0.23	0.23	0.02	0.02	0.00

and calculated form of vectors are shown for the better understanding. In Table 3 euclidean and cosine distances for these nodes are calculated. As it could be seen these distances are quite adequate and preserve information from CFT.

In order to make further clustering procedure more efficient some more steps could be done.

1. Normalization function could be used for ϵ_i. In the real world examples ϵ_1 usually higher than 0.3 and words with this similarity measure have nothing in common. On the contrary, words with similarity measure equal to 0.9 and higher are treated like synonyms. To take this into account following changes for each ϵ_i could be done:

$$\epsilon_i = min(max(\epsilon_i, \epsilon_{min}), \epsilon_{max}) \tag{3}$$

$$\epsilon_i = (\epsilon_i - \epsilon_{min})/(\epsilon_{max} - \epsilon_{min}) \tag{4}$$

2. Experiments have shown that in case of words similarity measure, based on word2vec model, following transformation allows to correct the scale.

$$\epsilon_i = (e^{2*\epsilon_i} - 1)/(e^2 - 1) \tag{5}$$

3. Center all the features by subtracting mean vector.
4. PCA [14] transformation could be done to reduce vectors' dimensionality. In the experimental part will be shown that features constructed according to the proposed algorithm demonstrate high variance ratio after PCA transformation.

Constructed features make a non-orthogonal basis of a new feature space, where each vector relates to a word. The last step is to construct features for sentences. To do this a membership relation should be defined. As long as each word feature has vector form any valid mathematics vector operations could be used. In further experiment were used following approach: for KPI k_i with words $w_{i1}, w_{i2}, ..., w_{in}$ feature vector were constructed as mean of related words vectors. Mean vector instead of simple sum is used as long as for clustering algorithm sentence "two repeated words" should be equal to "two repeated words".

4   Experiment Description and Result Evaluation

In order to comprehensively evaluate proposed features' advantages and drawbacks following clustering algorithms were used: K-means [26] with 23 clusters as parameter, Affinity Propagation [3], Mean Shift [4], Agglomerative Clustering [22] with average linkage with 23 clusters as parameter, HDBScan [9]. HDBscan was run with euclidean, manhattan and cosine distance functions, and for minimum cluster size were used values $10, 7, 5, 3$. Python library "scikit-learn" [19] were used in the experimental part, and other parameters of these methods also were ranged to get their best.

All experiments were performed on two datasets: "KPIs of General Education" with 1400 indicators, and "KPIs of Medicine Sphere" with 2500 indicators. These datasets were clustered by experts and the knowledge of the ground truth are given.

Following features were constructed:

1. "Simple vectors" - vectors described above: normalized vectors with components set to 1 for each CF Node related to the sentence and set to 0 for others.
2. "Simple vectors (PCA 200)" - "Simple vectors" reduced by using PCA to 200 dimentional vectors.
3. "Simple vectors (PCA 100)" - "Simple vectors" reduced by using PCA to 100 dimentional vectors.

4. "Simple vectors (PCA 50)" - "Simple vectors" reduced by using PCA to 50 dimentional vectors.
5. "Simple vectors (PCA 10)" - "Simple vectors" reduced by using PCA to 10 dimentional vectors.
6. "Simple vectors (PCA 3)" - "Simple vectors" reduced by using PCA to 3 dimentional vectors.
7. "H2V" - features proposed in this paper constructed by CFT transformation into vectors.
8. "H2V (PCA 200)" - "H2V" reduced by using PCA to 200 dimentional vectors.
9. "H2V (PCA 100)" - "H2V" reduced by using PCA to 100 dimentional vectors.
10. "H2V (PCA 50)" - "H2V" reduced by using PCA to 50 dimentional vectors.
11. "H2V (PCA 10)" - "H2V" reduced by using PCA to 10 dimentional vectors.
12. "H2V (PCA 3)" - "H2V" reduced by using PCA to 3 dimentional vectors.
13. "W2V" - 300 dimensional features constructed as mean vectors of words embedding obtained from word2vec model. In order to make "H2V" and "W2V" features comparable no modifications into generated CFT were done by experts. And no semantic templates or collocations were used in CFT, distinct words only. In other cases these features would have different information and comparison would not have any sense.
14. "W2V (PCA 200)" - "W2V" reduced by using PCA to 200 dimentional vectors.
15. "W2V (PCA 100)" - "W2V" reduced by using PCA to 100 dimentional vectors.
16. "W2V (PCA 50)" - "W2V" reduced by using PCA to 50 dimentional vectors.
17. "W2V (PCA 10)" - "W2V" reduced by using PCA to 10 dimentional vectors.
18. "W2V (PCA 3)" - "W2V" reduced by using PCA to 3 dimentional vectors.

There were 480 clustering experiments performed in total.

For result evaluation following metrics were chosen:

1. *Precision, Accuracy, Purity and Average Quality* - specially designed metrics which help not only evaluate clustering results but also understand its weak and string sides. These metrics are defined for each class as follows:

$$Precision_i = \frac{\max_j CM_{i,j}}{\sum_j CM_{i,j}} \tag{6}$$

$$Accuracy_i = \frac{\max_j CM_{i,j}}{\sum_k CM_{k,j_{max}}} \tag{7}$$

$$Purity_i = \frac{\max_j CM_{j,i}}{\sum_j CM_{j,i}}, \tag{8}$$

where CM $= [cm_{i,j} = (\omega_i \cap c_j), \omega_i$ - class with number i, c_j - cluster with number j].

Total *Precision*, *Accuracy* and *Purity* are calculated as average values. Also average value *Average Quality* $= (Precision + Accuracy + Purity)/3$ is used to rank results.

2. *Adjusted Rand Index (ARI)* - is a function that measures the similarity of the two assignments, ignoring permutations and with chance normalization [12]. Random (uniform) label assignments have a ARI score close to 0.0 for any value of clusters and samples (which is not the case for V-measure for instance). ARI has bounded range $[-1, 1]$: negative values are bad (independent labeling), similar clustering results have a positive ARI, 1.0 is the perfect match score. Also, no assumption is made on the cluster structure: can be used to compare clustering algorithms such as k-means which assumes isotropic blob shapes with results of spectral clustering algorithms which can find cluster with "folded" shapes.

Assuming that C is a ground truth class assignment and K the clustering, let us define a and b as:

- a, the number of pairs of elements that are in the same set in C and in the same set in K
- b, the number of pairs of elements that are in different sets in C and in different sets in K

$$RI = \frac{a + b}{C_2^{n_{samples}}}, \tag{9}$$

where $C_2^{n_{samples}}$ is the total number of possible pairs in the dataset (without ordering).

$$ARI = \frac{RI - E[RI]}{max(RI) - E[RI]}, \tag{10}$$

where E[RI] is mathematical expectation of random labeling.

3. *Silhouette Coefficient* - if the ground truth labels were not known, evaluation must be performed using the model itself [24]. The Silhouette Coefficient is an example of such an evaluation, where a higher Silhouette Coefficient score relates to a model with better defined clusters. The Silhouette Coefficient is defined for each sample and is composed of two scores: a - the mean distance between a sample and all other points in the same class; b - the mean distance between a sample and all other points in the next nearest cluster. The Silhouette Coefficient s for a single sample is then given as:

$$s = \frac{b - a}{max(a, b)} \tag{11}$$

The Silhouette Coefficient for a set of samples is given as the mean of the Silhouette Coefficient for each sample.

5   Analysis of Experiment Results

Both datasets contain 23 clusters determined by experts. Top 20 results according to *Average Quality* measure are presented in Tables 4 and 5. All results are in percentages. As long as algorithm HDBScan is a true clustering algorithm and always forms 'garbage cluster' with number -1, its evaluation metrics could not be compared to the others, so HDBscan top 10 results presented separately in Tables 6 and 7 with additional parameter 'percentage of garbage cluster'. The algorithm HDBScan has two main parameters: minimal cluster size and distance function. Minimal cluster size parameter equals to 7 and 5 respectively, and distance function parameter presented in the second column.

Table 4. Clustering results of "KPIs of Medicine Sphere" dataset

Features	Clustering algorithm	Prec.	Acc.	Prty	Avg Qlty	ARI	Silh. coeff.
Simp. vec. pca_10	Agg. clust.	86.5	57.0	79.4	68.2	41.9	47.0
h2v pca_10	Agg. clust.	85.8	43.9	92.9	68.4	50.3	48.6
w2v pca_3	K-means	67.5	60.5	79.3	69.9	53.3	49.7
Simp. vec. pca_10	K-means	76.1	68.8	73.6	71.2	55.4	50.9
w2v pca_3	Agg. clust.	80.8	55.7	87.3	71.5	57.6	51.1
Simp. vec. pca_200	K-means	74.7	65.3	79.7	72.5	67.0	51.5
w2v pca_10	K-means	78.8	66.3	80.0	73.1	67.2	51.7
Simp. vec. pca_100	K-means	90.7	56.6	91.2	73.9	70.4	52.2
Simp. vec.	K-means	75.8	68.1	80.7	74.4	71.1	52.8
Simp. vec. pca_50	K-means	81.0	67.7	85.9	76.8	71.5	52.9
w2v pca_10	Agg. clust.	79.6	70.5	85.3	77.9	72.2	53.7
h2v pca_10	K-means	79.9	71.8	87.5	79.6	72.4	53.9
w2v pca_50	K-means	86.3	70.5	89.5	80.0	73.1	55.0
w2v pca_100	K-means	80.1	72.1	88.0	80.1	73.3	55.6
h2v pca_200	Agg. clust.	82.1	72.6	89.3	81.0	74.2	55.9
w2v pca_200	K-means	80.0	75.3	91.6	83.5	74.9	56.4
h2v pca_50	K-means	85.3	79.1	88.0	83.5	78.5	60.8
w2v	K-means	86.0	77.6	89.7	83.7	79.2	63.0
h2v pca_100	K-means	85.5	78.2	90.1	84.2	80.0	63.7
h2v	K-means	85.5	79.3	93.5	86.4	81.8	66.5

Features constructed from presented transformation are much better than 'simple vectors' features and even slightly better than features constructed with word2vec model. Agglomerative clustering based on average linkage, K-means and HDBScan shows the best results. Average quality for HDBScan is generally

Table 5. Clustering results of "KPIs of General Education" dataset

Features	Clustering algorithm	Prec.	Acc.	Prty	Avg Qlty	ARI	Silh. coeff.
Simp. vec. pca_10	Agg. clust.	73.4	46.8	75.4	61.1	20.5	11.4
Simp. vec.	Agg. clust.	83.2	36.5	91.2	63.8	35.3	22.6
Simp. vec. pca_10	K-means	68.3	56.0	72.9	64.4	39.7	34.1
w2v pca_10	Agg. clust.	83.1	41.2	88.2	64.7	43.5	34.1
Simp. vec. pca_50	Agg. clust.	88.4	36.7	94.1	65.4	49.2	34.7
Simp. vec. pca_200	K-means	82.2	44.9	88.4	66.7	50.0	34.8
Simp. vec. pca_50	K-means	66.0	58.0	78.6	68.3	50.8	34.9
Simp. vec. pca_100	K-means	74.9	55.5	83.7	69.6	52.0	34.9
Simp. vec.	K-means	75.0	58.1	83.5	70.8	54.9	35.3
w2v pca_100	Agg. clust.	75.0	56.7	85.2	71.0	56.7	36.9
w2v pca_50	K-means	74.0	59.5	84.7	72.1	56.9	37.7
h2v pca_10	Agg. clust.	68.8	59.3	85.1	72.2	57.8	38.2
h2v pca_10	K-means	71.6	58.4	86.7	72.5	57.9	38.8
w2v pca_10	K-means	73.6	60.1	85.6	72.9	58.4	38.9
w2v	K-means	72.0	59.3	87.5	73.4	58.5	40.4
h2v pca_100	K-means	75.2	58.7	90.4	74.5	59.1	42.7
h2v	K-means	71.6	61.8	88.7	75.3	59.6	43.3
h2v pca_200	K-means	69.4	67.3	84.6	76.0	59.9	43.4
w2v pca_200	K-means	77.5	64.5	88.1	76.3	65.1	47.1
h2v pca_50	K-means	72.5	63.9	89.6	76.8	66.4	52.0

Table 6. Clustering results of "KPIs of Medicine Sphere" dataset (HDBScan)

Features	Measure	Prec.	Acc.	Prty	Avg Qlty	ARI	Silh. coeff.	Garb. coeff.
Simp. vec. pca_3	manhattan	43.6	88.1	84.6	86.4	29.3	88.4	19
Simp. vec. pca_3	euclidean	48.1	91.6	86.8	89.2	31.8	86.3	17
h2v pca_3	euclidean	47.6	93.7	86.0	89.9	30.8	87.1	18
Simp. vec. pca_10	euclidean	49.6	93.7	91.5	92.6	28.2	88.2	20
Simp. vec. pca_50	euclidean	50.2	93.8	92.1	93.0	27.4	89.4	21
Simp. vec. pca_50	manhattan	48.3	95.5	92.6	94.0	32.1	86.5	22
h2v pca_10	euclidean	51.0	96.6	94.9	95.7	30.0	87.8	19
h2v pca_10	manhattan	53.1	97.2	96.1	96.7	29.1	87.9	17
h2v pca_50	euclidean	49.7	98.6	96.6	97.6	27.8	89.3	19
h2v pca_50	manhattan	50.0	99.4	96.7	98.1	28.5	87.7	21

Table 7. Clustering results of "KPIs of General Education" dataset (HDBScan)

Features	Measure	Prec.	Acc.	Prty	Avg Qlty	ARI	Silh. coeff.	Garb. coeff.
Simp. vec.	manhattan	58.9	91.3	97.9	94.6	9.6	68.6	30
Simp. vec.	euclidean	58.2	95.1	97.4	96.2	9.6	72.2	28
Simp. vec. pca_200	euclidean	56.6	94.8	97.7	96.2	8.6	69.9	29
Simp. vec. pca_100	manhattan	57.3	94.8	98.6	96.7	9.3	71.8	30
Simp. vec. pca_100	euclidean	42.7	96.4	97.5	96.9	10.6	78.4	19
h2v pca_100	euclidean	54.2	96.1	98.2	97.2	8.5	71.0	25
h2v pca_200	euclidean	55.3	97.1	97.3	97.2	9.9	73.2	25
w2v pca_50	manhattan	41.7	96.6	97.9	97.2	8.4	74.4	24
h2v pca_50	manhattan	54.2	97.2	98.1	97.6	9.5	72.5	24
h2v	euclidean	55.4	97.3	98.1	97.7	9.1	70.9	25

higher but the cost for this is quite big 'garbage cluster' which is not labeled. There is a quite low rate for ARI in experiments with HSBScan because of 'garbage cluster' which is not a real cluster. So ARI is not applicable metric to HDBScan. It is especially worth noting that the simplest algorithms k-means shows acceptable results, which means that really big datasets could be clustered by using this approach.

Features with PCA transformation are on top of the list, in Table 8 explained variance ratio could be found. Even the worst results for 3 dimensional PCA transformation have 35.6% and more than 70% *average quality* and could be useful for visual 3D presentation of the clustering results.

Table 8. Average explained variance ratio in PCA for H2V features

Features	Explained variance ratio
h2v_pca_200	99.4%
h2v_pca_100	95.9%
h2v_pca_50	87.7%
h2v_pca_10	59.5%
h2v_pca_3	35.6%

6   Conclusions

In this paper an approach to feature construction from Clustering Feature Tree is presented. Proposed transformation allows to preserve information about feature hierarchy and reduce dimensionality of feature vectors. Experiment results with KPIs of Strategic Planning System of Russian Federation show that proposed approach is efficient in clustering tasks even after PCA transformation. Sufficient result for k-means algorithm proves that proposed features could be used for really large datasets.

Comparison with features constructed based on *word2vec* model shows that proposed approach have one significant advantage: CFT could combine machine learning algorithm results with expert knowledge, some CF Nodes could be manually added, deleted or changed, this means that CFT could contain more valuable information for specific task than general *word2vec* model.

Use cases of proposed transformation is not limited by sentence clustering: any fuzzy graph, and actually any weighted graph, topology, ontology or any other partially ordered set with weights, could be transformed into CFT and then into a set of feature vectors. Thus proposed approach could be used in many application spheres.

Acknowledgment. This study was supported Ministry of Education and Science of Russia in framework of project No 2.1182.2017/4.6 and Russian Foundation of base Research in framework of project No 16-47-732120 r_ofi_m.

References

1. Amorim, R.: Feature weighting for clustering: using K-means and the Minkowski. LAP Lambert Academic Publishing (2012)
2. Ball, G.H., Hall, David J.: Isodata: a method of data analysis and pattern classification, Stanford Research Institute, Menlo Park, United States. Office of Naval Re-search, Information Sciences Branch (1965)
3. Frey, B.J., Dueck, D.: Clustering by passing messages between data points. Science **315**, 972–976 (2007)
4. Comaniciu, D., Meer, P.: Mean shift: a robust approach toward feature space analysis. IEEE Trans. Pattern Anal. Mach. Intell. **24**, 603–619 (2002)
5. Dudarin, P., Pinkov, A., Yarushkina, N.: Methodology and the algorithm for clustering economic analytics object. Autom. Control. Process. **47**(1), 85–93 (2017)
6. Dudarin, P., Yarushkina, N.: Features construction from hierarchical classifier for short text fragments clustering. Fuzzy Syst. Soft Comput. **12**, 87–96 (2018). https://doi.org/10.26456/fssc26
7. Dudarin, P.V., Yarushkina, N.G.: Algorithm for constructing a hierarchical classifier of short text fragments based on the clustering of a fuzzy graph. Radio Eng. **2017**(6), 114–121 (2017)
8. Dudarin, P.V., Yarushkina, N.G.: An approach to fuzzy hierarchical clustering of short text fragments based on fuzzy graph clustering. In: Abraham, A., Kovalev, S., Tarassov, V., Snasel, V., Vasileva, M., Sukhanov, A. (eds.) IITI 2017. AISC, vol. 679, pp. 295–304. Springer, Cham (2018). https://doi.org/10.1007/978-3-319-68321-8_30
9. Ester M., Kriegel H. P., SanderJ., Xu X.: A density-based algorithm for discovering clusters in large spatial databases with noise. In: Proceedings of the 2nd International Conference on Knowledge Discovery and Data Mining, pp. 226–231. AAAI Press, Portland (1996)
10. Federal law "About strategic planning in Russian Federation" (2014). http://pravo.gov.ru/proxy/ips/?docbody=&nd=102354386

11. Han, X., Ma, J., Wu, Y., Cui, C.: A novel machine learning approach to rank web forum posts. Soft Comput. **18**(5), 941–959 (2014)
12. Hubert, L., Arabie, P.: Comparing partitions. J. Classif. **2**(1), 193–218 (1985). https://doi.org/10.1007/BF01908075
13. Jain, A.K., Murty, M.N., Flynn, P.J.: Data clustering: a review. ACM Comput. Surv. (CSUR) **31**(3), 264–323 (1999)
14. Jolliffe, I.T.: Principal Component Analysis, p. 487. Springer, Heidelberg (1986). https://doi.org/10.1007/b98835. ISBN 978-0-387-95442-4
15. Li, J., Wang, K., Xu, L.: Chameleon based on clustering feature tree and its application in customer segmentation. Ann. Oper. Res. **168**, 225 (2009). https://doi.org/10.1007/s10479-008-0368-4
16. Mansoori, E.G.: GACH: a grid based algorithm for hierarchical clustering of high-dimensional data. Soft Comput. **18**(5), 905–922 (2014)
17. Modha, D.S., Spangler, W.S.: Feature weighting in k-means clustering. Mach. Learn. **52**, 217 (2003). https://doi.org/10.1023/A:1024016609528
18. Mikolov T., Sutskever I., Chen K., Corrado G., Dean J.: Distributed representations of words and phrases and their compositionality. In: Proceedings of the 26th International Conference on Neural Information Processing Systems, 05–10 December, Lake Tahoe, Nevada, pp. 3111–3119 (2013)
19. Pedregosa, F.: Scikit-learn: machine learning in python. J. Mach. Learn. Res. **12**, 2825–2830 (2011)
20. Le, Q., Mikolov, T.: Distributed representations of sentences and documents. In: Proceedings of the 31st International Conference on Machine Learning, PMLR, vol. 32, no. 2, pp. 1188–1196 (2014)
21. Yeh, R.T., Bang, S.Y.: Fuzzy relation, fuzzy graphs and their applications to clustering analysis. In: Fuzzy Sets and their Applications to Cognitive and Decision Processes, pp. 125–149. Academic Press (1975). ISBN 9780127752600
22. Rokach, L., Maimon, O.: Clustering methods. In: Maimon, O., Rokach, L. (eds.) Data Mining and Knowledge Discovery Handbook. Springer, Boston (2005). https://doi.org/10.1007/0-387-25465-X_15
23. Rosenfeld, A.: Fuzzy graphs. In: Zadeh, L.A., Fu, K.S., Tanaka, K., Shimura, M. (eds.) Fuzzy Sets and Their Applications to Cognitive and Decision Processes, pp. 77–95. Academic Press, New York (1975)
24. Rousseeuw, P.J.: Silhouettes: a graphical aid to the interpretation and validation of cluster analysis. Comput. Appl. Math. **20**, 53–65 (1987). https://doi.org/10.1016/0377-0427(87)90125-7
25. Ruspini, E.H.: A new approach to clustering. Inform. Control **15**(1), 22–32 (1969)
26. Arthur, V., et al.: K-means++: the advantages of careful seeding. In: Proceedings of the Eighteenth Annual ACM-SIAM Symposium on Discrete Algorithms. Society for Industrial and Applied Mathematics (2007)
27. Blondel, V.D., Guillaume, J.-L., Lambiotte, R., Lefebvre, E.: Fast unfolding of communities in large networks. J. Stat. Mech. **2008**, P10008 (2008)
28. Zhang, J., Wang, Y., Feng, J.: A hybrid clustering algorithm based on PSO with dynamic crossover. Soft Comput. **18**(5), 961–979 (2014)
29. Zhang, T., Ramakrishnan, R., Livny, M.: BIRCH: an efficient data clustering method for very large databases. In: Proceedings of the 1996 ACM SIGMOD International Conference on Management of Data - SIGMOD 1996, pp. 103–114 (1996). https://doi.org/10.1145/233269.233324

Discrete Model of Asynchronous Multitransmitter Interactions in Biological Neural Networks

Oleg P. Kuznetsov[1]([✉]), Nikolay I. Bazenkov[1], Boris A. Boldyshev[1],
Liudmila Yu. Zhilyakova[1], Sergey G. Kulivets[1], and Ilya A. Chistopolsky[2]

[1] V. A. Trapeznikov Institute of Control Sciences of Russian Academy of Sciences,
65 Profsoyuznaya street, Moscow 117997, Russia
olpkuz@yandex.ru
[2] N. K. Koltzov Institute of Developmental Biology of Russian Academy of Sciences,
26 Vavilova street, Moscow 119334, Russia

Abstract. An asynchronous discrete model of nonsynaptic chemical interactions between neurons is proposed. The model significantly extends the previous work [5,6] by novel concepts that make it more biologically plausible. In the model, neurons interact by emitting neurotransmitters to the shared extracellular space (ECS). We introduce dynamics of membrane potentials that comprises two factors: the endogenous currents depending on the neurons of firing type, and the exogenous current, depending on the concentrations of neurotransmitters that the neuron is sensitive to. The firing type of a neuron is determined by the individual composition of endogenous currents. We consider three basic firing types: oscillatory, tonic and reactive. Each of them is essential for modeling central pattern generators, i.e. neural ensembles generating rhythmic activity in the absence of external stimuli. Variability of endogenous currents of different neurons leads to asynchronous neural interactions and significant fluctuations of phase durations in the activity patterns present in simple neural systems. An algorithm computing the behavior of the proposed model is provided.

Keywords: Asynchronous discrete model
Heterogeneous neuronal system · Extracellular space
Nonsynaptic interactions · Neurotransmitters

1 Introduction

The work continues the study of multitransmitter interactions in neural networks, initiated in [5,6]. The paper introduces new entities such as concentrations of neurotransmitters in the extracellular space, the rate of change in the membrane potential, and asynchronous interactions between neurons.

Modeling of neural networks can be viewed as two large disjoint directions. The first direction includes artificial neural networks (ANN), initially proposed

© Springer Nature Switzerland AG 2018
S. O. Kuznetsov et al. (Eds.): RCAI 2018, CCIS 934, pp. 190–205, 2018.
https://doi.org/10.1007/978-3-030-00617-4_18

as a machine learning technique for classification, pattern recognition etc. ANNs consist of simple threshold elements evolved from formal neurons of McCulloch-Pitts [16]. A review of classical artificial neural networks can be found in [12]. Currently multilayer neural networks known as deep learning architectures are extensively used for complex non-algorithmic and creative problems [11,15]. The described direction tends to maximal simplification of the neuron model and doesn't pretend to be biologically plausible. Complex network behavior is achieved by a large number of elements and connections between them.

The second direction focuses on biophysical processes occurring in a real neuron. These models describe in detail ionic currents flowing through ion channels in the neuron membrane. Specifically, dynamics of these currents when the action potential arises and propagates along the axon. These processes are described by systems of differential equations. Moreover, these models often require a large number of parameters to be selected empirically. The Hodgkin-Huxley model [13] and its modifications [10,17,19] is the most popular. In [14] the most successful models are compared by two factors: biological plausibility and computational efficiency. The author proposed another model suitable for simulation of large complex networks imitating cortical structures.

The model presented here describes the operation of simple networks with a small number of neurons characterized by individual properties. A typical example of such networks are central pattern generators [3,18] – neural ensembles that produce rhythmic motor output. The biological principles underlying the model are as follows:

- Interactions between neurons have a chemical basis; neurons release neurotransmitters and react to them not only through synapses, but also in the extrasynaptic way [4,9,21,22].
- Neurons are transmitter-specific. Each neuron: (a) in the active state releases a specific neurotransmitter [19,20]; (b) has a set of receptors selectively reacting to specific neurotransmitters [7,20]. There are excitatory and inhibitory receptors.
- Neurons communicate via a common extracellular space (ECS): each neuron is embedded into the common chemical environment mediating extrasynaptic interactions (volume transmission) [1,8].
- Neurons may have endogenous electrical activity of a certain type.

2   Modeling Objectives

There is a set of neurons in the extracellular space (ECS), which contains various neurotransmitters, either emitted by the neurons present in the ECS, or come from the outside. Neurons have receptors, and each receptor reacts to a specific transmitter. Receptors affect the membrane potential (MP) in a different way: the influence can have different sign (excitatory or inhibitory) and different strength (weight).

Neurons are heterogeneous and a neuron's type is determined by three basic properties:

- transmitters produced by the neuron (transmitter-specificity);
- the set of receptors which defines how the neuron reacts to different transmitters;
- the type of endogenous electrical activity – the ability to be active in the absence of external input.

We consider three basic types of endogenous activity:

1. Oscillatory neuron – periodically fires a burst of spikes when it is not inhibited.
2. Tonic neuron – holds a constant level of endogenous activity in the absence of inhibition.
3. Passive neuron – must be excited externally similar to the classic McCulloch-Pitts neuron.

A neuron is activated if its MP has reached the threshold value specific for each neuron. Activation occurs as a result of either endogenous activity or external input. Being active, the neuron releases one or more transmitters. The state of a neuron's activity (active/inactive) is called the *external state* of the neuron.

The neurons function in continuous time with discrete *events*. Examples of events: a neuron changes its state; a new transmitter has appeared in the ECS; the concentration of an existing transmitter has changed (including disappearance, see Sect. 3.3). The full list of events will be given below.

Events are represented as points on the continuous timescale which is divided by the events into discrete timesteps. The edges of the timesteps are called discrete moments enumerated as 0, 1, 2, ... The timesteps are also enumerated. Timestep t is the half-open interval $[t, t+1)$: the left bound is included in the timestep, the right bound belongs to the next timestep. No events occur inside the timestep. The timesteps have different length (duration) measured in continuous time units, expressed by positive numbers. The duration of the timestep t is denoted by $\tau(t)$. The position of the moment t on the timescale is denoted by $T(t)$, where

$$T(t) = \sum_{i=0}^{t-1} \tau(t). \tag{1}$$

3   The Model – Main Definitions

We introduce a heterogeneous neural system $\mathbf{S} = \langle \mathbf{N}, \mathbf{X(t)}, \mathbf{C}, \mathbf{T} \rangle$, where $\mathbf{N} = \{N_1, ..., N_n\}$ is the set of neurons, $\mathbf{X}$ is the extracellular space (ECS), $\mathbf{C} = \{c_1, ..., c_m\}$ is the set of neurotransmitters and $\mathbf{T}$ is the continuous timescale. At each timestep the neurons interact emitting the transmitters to the ECS and receiving them from it. Each neuron reacts to all transmitters present in the

ECS if it has appropriate receptors. There is no need to introduce a synapse as a separate entity: if neuron N_i produces a transmitter to which neurons N_k and N_l have receptors then there are causal links from N_i to N_k and N_l that can be interpreted as synapses.

3.1   Neuronal Inputs

Neuron N_i has a set of receptor slots, each slot is sensitive to some transmitter c_j and has weight $w_{ij} \in \mathbb{R}$. A slot represents all receptors to the transmitter c_j and its weight is the total effect from these receptors. The weight $w_{ij} = 0$ means that neuron N_i has no receptors to the transmitter c_j, $w_{ij} > 0$ means that this transmitter excites the neuron, $w_{ij} < 0$ means that the transmitter inhibits the neuron.

3.2   Neuronal Outputs

A neuron's activity is denoted by $y_i(t) \in \{0,1\}$: $y_i(t) = 1$ means the neuron is active at timestep t and $y_i(t) = 0$ means that the neuron is inactive. A neuron may be active or inactive at different timesteps. After activation a neuron emits to ECS a constant dose $d_{ij} \geq 0$ of transmitter c_j. The value of d_{ij} is the concentration of transmitter c_j produced by neuron N_{ij}. If a neuron doesn't produce transmitter c_j then $d_{ij} = 0$. During a timestep d_{ij} remains constant.

3.3   Extracellular Space

The *external state* of the ECS at timestep t is a vector $X(t) = (x_1(t), \dots, x_m(t))$, where $x_j(t)$ is the total concentration of transmitter c_j in the ECS during the timestep t. We introduce the following assumptions:

1. one dose of any transmitter is sufficient to influence all receptors sensitive to it;
2. dose d_{ij} of transmitter c_j exists in the ECS while neuron N_i is active, plus some duration τ_{c_j} independent of N_i.

Let us introduce a Boolean variable (a predicate) $I_{d_{ij}}(t)$:

$$I_{d_{ij}} = \begin{cases} 1, \text{if } d_{ij} \text{ exists at timestep } t \text{ and } y_i(t) = 0 \\ 0, \text{otherwise.} \end{cases} \tag{2}$$

If no transmitters come from outside, concentration $x_j(t)$ is defined by the equation

$$x_j(t) = \sum_{i=1}^{n} d_{ij}(y_i(t) + I_{d_{ij}}(t)) \tag{3}$$

An appearance of a new transmitter is a separate event only when the transmitter comes from the outside. Otherwise it is related to another event – activation of a neuron. A disappearance of a transmitter is always a separate event.

In addition, the concentration of a transmitter discretely changes when the transmitter was emitted into the ECS by several neurons: when a dose produced by one neuron disappears, other doses of the same transmitter may persist. This is an event since the concentration of a transmitter changes and therefore the influence to the neurons changes as well. Inside a timestep, the state of the ECS does not change.

An external state of the system at timestep t is the vector of the external states of all the neurons and the ECS:

$$Z(t) = (Y(t), X(t)) = (y_1(t), \ldots, y_n(t), x_1(t), \ldots, x_m(t)).$$

3.4   Membrane Potential

The membrane potential (MP) $U_i(t)$ of neuron N_i varies in the range $U_i^{min} \leq U_i(t) \leq U_i^{max}$. In the absence of external inputs the MP varies in the range $U_i^0 \leq U_i(t) \leq U_i^{max}$, $U_i^0 \geq U_i^{min}$. A neuron is active if its membrane potential is not below the threshold $P_i \leq U_i^{max}$:

$$y_i(t) = \begin{cases} 1, \text{if } U_i(t) \geq P_i, \\ 0, \text{otherwise.} \end{cases} \tag{4}$$

The values of U_i^{min}, U_i^{max}, U_i^0 and P_i are individual for every neuron. Inside a timestep the dynamics of the membrane potential is driven by the constant *total current* $v_i(t)$:

$$v_i(t) = s_i(t) + v_{ien}^{\alpha}(t), \tag{5}$$

where $s_i(t)$ is the *exogenous current* proportional to the external inputs:

$$s_i(t) = h \sum_{j=1}^{m} w_{ij} x_j(t), \tag{6}$$

v_{ien}^{α} is the endogenous current, α is the parameter which depends on the neuron's activity type (each neuron type is characterized by a specific set of endogenous currents) and the current phase of the endogenous dynamics. We assume $h = 1$ in the rest of the paper.

Subsection 3.5 explains behavior of MP for different neuron types. All possible values of α are given in Table 2.

3.5   Neural Activity Types

All neurons in the model belong to one of the following types of endogenous activity: oscillatory, tonic, and reactive. Each type is characterized by an individual structure of endogenous currents.

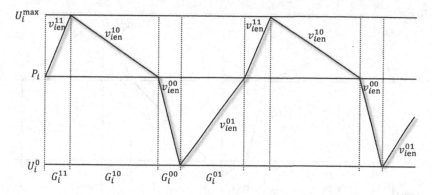

Fig. 1. Membrane potential dynamics of an oscillatory neuron in the autonomous mode

Oscillator. Oscillatory, or bursting, neuron N_i periodically fires a burst of spikes. In the absence of external inputs, the burst duration $\tau_i^1 > 0$ and the quiescent period $\tau_i^0 > 0$ are constant. They are shaped by endogenous currents active in different phases of the oscillatory cycle. Figure 1 illustrates the four phases of the cycle. A phase is denoted by G^{kl}, $k, l \in \{0, 1\}$ where $k = y_i(t)$ and $l = 1$ if the MP is increasing during the phase and $l = 0$ otherwise.

The burst begins when the MP reaches threshold P_i, then U_i rises from P_i to U_i^{max}. This is the phase G^{11} and the corresponding endogenous current (*endogenous growth current*) is v_{ien}^{11}. The process transits to discharge phase G^{10}: the MP declines from U_i^{max} to P_i with *endogenous discharge current* v_{ien}^{10}. Then the reset phase begins when the MP falls to U_i^0 driven by the *endogenous reset current* v_{ien}^{00}. The reset phase is followed by the recharge phase G^{01} when the membrane potential is driven from U_i^0 up to P_i by the *endogenous recharge current* v_{ien}^{01}. The currents v_{ien}^{01} and v_{ien}^{11} are positive and $v_{ien}^{10}, v_{ien}^{00}$ are negative.

The burst duration (active phase) τ_i^1 equals

$$\tau_i^1 = \frac{U_i^{max} - P_i}{v_{ien}^{11}} + \frac{P_i - U_i^{max}}{v_{ien}^{10}}, \tag{7}$$

and the quiescent period τ_i^0 equals

$$\tau_i^0 = \frac{U_i^0 - P_i}{v_{ien}^{00}} + \frac{P_i - U_i^0}{v_{ien}^{01}}. \tag{8}$$

External currents may change the burst and quiescent durations and the whole dynamics get more complex. An excitatory input reduces the recharge phase and prolongs the discharge phase. An inhibitory input will slow down the recharge and speed up the discharge phases. An inhibition may shift the MP below U_i^0, that situation is called G_i^{min} and does not appear in the autonomous mode. Moreover, a strong enough input may change the sign of the total rate $v_i(t)$. The extended equations for total rate are provided at the end of the section.

Figure 2 illustrates the oscillatory MP dynamics under the external inhibition. The vertical lines marks the events changing the behavior of the neuron.

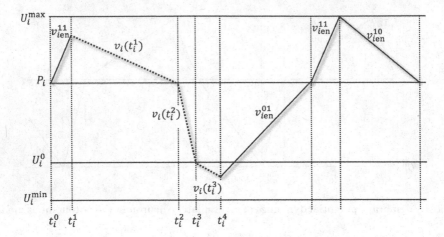

Fig. 2. Membrane potential dynamics of an oscillatory neuron with inhibitory input from timestep t_i^1 to t_i^4.

There is no external input at timestep t_i^0 and the MP grows with endogenous rate v_{ieb}^{11}. At the moment t_i^1 another neuron was activated and emitted a neurotransmitter causing a strong inhibition to N_i. The total rate $v_i(t_i^1)$ became negative and the MP began to decline. As a result, at the moment t_i^2 neuron N_i was turned off, and its MP continued its fall with the total rate $v_i(t_i^2)$. When U_i^0 is reached at the moment t_i^3, the endogenous current v_{ien}^{01} was switched on, and the total MP rate decreased in absolute value, but remained negative, so the fall continued. At the moment t_i^4 the inhibitory transmitter ceased to exist, and the MP began to grow at the rate v_{ien}^{01}.

Now we introduce a strict definition of an *event*. An *event* is either: (1) a change of a neuron's activity, or (2) a change of a MP rate of a neuron, or (3) a decrease in concentration of a transmitter, e.g. a disappearance of a dose d_{ij}.

A switch to another endogenous current is not necessarily associated with a change of the neuron's activity state. As Figs. 1 and 2 show for the oscillator, there are ranges of the MP where the activity state is the same, but the endogenous currents are different. In the range $J_i^{max} = [P_i, U_i^{max}]$, the currents are v_{ien}^{10} or v_{ien}^{11}, in the range $J_i^0 = [U_i^0, P_i)$ – currents v_{ien}^{00} or v_{ien}^{01}. In the range $J_i^{min} = [U_i^{min}, U_i^0)$ there is only current v_{ien}^{01}.

Tonic Neuron. A tonic neuron N_i is permanently active when it is not inhibited, and by default $U_i = U_i^{max}$. Thus, there are no discharge v_{ien}^{10} and reset v_{ien}^{00} endogenous currents. When the inhibition is removed, it tends to reach U_i^{max}, i.e. its endogenous current is always positive ($l = 1$), however it depends on whether the neuron is active at time t ($y_i(t) = 1$) or not. If $y_i(t) = 1$ ($U_i \geq P_i$), then the MP is driven by the endogenous current v_{ien}^{11}. If $y_i(t) = 0$ the MP rate is v_{ien}^{01}, this situation is denoted by the indicator $I_i^{min}(t)$.

Reactive Neuron. A reactive neuron N_i stays inactive until it is excited externally. The default MP is $U_i = U_i^0$. Therefore, it does not have recharge v_{ien}^{01} and growth v_{ien}^{11} endogenous currents. When the excitation is removed at time t, it tends to reach U_i^0, but its endogenous current depends on what range the neuron's MP belongs to at the moment. If $y_i(t) = 1$, the endogenous current is v_{ien}^{10}. In the range J_i^0, the current is v_{ien}^{00}. The situation of $U_i = U_i^0$ and no inhibition is applied, is a special phase G_i^{0*}, specific for the reactive neuron only. It is a point at the boundary of the ranges J_i^0 and J_i^{min}. At the point the endogenous current is $v_{ien}^{0*} = 0$. If, as a result of inhibition, the MP of a reactive neuron is below U_i^0, then a positive endogenous current $-v_{ien}^{00}$ switches on, tending to return the MP to U_i^0.

3.6 Summary

The described parameters can be divided into three classes: absolutely static (remain stable during the model functioning), autonomously static (stable in the absence of external inputs) and dynamic. Absolutely static parameters include: the number of receptors, weights, thresholds, etc. (see Table 1). Autonomously static are the burst duration and the recharge period, i.e. the time required by the membrane potential to reach the threshold P_i from U_i^0. Dynamic parameters are the variables $y_i(t)$, $U_i(t)$, $v_i(t)$, etc. (see Sect. 4.1).

The static parameters are listed in the Table 1.

Table 1. Static parameters of the model

	P_i	U_i^{max}	U_i^0	U_i^{min}	v_{ien}^{00}	v_{ien}^{01}	v_{ien}^{10}	v_{ien}^{11}	c_1	...	c_m	w_{i1}	...	w_{im}
N_1	P_1	U_1^{max}	U_1^0	U_1^{min}	v_{1en}^{00}	v_{1en}^{01}	v_{1en}^{10}	v_{1en}^{11}	d_{11}	...	d_{1m}	w_{11}	...	w_{1m}
...	...	...	...	...	...	...	...	...	...	...	...	...	...	...
N_n	P_n	U_n^{max}	U_n^0	U_n^{min}	v_{nen}^{00}	v_{nen}^{01}	v_{nen}^{10}	v_{nen}^{11}	d_{n1}	...	d_{nm}	w_{n1}	...	w_{nm}
τ_{c_j}									τ_{c_1}	...	τ_{c_m}			

We introduce a notion of a *situation*, which generalizes the notion of a phase to every neural type and the case where external inputs are present.

Situation $G_i^\alpha(t)$ is the pair $(y_i(t), v_{ien}^\alpha)$, where $\alpha \in \{00, 01, 10, 11, 0*, min\}$. A transition from one situation to another is associated with a switch between the endogenous currents or a change in the neuron's activity. In any case, it is triggered by an event.

The fact that neuron N_i is in situation $G_i^\alpha(t)$ is represented as a predicate (the situation indicator) $I_i^\alpha(t)$: $I_i^\alpha(t) = 1$ if $U_i(t)$ is in the situation G_i^α; otherwise $I_i^\alpha(t) = 0$. Obviously, for any t, one and only one of these indicators equals to 1. The notations G_i^α are called the *names of situations*.

Not every combination of $y_i(t)$, v_{ien}^α is feasible. Different neural types are associated with different sets of endogenous currents and possible situations listed in Table 2.

Table 2. Situations for different neural types

Oscillatory neuron	Tonic neuron	Reactive neuron
$G_i^{11}(t) = (1, v_{ien}^{11})$	$G_i^{11}(t) = (1, v_{ien}^{11})$	$G_i^{10}(t) = (1, v_{ien}^{10})$
$G_i^{10}(t) = (1, v_{ien}^{10})$	$G_i^{01}(t) = (0, v_{ien}^{01})$	$G_i^{00}(t) = (0, v_{ien}^{00})$
$G_i^{00}(t) = (0, v_{ien}^{00})$		$G_i^{min}(t) = (0, -v_{ien}^{00})$
$G_i^{01}(t) = (0, v_{ien}^{01})$		$G_i^{0*}(t) = (0,0)$ – the rest state $U_i = U_i^0$

Moreover, there are three situations $G_{d_{ij}}$ for neurotransmitter doses: $y_i(t) = 1$ (d_{ij} exists); $y_i(t) = 0$, but d_{ij} still exists; $y_i(t) = 0$ and d_{ij} does not exist. The first is equivalent to one of the neural situations. The second is indicated by $I_{d_{ij}}$, introduced in the Eq. (2). The third situation is denoted as $G_{d_{ij}0}$.

The situation indicators can be used to transform the total MP rate Eq. (5) to equations, specific for each neural type.

Oscillatory neuron:

$$v_i(t) = s_i(t) + \sum_{k,l} v_{ien}^{kl} I_i^{kl}(t). \tag{9}$$

Tonic neuron:

$$v_i(t) = s_i(t) + \sum_{k} v_{ien}^{k1} I_i^{k1}(t). \tag{10}$$

Reactive neuron:

$$v_i(t) = s_i(t) + \sum_{k} v_{ien}^{k0} I_i^{k0}(t) - v_i^{00} I_i^{min}(t). \tag{11}$$

4 Model Dynamics: Simulation

4.1 Dynamic Parameters and the Internal System State

An algorithm implementing the proposed model should compute a sequence of the external states at the moments $0, 1, \ldots, t, t+1, \ldots$. Thus, it should compute the state at the moment $t + 1$, given the state at the moment t. Because the model is asynchronous, before the next state is computed, one should compute the moment $t + 1$ itself, i.e. the duration $\tau(t)$ of the timestep $[t, t + 1)$. The dynamic parameters are also updated at each timestep.

Here we introduce several new dynamic parameters in addition to $y_i(t)$, $x_j(t)$, $U_i(t)$, $s_i(t)$, $v_i(t)$, I_i^α.

Residual Potential. $\Delta U_i(t)$ is the "distance" to the *nearest event* associated with neuron N_i. The nearest event is determined not only by the current situation, but also by the sign of the total rate (5) and the neuron's type. All possible cases are specified by *transition tables*, separate for each neural type. Table 3 shows the transition table for the oscillatory neuron. The tables for other neural types are constructed in the same way. Each row describes a transition from the situation in the second column, depending on the rate of the MP (third column). The nearest event is given in the fourth column and the formulas for residual potentials are listed in the fifth. Sign '∞' means that the MP will not change until the next event. The case $v_i(t) = 0$ means $\Delta U_i(t) = \Delta U_i(t-1)$, so it is not included in the table.

Table 3. Transition table for the oscillatory neuron

	Situation	Sign of $v_i(t)$	Nearest event	$\Delta U_i(t)$
1	$G_i^{00} = (0, v_{ien}^{00})$	+	$y_i = 1;\ I_i^{11} = 1$	$P_i - U_i(t)$
2	$G_i^{00} = (0, v_{ien}^{00})$	−	U_i^0 is reached; $I_i^{01} = 1$	$U_i(t) - U_i^0$
3	$G_i^{01} = (0, v_{ien}^{01})$	+	$y_i = 1;\ I_i^{11} = 1$	$P_i - U_i(t)$
4	$G_i^{01} = (0, v_{ien}^{01})$	−	U_i^{min} is reached; $I_i^{01} = 1$ – the situation doesn't change, no events	∞
5	$G_i^{11} = (1, v_{ien}^{11})$	+	U_i^{max} is reached; $I_i^{10} = 1$	$U_i^{max} - U_i(t)$
6	$G_i^{11} = (1, v_{ien}^{11})$	−	$y_i = 0;\ I_i^{00} = 1,\ I_{d_{ij}} = 1$	$U_i(t) - P_i$
7	$G_i^{10} = (1, v_{ien}^{10})$	+	U_i^{max} is reached; $I_i^{10} = 1$ – the situation doesn't change, no events	∞
8	$G_i^{10} = (1, v_{ien}^{10})$	−	$y_i = 0;\ I_i^{00} = 1,\ I_{d_{ij}} = 1$	$U_i(t) - P_i$

The nearest event to be realized is the event with the minimal residual time.
The residual time $\tau_{ri}(t)$ of neuron N_i is the time to the nearest event given the current values of the MP $U_i(t)$ and the total rate $v_i(t)$.

$$\tau_{ri}(t) = \begin{cases} \frac{\Delta U_i(t)}{|v_i(t)|}, & \text{if } v_i(t) \neq 0 \text{ and } \Delta U_i(t) < \infty; \\ \infty, \text{otherwise} \end{cases} \tag{12}$$

The residual time $\tau_{d_{ij}}(t)$ of a dose d_{ij} for transmitter c_j for any $t \neq 0$ is defined as follows:

$$\tau_{d_{ij}}(t) = \begin{cases} \tau_{ri}(t) + \tau_{c_j}, & \text{if } y_i(t) = 1; \\ \tau_{c_j}, & \text{if } y_i(t) = 0 \text{ and } y_i(t-1) = 1; \\ \tau_{ri}(t), & \text{if } y_i(t) = 0 \text{ and } \tau(t-1) = \tau_{d_{ij}}(t-1); \\ \tau_{d_{ij}}(t-1) - \tau(t-1), & \text{otherwise.} \end{cases} \tag{13}$$

If in any of these rows $\tau_{ri}(t) = \infty$, then $\tau_{d_{ij}}(t) = \infty$ for the row.

Let us explain Eq. (13). The first row corresponds to the second point from Subsect. 3.3. The second row correspond to the case when neuron N_i turned off at the moment t. The third row describes the case when a dose d_{ij} ceased to exist at the moment t; the dose will appear again when neuron N_i turns on.

The fourth row describes the last case when at the moment $t - 1$ an event, not associated with N_i and d_{ij}, occurs after N_i turned off and before d_{ij} disappeared.

The equation for $\tau_{d_{ij}}(0)$ is more compact:

$$\tau_{d_{ij}}(0) = \begin{cases} \tau_{ri}(0) + \tau_{c_j}, & \text{if } y_i(0) = 1; \\ \infty, & \text{if } y_i(t) = 0. \end{cases} \tag{14}$$

4.2   Algorithm of the Neural System Functioning

The algorithm's input are the static parameters specified by Table 1 and the initial MP values $U_1(0), \ldots, U_n(0)$. They completely determine the external state $Z(0)$ and the situations for every element of the system. Thus, call the vector $Q(0) = (U_1(0), \ldots, U_n(0))$ the internal system state.

At the moment t given $Q(0)$ the algorithm performs the following steps:

1. Compute the membrane potentials $U_i(t)$:

$$U_i(t) = U_i(t - 1) + \tau(t - 1)v_i(t - 1).$$

Here
   - if $U_i(t - 1) + \tau(t - 1)v_i(t - 1) > U_i^{max}$, set $U_i(t) = U_i^{max}$;
   - if $U_i(t - 1) + \tau(t - 1)v_i(t - 1) < U_i^{max}$, set $U_i(t) = U_i^{min}$.
2. Compute the neural inputs, Eq. (6).
3. Compute the total MP rates, Eqs. (9)–(11).
4. Compute the residual potentials by the transition table.
5. Compute the residual time by Eqs. (12) and (14).
6. Compute the duration $\tau(t)$ of the timestep $[t, t + 1]$:

$$\tau(t) = \min_{i,j}(\tau_{ri}(t), \tau_{d_{ij}}(t)).$$

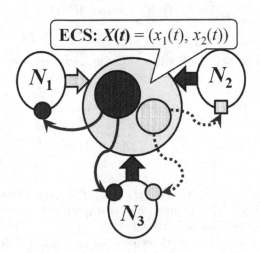

Fig. 3. Example. Three neurons with common ECS.

7. Find the nearest events:
   - If $\tau(t) = \tau_{rg}(t)$ then the event is the transition to another situation of neuron N_g:
     For the current indicator $I_g(t)$ and the sign of $v_g(t)$, find the next situation in the transition table.
   - If $\tau(t) = \tau_{d_{gh}}(t)$ then the event is that dose d_{gh} is removed from the ECS.

The algorithm is implemented as a software tool.

5   Example: An Ensemble of Three Neurons

Consider a neural ensemble consisting of three neurons (Fig. 3) with the following types: N_1 – oscillatory, N_2 – reactive, N_3 – oscillatory. Neuron N_1 produces transmitter c_1, N_2 and N_3 produce transmitter c_2. In the figure the circles denote inhibitory receptors, the square denotes an excitatory receptor.

The static parameters are listed in Table 4, d_{ij}, $i = 1, \ldots, n$, $j = 1, \ldots, m$ lies at the intersection of the row N_i and the column c_j.

The transition tables for N_1 and N_3 are derived by substitution of the parameters of Table 4 to the Table 3. The result is given in Table 5. The table for neuron N_2 is derived by the transition table of the reactive neuron which is not provided here.

The initial state (membrane potentials): $Q(0) = (0.9, 0, 0)$.

The initial external state: $Z(0) = (1, 0, 0, 0.7, 0)$.

Situations:
$$G_i^{10} = (y_1(0) = 1, -0.2), \quad G_2^{00} = (y_2(0) = 0, -0.2), \quad G_3^{01} = (y_3(0) = 0, 0.8).$$

The simulation protocol of the first 10 timesteps for the given parameters and initial conditions are provided in Table 6. Figure 4 shows the membrane potential dynamics for every neuron.

Note that the sequence of the external states contains identical timesteps: 2–4, 5–6, 7–8, 9–10. An external observer could not distinguish these timesteps, so they can be combined into a single timestep which duration is the sum of the durations of the original timesteps. The enlarged timesteps are listed in the Table 7.

The combined external sequence is called *a rhythm*. Rhythms are useful to describe behavioral patterns because an external observer can see only them.

Table 4. Static parameters for the system in Fig. 3

	P_i	U_i^{max}	U_i^0	U_i^{min}	v_{ien}^{00}	v_{ien}^{01}	v_{ien}^{10}	v_{ien}^{11}	c_1	c_2	w_{i1}	w_{i2}
N_1	0.6	0.9	0	−0.2	−0.6	0.85	−0.2	0.95	0.7	—	0	−1
N_2	0.6	0.7	0	−0.2	−0.2	0	−0.2	0	—	0.6	1	0
N_3	0.4	0.6	0	−0.2	−0.2	0.8	−0.2	0.9	—	0.7	−1	−1
τ_{c_j}									0.1	0.1		

Table 5. Transition table for neuron N_1

	Situation	Sign of $v_1(t)$	Nearest event	$\Delta U_1(t)$
1	$G_1^{00} = (0, -0.6)$	+	$y_1 = 1$; $I_1^{11} = 1$	$0.6 - U_1(t)$
2	$G_1^{00} = (0, -0.6)$	−	U_1^0 is reached; $I_1^{01} = 1$	$U_1(t) - 0$
3	$G_1^{01} = (0, 0.85)$	+	$y_1 = 1$; $I_1^{11} = 1$	$0.6 - U_i(t)$
4	$G_1^{01} = (0, 0.85)$	−	U_1^{min} is reached; $I_1^{01} = 1$ – the situation does not change, no events	∞
5	$G_i^{11} = (1, 0.95)$	+	U_1^{max} is reached; $I_1^{10} = 1$	$0.9 - U_1(t)$
6	$G_i^{11} = (1, 0.95)$	−	$y_1 = 0$; $I_1^{00} = 1$, $I_{d_{11}} = 1$	$U_1(t) - 0.6$
7	$G_i^{10} = (1, -0.2)$	+	U_1^{max} is reached; $I_1^{10} = 1$ – the situation doesn't change, no events	∞
8	$G_i^{10} = (1, -0.3)$	−	$y_1 = 0$; $I_1^{00} = 1$, $I_{d_{11}} = 1$	$U_1(t) - 0.6$

Table 6. Dynamics of parameters for the simulated example

t	0	1	2	3	4	5	6	7	8	9	10
$Y(t)$	100	110	010	010	010	000	000	001	001	000	000
$U_1(t)$	0.9	0.66	0.6	0.48	0	0.0094	0.0344	0.31	0.46	0.4933	0.5083
$U_2(t)$	0	0.6	0.6375	0.6875	0.6075	0.6	0.58	0.515	0.315	0.2706	0.2506
$U_3(t)$	0	0.12	0.0825	0.0325	0.1125	0.12	0.14	0.4	0.6	0.4	0.38
$G_1^\alpha(t)$	G_1^{10}	G_1^{10}	G_1^{00}	G_1^{00}	G_1^{01}	G_1^{01}	G_1^{01}	G_1^{01}	G_1^{01}	G_1^{01}	G_1^{01}
$G_2^\alpha(t)$	G_2^{00}	G_2^{10}	G_2^{10}	G_2^{10}	G_2^{10}	G_2^{00}	G_2^{00}	G_2^{00}	G_2^{00}	G_2^{00}	G_2^{00}
$G_3^\alpha(t)$	G_3^{01}	G_3^{01}	G_3^{01}	G_3^{01}	G_3^{01}	G_3^{01}	G_3^{01}	G_3^{11}	G_3^{10}	G_3^{00}	G_3^{00}
$\tau(t)$	1.2	0.075	0.1	0.4	0.0375	0.1	0.325	1	0.2222	0.1	0.1079
v_1	−0.2	−0.8	−1.2	−1.2	0.25	0.25	0.85	0.15	0.15	0.15	0.85
v_2	0.5	0.5	0.5	−0.2	−0.2	−0.2	−0.2	−0.2	−0.2	−0.2	−0.2
v_3	0.1	−0.5	−0.5	0.2	0.2	0.2	0.8	0.2	−0.9	−0.9	−0.2
d_{11}	0.7	0.7	0.7	0	0	0	0	0	0	0	0
d_{22}	0	0.6	0.6	0.6	0.6	0.6	0	0	0	0	0
$d32$	0	0	0	0	0	0	0	0.7	0.7	0.7	0

Table 7. Combined timesteps

Timesteps	0	1	2–4	5–6	7–8	9–10	11
Rhythm	100	110	010	000	001	000	100
Duration	1.2	0.075	0.5375	0.425	1.2222	0.2079	

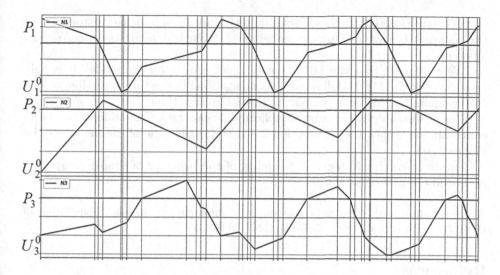

Fig. 4. Membrane potentials of the interacting neurons.

6   Discussion

The well-established models of threshold networks [2,12,23] are synchronous: all elements have the same time parameters. Therefore, in these models there is no need to consider the membrane potential and its dynamics: one can assume that the activation follows the moment when the input reaches the threshold with a delay, which is the same for all elements. Moreover, the external state $Y(t)$ uniquely determines $Y(t+1)$, and in the autonomous mode the network functions like a finite state machine: the sequence $Y(0), \ldots, Y(t), \ldots$, becomes periodic at some t'. The lengths of the period and the subperiod do not exceed $2n$, where n is the number of elements in the network.

Our model is different. Table 6 shows that the external state $Y(t)$ is insufficient to determine $Y(t+1)$. Thus, the question, whether the model is a finite state machine, is still open.

7   Conclusion

Here we summarize the main assumptions adopted in the model. These assumptions greatly simplify the biological reality.

1. The dose d_{ij} of a transmitter c_j, produced by a neuron N_i, remains constant while the neuron is active plus the residual lifetime of τ_{c_j}. Then this dose disappears immediately.
2. A single dose of any transmitter is sufficient to influence all receptors sensitive to it.
3. During each timestep the membrane potentials change with constant rates, i.e. linearly.

The question is do these assumptions lead to significant distortions of the external behavior. Further modeling of biological neural systems should answer it.

We emphasize that the MP linearity is not crucial for the model. It is only necessary for the MP to follow some computationally tractable function, so one can compute the MP at any point of the continuous time.

The first assumption is more complicated, because if we introduce a "smooth" concentration decay, the assumption 2 and the whole logic of the model will be violated.

Future research will focus at modeling of the behavior of simple neural systems and the neuromodulatory effects.

Acknowledgement. The work was partially supported by Russian Foundation of Basic Research (projects no. 17-07-00541, 17-29-07029).

References

1. Agnati, L.F., Guidolin, D., Guescini, M., Genedani, S., Fuxe, K.: Understanding wiring and volume transmission. Brain Res. Rev. **64**(1), 137–159 (2010). https://doi.org/10.1016/j.brainresrev.2010.03.003, http://www.sciencedirect.com/science/article/pii/S0165017310000214
2. Amari, S.I.: Learning patterns and pattern sequences by self-organizing nets of threshold elements. IEEE Trans. Comput. **C–21**(11), 1197–1206 (1972). https://doi.org/10.1109/T-C.1972.223477
3. Balaban, P., et al.: Tsentral'nyye generatory patterna (CPGs). Zhurn. vyssh. nerv. deyat. **63**(5), 1–21 (2013)
4. Bargmann, C.I.: Beyond the connectome: how neuromodulators shape neural circuits. BioEssays: News Rev. Mol. Cell. Dev. Biol. **34**(6), 458–465 (2012). https://doi.org/10.1002/bies.201100185
5. Bazenkov, N., Dyakonova, V., Kuznetsov, O., Sakharov, D., Vorontsov, D., Zhilyakova, L.: Discrete modeling of multi-transmitter neural networks with neuronal competition. In: Samsonovich, A.V., Klimov, V.V. (eds.) BICA 2017. AISC, vol. 636, pp. 10–16. Springer, Cham (2018). https://doi.org/10.1007/978-3-319-63940-6_2
6. Bazenkov, N., et al.: Diskretnoe modelirovanie mezhneironnyh vzaimodeistvii v multitransmitternih setyah [Discrete modeling of neuronal interactions in multi-neurotransmitter networks]. Iskusstvenny Intellekt i Prinyatie Reshenii [Artif. Intell. Decis. Making] **2**, 55–73 (2017)
7. Brezina, V.: Beyond the wiring diagram: signalling through complex neuro modulator networks. Philos. Trans. R. Soc. Lond. B: Biol. Sci. **365**(1551), 2363–2374 (2010). https://doi.org/10.1098/rstb.2010.0105, http://rstb.royalsocietypublishing.org/content/365/1551/2363
8. Dyakonova, V.: Neyrotransmitternyye mekhanizmy kontekst-zavisimogo povedeniya. Zhurn. vyssh. nerv. deyat. **62**(6), 1–17 (2012)
9. Dyakonova, V.E., Dyakonova, T.L.: Coordination of rhythm-generating units via no and extra synaptic neurotransmitter release. J. Comput. Physiol. A **196**(8), 529–541 (2010). https://doi.org/10.1007/s00359-010-0541-5

10. FitzHugh, R.: Mathematical models of excitation and propagation in nerve. In: Schwan, H. (ed.) Biological Engineering, Chap. 1, pp. 1–85. McGraw Hill Book Co., New York (1969)
11. Goodfellow, I., Bengio, Y., Courville, A.: Deep Learning. MIT Press, Cambridge (2016)
12. Haykin, S.: Neural Networks: A Comprehensive Foundation, 3rd edn. Prentice-Hall Inc., Upper Saddle River (2007)
13. Hodgkin, A.L., Huxley, A.F.: A quantitative description of membrane current and its applications to conduction and excitation in nerve. J. Physiol. (Lond.) **116**, 500–544 (1952)
14. Izhikevich, E.M.: Which model to use for cortical spiking neurons? IEEE Trans. Neural Netw. **15**(5), 1063–1070 (2004)
15. LeCun, Y., Bengio, Y., Hinton, G.: Deep learning. Nature **521**, 436–444 (2015). https://doi.org/10.1038/nature14539
16. McCulloch, W., Pitts, W.: A logical calculus of the ideas immanent in nervous activity. Bull. Math. Biophys. **5**, 115–133 (1943)
17. Morris, C., Lecar, H.: Voltage oscillations in the barnacle giant muscle fiber. Biophys. J. **35**(1), 193–213 (1981)
18. Mulloney, B., Smarandache, C.: Fifty years of CPGs: two neuroethological papers that shaped the course of neuroscience. Front. Behav. Neurosci. **4**(45), 1–8 (2010). https://doi.org/10.3389/fnbeh.2010.00045
19. Nagumo, J., Arimoto, S., Yoshizawa, S.: An active pulse transmission line simulating nerve axon. Proc. IRE **50**(10), 2061–2070 (1962). https://doi.org/10.1109/JRPROC.1962.288235
20. Sakharov, D.: Biologicheskiy substrat generatsii povedencheskikh aktov. Zhurn. obshch. biologii. **73**(5), 334–348 (2012)
21. Sem'yanov, A.V.: Diffusional extrasynaptic neurotransmission via glutamate and gaba. Neurosci. Behav. Physiol. **35**(3), 253–266 (2005). https://doi.org/10.1007/s11055-005-0051-z
22. Vizi, E., Kiss, J.P., Lendvai, B.: Non synaptic communication in the central nervous system. Neurochem. Int. **45**(4), 443–451 (2004). https://doi.org/10.1016/j.neuint.2003.11.016, http://www.sciencedirect.com/science/article/pii/S0197018603002493. Role of Non-synaptic Communication in Information Processing
23. Wang, R.S., Albert, R.: Effects of community structure on the dynamics of random threshold networks. Phys. Rev. E **87**, 012810 (2013). https://doi.org/10.1103/PhysRevE.87.012810

eLIAN: Enhanced Algorithm
for Angle-Constrained Path Finding

Anton Andreychuk[1](✉), Natalia Soboleva[2], and Konstantin Yakovlev[2,3,4]

[1] Peoples' Friendship University of Russia, Moscow, Russia
andreychuk@mail.com
[2] National Research University Higher School of Economics, Moscow, Russia
nasoboleva@edu.hse.ru, kyakovlev@hse.ru
[3] Federal Research Center "Computer Science and Control" of Russian Academy
of Sciences, Moscow, Russia
yakovlev@isa.ru
[4] Moscow Institute of Physics and Technology, Dolgoprudny, Russia
yakovlev.ks@mipt.ru

Abstract. Problem of finding 2D paths of special shape, e.g. paths comprised of line segments having the property that the angle between any two consecutive segments does not exceed the predefined threshold, is considered in the paper. This problem is harder to solve than the one when shortest paths of any shape are sought, since the planer's search space is substantially bigger as multiple search nodes corresponding to the same location need to be considered. One way to reduce the search effort is to fix the length of the path's segment and to prune the nodes that violate the imposed constraint. This leads to incompleteness and to the sensitivity of the's performance to chosen parameter value. In this work we introduce a novel technique that reduces this sensitivity by automatically adjusting the length of the path's segment on-the-fly, e.g. during the search. Embedding this technique into the known grid-based angle-constrained path finding algorithm LIAN, leads to notable increase of the planner's effectiveness, e.g. success rate, while keeping efficiency, e.g. runtime, overhead at reasonable level. Experimental evaluation shows that LIAN with the suggested enhancements, dubbed eLIAN, solves up to 20% of tasks more compared to the predecessor. Meanwhile, the solution quality of eLIAN is nearly the same as the one of LIAN.

Keywords: Path planning · Path finding · Grid · Angle-constrained
LIAN

1 Introduction

Path finding is a vital capability of any intelligent agent operating in real or simulated physical environment. In case this environment is known a priory and remains static, grids are commonly used to explicitly represent it as they are simple yet informative and easy-to-update graph models. Grids appear naturally in

S. O. Kuznetsov et al. (Eds.): RCAI 2018, CCIS 934, pp. 206–217, 2018.
https://doi.org/10.1007/978-3-030-00617-4_19

game development [11] and are widely used in robotics [4,12]. When agent's environment is represented by a grid, heuristic search algorithms, such as A* [7] and its modifications, are typically used for path planning. The path planning process is carried out in the state-space, where states are induced by grid elements, and the function of generating successors defines which movements on the grid are allowed or not [15].

Numerous works on grid-based path finding allow agent to move only between the cardinally adjacent grid elements, e.g cells or corners (sometimes, diagonal moves are allowed as well), see [1,6,10] for example. This often results in finding paths, containing multiple sharp turns (see Fig. 1a), which might be undesirable for some applications. To mitigate this issue a few techniques have been proposed. First, smoothing can be carried out as the post-processing step [1], i.e. when the path has already been planned. Second, smoothing can be interleaved with the state-space exploration resulting in so-called any-angle behavior (see Fig. 1b). Originally any-angle algorithms, like Theta* [3], Lazy Theta* [9] were lacking optimality, by recently ANYA* [5] was introduced which guarantees finding shortest any-angle paths between the given grid cells.

Any-angle algorithms allow agent to move in any arbitrary direction, but the resulting paths still might contain sharp turns (see Fig. 1b). [13] addresses this issue by keeping the number of turns to a possible minimum. In [8] a modification of Theta* was proposed, that prunes away the nodes that lead to the turns violating the given kinematic constraints of the agent. Finally, in [14] LIAN algorithm was proposed that seeks for the so-called angle-constrained paths, e.g. the paths comprised of line segments (with the endpoints tied to the grid elements), having the property that the angle between any two consecutive segments does not exceed the predefined threshold (see Fig. 1c). In this work we focus on these sort of paths, e.g. angle-constrained paths.

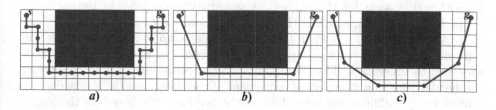

Fig. 1. Different types of grid paths. (a) A*-path composed of the moves between cardinally adjacent cells; (b) any-angle path; (c) angle-constrained path

Finding angle-constrained paths is a much harder problem to solve compared to finding shortest paths, as the algorithm has to keep numerous nodes corresponding to the same grid element because the latter can be approached from different directions, rather than keeping only one node and re-writing its g-value (cost of the best path known so far). In order to keep the number of nodes in the search space to a reasonable minimum LIAN considers only states that correspond to the moves of the fixed length, which is the parameter of the algorithm.

This, obviously, leads to incompleteness and to the sensitivity of the planner's performance to the chosen parameter value. In this work, we mitigate this issue by introducing an original technique that adjusts the parameter value on-the-fly while the algorithm performs the search. This does not make the algorithm complete but significantly increases the chance of finding the solution. As shown experimentally, success rate of the enhanced planner, dubbed eLIAN (enhanced LIAN) exceeds the one of the original algorithm up to 20%. Meanwhile, the solution quality of eLIAN and LIAN is nearly the same.

2   Background

Problem Statement. Consider a grid composed of blocked and unblocked cells and a point-agent that is allowed to move from one unblocked cell to the other following the straight line segment connecting the centers of those cells. The move is considered feasible if there exists line-of-sight between the endpoints of the move, in other words – if the corresponding segment of the straight line does not intersect any blocked cell. In practice the well-known in computer graphics algorithm from [2] adopting slight modifications can be used to efficiently perform line-of-sight checks.

Consider now a path, π, composed of feasible moves, e_i, and the value $\alpha_m(\pi) = max(|\alpha(e_1, e_2)|, |\alpha(e_2, e_3)|, \ldots, |\alpha(e_{v-1}, e_v)|)$, where $\alpha(e_j, e_{j+1})$ is the angle between two consecutive line segments representing the moves. The task of the planner is formalized as follows. Given the start and the goal cell, as well as the constraint $\alpha_{MAX} \in [0°; 180°]$, find a feasible path π, connecting them, such that $\alpha_m(\pi) \leq \alpha_{MAX}$.

LIAN Algorithm. LIAN (abbreviation for "limited angle") is a heuristic search algorithm that seeks for the angle-constrained path composed of the segments of the fixed length, Δ, which is the input parameter of the algorithm.

LIAN search node **s** is identified by the tuple $[s, bp(\mathbf{s})]$, where s is the grid cell and $bp(\mathbf{s})$ is the back-pointer to the predecessor of **s**, e.g. the node which was used to generate **s**. The back-pointer to the predecessor allows to identify the heading, i.e. the direction of movement of the agent in case if it arrives to s from $bp(\mathbf{s})$. Additionally such data as the g-value – the length of the angle-constrained path from the start node to the current node and the h-value – approximate length of the path to the goal is also associated with the node.

LIAN explores the search-space in the same way A* does. On each step it chooses the most promising node, e.g. the one minimizing f-value ($f(\boldsymbol{s}) = g(\boldsymbol{s}) + h(\boldsymbol{s})$) from *OPEN* (set of previously generated nodes which are the candidates for further processing) and expands it. Expanding includes removing the node from *OPEN*, adding it to *CLOSED* (set of already processed nodes), generating valid successors and adding them to *OPEN*. Generating successors is a multi-step procedure including: (a) estimating the cells that lie within Δ-distance, (b) pruning away the cells that result in a move violating given angle constraint; (c) pruning away the cells that violate line-of-sight constraint; (d) pruning away the cells that result in generating nodes that have already been visited before. To find the cells lying within Δ-distance midpoint circle algorithm is used that builds discrete approximation of a circumference centered in a given cell and of a given Δ-radius. Such circumferences are depicted on Fig. 2. As one can see, the number of cells that satisfy angle-constraints grows with Δ.

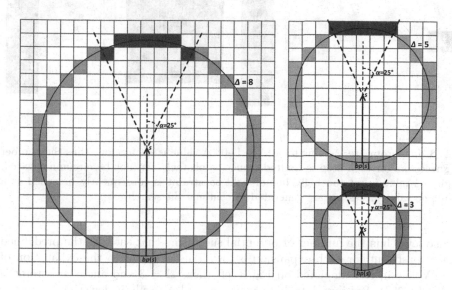

Fig. 2. The process of generating successors. The cells identified by midpoint circle algorithm are marked by light-gray. Among them only those cells are considered that satisfy the angle constraint (marked by dark-gray).

Algorithm stops when either the goal node is retrieved from *OPEN* or the *OPEN* is exhausted. In the first case the sought angle-constrained path can be re-constructed using the back-pointers, in the second one algorithm returns failure to find the path.

For the sake of space LIAN pseudocode is omitted but one can consult eLIAN code given in Algorithm 1 and Algorithm 2. LIAN code is the same stating that $\Delta_{max} = \Delta_{min} = \Delta$ for LIAN. Thus, when expanding a node, condition on line 11 is always false for LIAN and lines 12–13 are skipped. Condition on line 17 is also always false thus line 18 is skipped as well.

Sensitivity to the Input Parameter. As mentioned before, in general LIAN provides no guarantees to find a path even though it exists. One can only claim that if the angle-constrained path comprised of Δ-segments exists, the algorithm will find it [14]. In practice quite often LIAN fails to find a solution due to inappropriately chosen Δ-value.

On Fig. 3 two scenarios are depicted when setting the Δ-value too low/too high leads to failure. When Δ is low the number of cells lying within Δ-distance

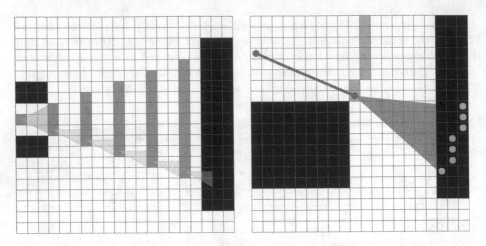

Fig. 3. Inappropriate Δ-value for LIAN. Left: Δ-value is set too low, thus the number of successors on each step is limited, thus the obstacle can not be circumnavigated. Right: Δ-value is set too high, line-of-sight constraint is consequently breached as a result no valid successors are generated to continue the search.

is also low, thus the number of potential successors that satisfies the predefined angle constraint for each expansion is low. This can lead to the exhaustion of *OPEN*, like in case depicted on Fig. 3 left (here $\Delta = 3$ and angle constraint is set to be 25°). Setting Δ to high values can also result in lowering down the number of successors but now it is mainly due to the violation of line-of-sight constraints in certain cases – see Fig. 3 right.

3   eLIAN

To mitigate the influence of the parameter Δ on the effectiveness of the algorithm we suggest adjusting the former on-the-fly thus adapting LIAN behavior to given path finding instance, e.g. to the configuration of the obstacles and to the particular locations of start and goal.

In both cases described above and shown on Fig. 3 algorithm would not have failed to find the solution if Δ was either higher (the first case), or lower (the second case). Thus rather than fixing Δ-value we suggest setting the upper and the lower limit, Δ_{max} and Δ_{min}, and allow Δ to take values from the range $[\Delta_{min}, \Delta_{max}]$.

Introducing the Δ range leads to the following questions: (a) what is the initial value of Δ from this range; (b) what is the procedure of adjusting Δ; (c) how is it embedded to the search algorithm. We suggest the following answers to these questions.

The initial value equals Δ_{max}. The rationale behind this is that setting Δ to high values potentially leads to large number of successors, except the cases (discussed above), which are to be handled by the next adjusting procedure.

Algorithm 1. eLIAN–Main-Loop

1 $bp(\textbf{start}):=\emptyset$; $g(\textbf{start}):=0$;
2 $\Delta(\textbf{start}) := \Delta_{max}$;
3 $OPEN$.push(**start**); $CLOSED:=\emptyset$;
4 **while** $OPEN$ *is not empty* **do**
5      $\textbf{s}:=\text{argmin}_{\textbf{s}\in OPEN} f(\textbf{s})$;
6      $OPEN$.remove(**s**);
7   **if** $s = goal$ **then**
8   **return** $getPathFromParentPointers(\textbf{s})$;
9      $CLOSED$.push(**s**);
10     $Expand(\textbf{s}, \alpha_m, \Delta_{min})$;
11 **return** "path not found";

In case no successors are generated at some step when Δ_{max} is in use it should be lowered down by multiplying it by a factor of $k \in (0; 1)$, which becomes the algorithm's parameter. Now, in order to keep track of Δ-values they should be associated with the search nodes like g-values and h-values. We will refer to the exact Δ-value of the node as to $\Delta(\textbf{s})$. After fixing $\Delta(\textbf{s})$ for the current node **s**, the latter is inserted back to $OPEN$, thus in case of re-expansion another set of successors will be generated for the node and possibly this set will not be empty. In case it is, $\Delta(\textbf{s})$ is decremented again. Lowering down the Δ-value continues up to a point when $k \cdot \Delta(\textbf{s}) < \Delta_{min}$. If this condition holds, e.g. the Δ-value can not be lowered down any more as it goes out of the range, the search node is discarded for good.

To provide additional flexibility to the algorithm we suggest not only decrementing Δ-value but also incrementing it after a few consecutive successful attempts of successors' generation. Number of such attempts can be seen as another input parameter. As a result the algorithm tries to keep Δ-value as high as possible and lowers it down only when needed.

We name the resultant algorithm as eLIAN (stands for "enhanced LIAN"). Pseudocode of eLIAN is presented as Algorithm 1 (main loop) and Algorithm 2 (node expansion). Please note that the code increments Δ-value after 2 consecutive successful expansions (see line 17 of the Algorithm 2), but it can be easily modified for another threshold.

An example of eLIAN trace is presented in Fig. 4. The initial Δ-value is 8 and it is too high for the given task due to rather small passages between the obstacles. Regular LIAN algorithm would have failed to solve this task as it would not be able to generate valid successors that bypass the obstacles. In contrast, eLIAN decreased the Δ-value to 4 after two steps and successfully found a sequence of small sections going through the passage. After the obstacle was successfully bypassed Δ-value was increased back to 8 leading to more straightforward and rapid movement towards the goal.

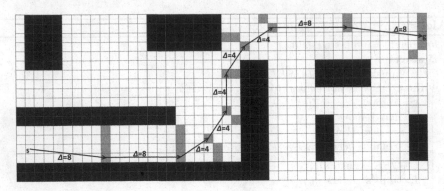

Fig. 4. An example of how eLIAN works. Possible Δ-values are 8 and 4. Angle constraint is $\alpha = 35°$

The following properties of eLIAN can be inferred.

Property 1. eLIAN is correct, e.g. it terminates after the finite number of steps.

Sketch of proof. Algorithm is performing the search until the *OPEN* list is empty (or until the goal node is retrieved from it). *OPEN* contains only elements that correspond to the grid cells the total number of which is finite. As the algorithm operates only by Δ-sections the number of potential parents of the cell per each Δ is also finite. eLIAN operates finite number of Δ-values in range $[\Delta_{min}; \Delta_{max}]$. Thus the number of all nodes possibly considered by eLIAN is finite.

At the same time when a new node is generated the algorithm checks whether this node (the node defined by the same cell and the same parent) has been processed before already (lines 7–9 of Algorithm 2). And in case the *CLOSED* already contains such node it is pruned and is not added to *OPEN*. Taking into account the fact that on each step an element is removed from *OPEN* (line 6 of Algorithm 1) and added to *CLOSED* one can infer that sooner or later either the goal node will be retrieved or there will not be any node that could be added to *OPEN* and it will become empty. In both cases (lines 7, 11 of Algorithm 1) algorithm terminates.

Property 2. If there exists an angle-constrained path comprised of Δ_{max}-segments, eLIAN will find a solution.

Sketch of proof. As the path composed of Δ_{max}-segments exists, eLIAN will always maintain at least one node lying on that path as the first expansion happens with Δ_{max}. Thus eLIAN always has an option to expand this node and to generate another one lying on Δ_{max}-path. Thus, sooner or later either this path will be found or eLIAN will find another path comprised of the segments of alternating lengths (due to the expansions of nodes with different Δ-values that have lower f-values).

Algorithm 2. Expand(**s**, α_m, Δ_{min})

---

1 $SUCC := getDeltaSuccessors(\mathbf{s},\Delta(\mathbf{s}))$;
2 if $dist(s, goal) < \Delta(\mathbf{s})$ then
3 ⎣ $SUCC$.push(**goal**);

4 for each $\mathbf{s}' \in SUCC$ do
5 | if $lineOfSight(s, s')$=false or
   | $angleConstraint(bp(s),s,s')$=false then
6 | ⎣ $SUCC$.remove(**s**′);
7 | for each $\mathbf{s}'' \in CLOSED$ do
8 | | if $s' = s''$ and $bp(s')$=$bp(s'')$ then
9 | | ⎣ $SUCC$.remove(**s**′);

10 if $SUCC = \emptyset$ then
11 | if $\Delta(\mathbf{s}) > \Delta_{min}$ then
12 | | $\Delta(\mathbf{s}):= k\cdot \Delta(\mathbf{s})$; $//k \in (0; 1)$ is a parameter
13 | ⎣ $OPEN$.push(**s**);

14 else
15 | for each $\mathbf{s}' \in SUCC$ do
16 | | $g(\mathbf{s}'):=g(\mathbf{s})+dist(s, s')$;
17 | | if $\Delta(\mathbf{s})=\Delta(bp(\mathbf{s}))$ and $\Delta(\mathbf{s})<\Delta_{max}$ then
18 | | ⎣ $\Delta(\mathbf{s}'):=\Delta(\mathbf{s})/k$;
19 | ⎣ $OPEN$.push(**s**′);

---

4 Experimental Evaluation

To empirically evaluate the suggested algorithm, eLIAN, and to compare it with the predecessor, LIAN, we used the well-known in the community benchmark sets from N. Sturtevant's collection [11]: Baldur's Gate (*BG*) and Warcraft III (*WIII*). Both collections contain maps used in computer games. *BG* maps (75 in total) represent mostly indoor environments while *WIII* maps (36 in total) – imaginary outdoor environments. Benchmark repository contains path finding instances organized into the so-called buckets. Instances belonging to the bucket have nearly the same solution cost (path length as found by A*). We used instances that belong to the most "tough" buckets, e.g. the instances with the highest A* solution costs. There were taken 14 tasks per map for BG and 30 tasks for WIII to get more than 1000 tasks per each collection. We have also used a collection of maps that represent urban outdoor environment. These maps were retrieved from OpenStreetMaps[1]. 100 maps of 1.2×1.2 km fragments were converted to 501×501 grids (thus each grid cells corresponds to 2.7×2.7 area). 10 tasks per map were generated in such a way that a distance between start and goal is greater or equal to 1 km (400 cells). An example of a map fragment

---

[1] http://www.openstreetmaps.org/export.

Fig. 5. Left: Map fragment of an urban environment retrieved from OpenStreetMaps. Right: An example of a task, solved by eLIAN on the corresponding grid.

that was used for the experimental evaluation and the corresponding grid are presented in Fig. 5.

Experimental evaluation was carried out on Windows operated PC with Intel Core 2 Duo Q8300 @2.5 Ghz processor and 2 GB of RAM. For the sake of fair comparison both algorithms were coded from scratch using the same programming techniques and data structures.

It is known from previous research that finding angle-constrained paths might take time, so inflated heuristic was used to speed-up the search. Heuristic weight was set to 2. Besides, 5 min time limit was introduced. In case the algorithm did not come up with the solution within this time, the result of the run was considered to be failure.

The following algorithms were evaluated: LIAN-20, e.g. the original algorithm with $\Delta = 20$; eLIAN-20-10, e.g. $\Delta_{max} = 20$, $\Delta_{min} = 10$; eLIAN-20-5, e.g. $\Delta_{max} = 20$, $\Delta_{min} = 5$. For eLIAN we used $k = 0.5$, e.g. Δ-value was cut in half in case no valid successors were generated. This means that Δ-value alternated between 10 and 20 for eLIAN-20-10 and between 5 and 10 and 20 for eLIAN-20-5. The angle constraint was set to be 20°, 25° and 30°.

Table 1. Success rate of LIAN and eLIAN

	City Maps			Baldur's Gate			Warcraft III		
	20°	25°	30°	20°	25°	30°	20°	25°	30°
LIAN-20	84.2%	89.0%	91.8%	61.24%	71.71%	81.05%	78.33%	82.22%	85.74%
eLIAN-20-10	89.5%	93.5%	95.2%	74.38%	84.76%	88.0%	83.98%	88.33%	91.76%
	(+5.3)	(+4.5)	(+3.4)	(+13.14)	(+13.05)	(+6.95)	(+5.65)	(+6.11)	(+6.02)
eLIAN-20-5	92.4%	95.4%	97.0%	82.38%	88.29%	87.52%	87.5%	92.59%	93.61%
	(+8.2)	(+6.4)	(+5.2)	(+21.14)	(+16.57)	(+6.48)	(+9.17)	(+10.37)	(+7.87)

Table 2. Normalized number of tasks that were not solved by LIAN but solved with eLIAN

	City Maps			Baldur's Gate			Warcraft III		
	20°	25°	30°	20°	25°	30°	20°	25°	30°
eLIAN-20–10	27.8%	35%	37.2%	34.7%	49.1%	44.7%	17.5%	22.6%	32%
eLIAN-20–5	43.3%	46.7%	60.5%	56.6%	73%	76.7%	34.5%	51.6%	50.5%

Success rates of the algorithms are reported in Table 1. As one can see eLIAN always solves more tasks that the predecessor, no matter which collection or angle constraint is used. The most notable difference is predictably between LIAN-20 and eLIAN-20-5. Suggested technique of dynamic Δ-value adjusting leads to solving up to 20% more tasks (see Baldur's Gate-20°) and thus substantially increases the effectiveness of the algorithm. The normalized number of instances that were not solved by LIAN but solved by eLIAN is presented in Table 2. In absolute numbers LIAN failed to solve 659 tasks for $\Delta = 20°$, 451 - for $\Delta = 25°$ and 299 - for $\Delta = 30°$. Analyzing the figures, one can claim that on average eLIAN is able to successfully handle (within 5 min time cap) half of the tasks that LIAN is unable to solve.

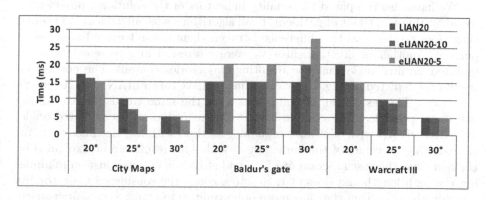

Fig. 6. Averaged runtime of LIAN and eLIAN

To conduct a fair comparison of the efficiency indicators, e.g. the resultant runtimes, we averaged those values across the intstances that were solved by all three algorithms: LIAN-20, eLIAN-20-10 and eLIAN-20-5. Instead of arithmetic mean value, we report the median, since there is a big discrepancy in the algorithms' runtimes (up to one order of magnitude). The results are depicted on Fig. 6. As one might see on City Maps and Warcraft III collections eLIAN works faster than the predecessor, while on Baldur's Gate eLIAN's runtime is equal or higher than the one of LIAN. This might be explained by the difference in types of environments. City Maps and Warcraft III collections consist of the maps representing outdoor environments populated with stand-alone and rather small obstacles, while most maps in Baldur's Gate collection represent indoor environments with rooms, corridors and walls. Due to the increased branching factor (e.g. re-expnaing the nodes with attempting different $Delta$-values) in case of getting stuck into obstacles represented by walls, eLIAN has to spend more time circumnavigating them in comparison with LIAN.

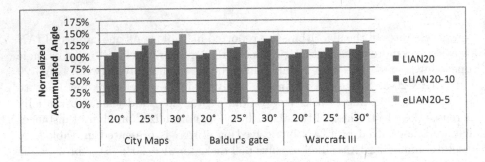

Fig. 7. Normalized accumulated turning angle

We have also compared the quality indicators of the solutions, provided by the algorithms. In terms of pathlength, all algorithms with all values of Δ showed very close results, and the difference between them doesn't exceed a couple of percent. Another quality indicator we were interested in represents the accumulated turning angle that the resulting trajectories contain. The normalized results are depicted on Fig. 7. As a baseline we have taken LIAN with 20°- angle constraint. The results on all collections show the same trends: increasing the maximum possible turning angle leads to higher values of accumulated angle. Moreover, eLIAN with allowance of making sections with less size finds trajectories with higher values of this indicator as well. This result can be explained by the fact, that the cost function of LIAN and eLIAN is targeted just on minimizing the path length and doesn't take into account the considered indicator. As a result, the algorithm that has more opportunities to explore the search-space, e.g. eLIAN, finds paths that contain more turns.

5   Conclusions

We have considered a problem of grid-based angle-constrained path planning and presented a novel technique that significantly boosts the performance of the state-of-the-art planner tailored to solve those type of problems. As shown experimentally, success rate of the enhanced algorithm is notably higher (up to 20%) while solution quality is nearly the same. Obvious and appealing direction of future research is elaborating on further enhancements to make the algorithm complete.

Acknowledgments. The work was partially supported by the "RUDN University Program 5–100" and by the special program of the presidium of Russian Academy of Sciences.

References

1. Botea, A., Müller, M., Schaeffer, J.: Near optimal hierarchical path-finding. J. Game Dev. **1**(1), 7–28 (2004)
2. Bresenham, J.E.: Algorithm for computer control of a digital plotter. IBM Syst. J. **4**(1), 25–30 (1965)
3. Daniel, K., Nash, A., Koenig, S., Felner, A.: Theta*: any-angle path planning on grids. J. Artif. Intell. Res. **39**, 533–579 (2010)
4. Elfes, A.: Using occupancy grids for mobile robot perception and navigation. Computer **22**(6), 46–57 (1989)
5. Harabor, D., Grastien, A., Oz, D., Aksakalli, V.: Optimal any-angle pathfinding in practice. J. Artif. Intell. Res. **56**, 89–118 (2016)
6. Harabor, D.D., Grastien, A.: Online graph pruning for pathfinding on grid maps. In: Proceedings of The 25th AAAI Conference on Artificial Intelligence (AAAI-2011), pp. 1114–1119 (2011)
7. Hart, P.E., Nilsson, N.J., Raphael, B.: A formal basis for the heuristic determination of minimum cost paths. IEEE Trans. Syst. Sci. Cybern. **4**(2), 100–107 (1968)
8. Kim, H., Kim, D., Shin, J.U., Kim, H., Myung, H.: Angular rate-constrained path planning algorithm for unmanned surface vehicles. Ocean Eng. **84**, 37–44 (2014)
9. Nash, A., Koenig, S., Tovey, C.: Lazy theta*: any-angle path planning and path length analysis in 3D. In: Proceedings of the 24th AAAI Conference on Artificial Intelligence (AAAI-2010), pp. 147–154. AAAI Press (2010)
10. Silver, D.: Cooperative pathfinding. In: Proceedings of The 1st Conference on Artificial Intelligence and Interactive Digital Entertainment (AIIDE-2005), pp. 117–122 (2005)
11. Sturtevant, N.R.: Benchmarks for grid-based pathfinding. IEEE Trans. Comput. Intell. AI Games **4**(2), 144–148 (2012)
12. Thrun, S.: Learning occupancy grid maps with forward sensor models. Auton. Robots **15**(2), 111–127 (2003)
13. Xu, H., Shu, L., Huang, M.: Planning paths with fewer turns on grid maps. In: Proceedings of The 6th Annual Symposium on Combinatorial Search (SoCS-2013), pp. 193–201 (2013)
14. Yakovlev, K., Baskin, E., Hramoin, I.: Grid-based angle-constrained path planning. In: Hölldobler, S., Krötzsch, M., Peñaloza, R., Rudolph, S. (eds.) KI 2015. LNCS (LNAI), vol. 9324, pp. 208–221. Springer, Cham (2015). https://doi.org/10.1007/978-3-319-24489-1_16
15. Yap, P.: Grid-based path-finding. In: Cohen, R., Spencer, B. (eds.) AI 2002. LNCS (LNAI), vol. 2338, pp. 44–55. Springer, Heidelberg (2002). https://doi.org/10.1007/3-540-47922-8_4

Mobile Robotic Table with Artificial Intelligence Applied to the Separate and Classified Positioning of Objects for Computer-Integrated Manufacturing

Héctor C. Terán[✉], Oscar Arteaga, Guido R. Torres,
A. Eduardo Cárdenas, R. Marcelo Ortiz, Miguel A. Carvajal,
and O. Kevin Pérez

Universidad de las Fuerzas Armadas ESPE, Sangolquí, Ecuador
{hcteran, obarteaga, grtorres, mecardenas2, amortiz8,
macarvajal, pokevin}@espe.edu.ec

Abstract. The article is about the construction of a robotic table for the separation and classification of objects for computer integrated manufacturing, consisting of a surface with gears distributed in modules with two gears each for x and y respectively, which allow the displacement of objects and their classification. The computer vision algorithms used to determine the position of the object are detailed. In addition, the Kd, Ki, Kp constants are auto tuned through a neural network to locate and track the objects transported. The design and construction features as well as the mechanical and electronic system are integrated into a mechatronic system. Finally, for the operation of the robotic table was tested with objects of different weights, the results showed the usefulness of the separation and classification system for the tracking of preset trajectories.

Keywords: Separation · Classification · Artificial intelligence

1 Introduction

The evolution of robotics has made the processes faster and faster, mainly in the manufacturing industry where large production capacity is required without neglecting quality, such as separation and sorting systems on raw material conveyor belts for the supply of numerical control machines and their respective palletizing and distribution [1–3].

In recent years, research in artificial intelligence has developed algorithms for automatic adjustment through learning layers that can solve assigned objectives [4, 5]. For example, a machine learning network can decide, which the best mathematical model to be applied is, in this paper case it is used for an auto tuning of the constants of a PID algorithm to track the trajectory of the object on the robotic table, to meet the high demand for speed of current production lines.

© Springer Nature Switzerland AG 2018
S. O. Kuznetsov et al. (Eds.): RCAI 2018, CCIS 934, pp. 218–229, 2018.
https://doi.org/10.1007/978-3-030-00617-4_20

2 Design and Construction

2.1 Structure

The construction of the table is directly related to the mechanical and electronic components, which are distributed in such a way that they allow the movement of the transported object depending on the position obtained from the artificial vision algorithm. The Fig. 1 shows a diagram of how the components interact with each other.

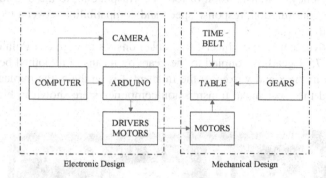

Electronic Design   Mechanical Design

Fig. 1. Electronic and mechanical schematic of the robotic table.

2.2 Mechanical Design

One of the most critical steps for the development of the complete system are the actuators and controllers, the general design for each of the components was used in the construction of the table for which a 3D design software was used, based on a flexible production according to the requirements of the industry 4.0 taking into account aspects of modularity and manufacturing. Figure 2 shows a three-dimensional design of the structure to be implemented, which provides a robust and high precision electronic and mechanical system thanks to the implementation of the latest technology algorithm.

Fig. 2. Computer-aided design of the robotic table

The mechanical design is based on the different movement needs to provide the transfer of objects to the left, right, front and back, the first aspect of the construction is the frame that will support the structure. The frame of the table is made of steel with a dimension of 760 × 760 mm which makes the table a prototype with a high degree of usability.

The mechanical transmission system consists of gears and pulleys to transmit the movement of the motors. The present work is related to the design and manufacture of the parts in numerical control machines, which allows the motion system to have the maximum efficiency of transmission of forces corresponding to the transmission ratio, torque, energy consumption, response speed and compatibility between the surfaces of the table and the frame.

For the movement of the objects, the table consists of eighteen Polulo 25DX54L motors with a 75:1 gearbox coupled to the gears by means of a toothed belt to control the movement in the x and y axes, at the moment it is on them, the implementation of the motors and the mechanical transmission components are shown in Fig. 3.

Fig. 3. Mechanical transmission system

The Fig. 4 shows the main part, which consists of a 200 × 200 mm square with a double rack that is coupled with the gears in the robotic table, allowing the movement of the objects on the surface, this piece was built in acrylic to reduce friction.

Fig. 4. Double zipper base

2.3 Electronic Design

The electronic system is made up of different parts for data acquisition and motor control. The camera is the main device for obtaining images, which after a sophisticated study of computer vision algorithms provides the required position of an object on the table. This data is interpreted by computer. The computer has the following features: Intel i7, 8 GB of RAM and 2 GB of video memory, with the Windows Operating System and the QT software, which is responsible for the algorithms of artificial vision, the logic of control of the direction of rotation of the motors and the auto tuning of the PID controller for the path of the objects (Fig. 5).

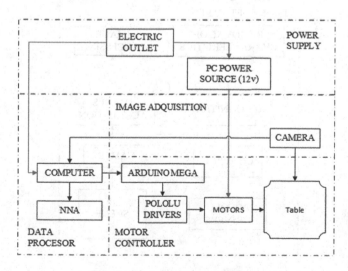

Fig. 5. Electronic system diagram

An Arduino Mega card receives information through the serial port to control the engines, the Pololu TB6612FNG driver with two-channel jumper for each engine was used.

2.4 Artificial Vision

An artificial vision system was used to obtain the data, which is of vital importance since it allows us to understand the image processing.

The objective of our artificial vision application is to implement a system which is used to identify the pieces placed on the robotic table, to later find their exact location and perform the tasks of separation and classification [6].

To determine the position, the implemented system has a machine vision system, where from a digitized image it is possible to determine the coordinates of the position and orientation in the image plane which are then transmitted to the workspace of the robotic table.

The implemented artificial vision is able to provide the following five stages: data acquisition, pre-processing, feature extraction, classification and operation of the robotic table. Once a part is recognized, the control signal is sent to the robotic table for separation and classification according to requirements previously set by the operator [7].

By means of artificial vision, the image provided by the camera is transformed into a representation of the elements present in the robotic table. The representation has the necessary information for the robotic table to make the necessary movements for the correct fulfillment of the configured task. In order to obtain the necessary information, these stages are presented: The scene, image capture (computer acquisition), image processing (enhancement of geometric features), image segmentation, feature extraction, object recognition and signals for the robotic table. The system schematic is shown in Fig. 6.

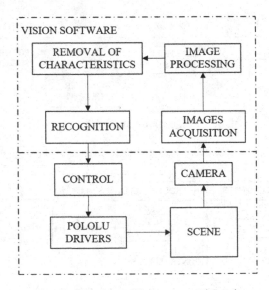

Fig. 6. Computer vision system schematic

Lighting: It is done by means of a fluorescent lamp to illuminate the objects.

Video camera: An analog CCD video camera with NTSC format is used.

Computer: Laptop with an Intel Core i7 processor with a dedicated 4 GB Nvidia video card.

Software: The program was implemented through the OpenCv library in Matlab.

In order for the computer vision system to perform an automatic movement of the robotic table, a decision making system was implemented, which allows the table movement to be executed when a specific part has been detected.

The position of the part is analyzed within a decision tree, with the objective of obtaining the movement of the robotic table for each sample provided with the camera, and then a state machine is made for the feedback of the system with the coordinates of the object, which are used as data for the decision tree.

When the parameters for the movement of the robotic table are already available, a string of characters is sent through the serial port via the RS232 protocol, which corresponds to the coordinates and then the execution of the movement of the motors to achieve the separation and classification of the parts according to the operator's requirements. Figure 7 corresponds to the state machine schematic.

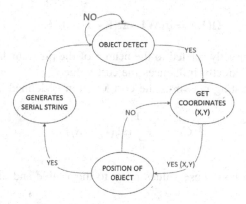

Fig. 7. Diagram of the state machine

Once the serial string corresponding to the coordinates of each object has been defined, it is received by the Arduino, which interprets each command coming from the vision system and control the drivers that modulate the PWM signal required for each motor.

2.5 Artificial Intelligence

For the trajectory control of the object of interest, an Integral Derivative Proportional PID algorithm was used. This controller is widely used in the industry due to its low computational cost and high simplicity; there are many methods to obtain the values of KD, KP, KI and KI constants, some of which take a mathematical modeling process that requires an expert mathematician. The other method is the heuristic tuning required by an expert in the process, however, we will never be sure if the response of the system can be improved or not. In this article we propose the use of an artificial neuronal network with regressive propagation to find the values of the constants, the controller has as inputs the x and y coordinates of the object obtained from the image processing described in the previous section and as output the speed of each motor.

The advantages of using a neural network over traditional tuning methods are to dispense with the mathematical model and to add robustness to the motion control system, because in a classic PID there are differences between the model and reality due to variables that are not considered physical parameters such as air friction and moments of inertia [8, 9].

The regressive propagation algorithm was used to define the weights of the connection between each perceptrons, for which we took a simple network with the weights w1 where the activation of an omega neuronal is defined by the cost function for a training iteration is shown in the equation C_o.

$$C_o = \left(\Omega(k) - X_{rbf} \right)^2 \tag{1}$$

Where the last activation $\Omega (k)$ is calculated with the weight w1 which is to be determined multiplied by the previous activation $\Omega (k - 1)$ plus a bias value b(k) evaluated by a sigmoid function the result is shown in equation $\Omega (k)$.

$$\Omega(\mathbf{K}) = \sigma(\mathbf{W1} * \Omega(\mathbf{K-1}) + \mathbf{b}(\mathbf{K})) \tag{2}$$

Each of these is directly related to the neuron of the previous layer and determines the sensitivity which directly influences the cost value C_o.

Finally, for all existing networks the cost function is defined by.

$$C_O = \sum_{j=0}^{nk-1} \left(\Omega_j(k) - X_j \right)^2 \tag{3}$$

The Fig. 8 shows how these values relate to the visible and hidden layers.

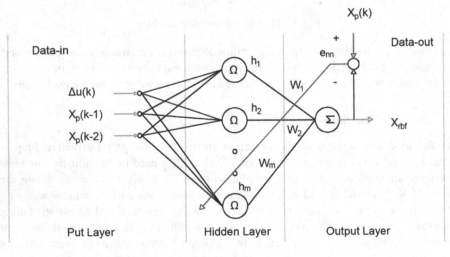

Fig. 8. Artificial neural network (ANN) layer

3   Experiments and Results

For trajectory control and structure prediction, the optimal artificial neural network was analyzed, and as a starting point the parameters in training ratios, their validation and the set of tests (a:b:c) were related, i.e. 100:25:25, 150:15:15 and 200:5:5. Connection weights were used with 4 randomized trials varying the parameters to find the ratio with the minimum square error (MSE), it is necessary that some interactions be executed to converge and obtain the required learning [10–12].

The accuracy was modified through the hidden nodes for the 200:5:5 ratios with 4 neurons in the hidden layer, this was obtained by trial and error. The hyperbolic function f(x) was used for the hidden layer with a linear transfer function applied to each output layer within −1 and 1 interval. The evaluation was based on connection weights with a comparison of the minimum square error (MSE) obtained by comparing the expected results with those of the prediction using the equation (MSE).

$$f(x) = \frac{1}{(1 + \exp(x))} \tag{4}$$

$$MSE = \frac{\sum_{m-1}^{m} (y_m - d_m)^2}{M}, \tag{5}$$

where m is the sample, $(y_m - d_m)$ are the network outputs and M is the cluster size studied.

The value for the deviation was determined by applying a correlation coefficient R with equation (R) to the interactions.

$$R = \sqrt{\frac{\sum_{m-1}^{M} (y_m - y^-)^2 - \sum_{m-1}^{M} (y_m - y_m)^2}{\sum_{m-1}^{M} (y_1 - y^-)^2}} \tag{6}$$

Where m are the measured dependent variables, y_m are the adjusted dependent variables for the independent variable, $\sum_{m-1}^{M} (y_m - y^-)^2$ is the sum total of squares, $\sum_{m-1}^{M} (y_m - y_m)^2$ is the sum total of remaining squares.

Starting from the prediction with three experimental models, a model with recommended proportions was chosen for validation with 1000 tests, being the critical point where the learning transition stabilizes at 66 tests with a minimum quadratic error (MSE) of 0.00181818 using 4 neurons (See Fig. 9).

The training correlation R is 0.98853 with a validation value R = 0.99947 subjected to tests reaches R = 0.99961, a differential error of 0.5% is acquired in the

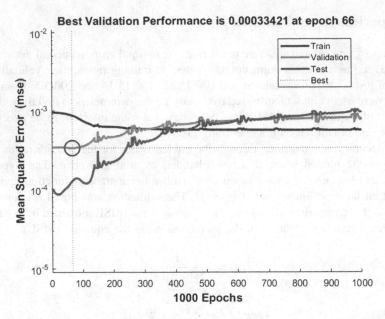

Fig. 9. Validation of tests with quadratic mean error to 4 neurons in a ratio (200:5:5)

transitions of each path so the dependent variable is validated by the independent variable Fig. 10 as a result the sectioned point path represents the ideal fit to the model.

In Fig. 11 we can see the slope in relation to the 1000 interactions made, the results allow us to analyze that the learning was executed in random incremental periods starting from 66 interactions and ending in 934 interactions to reach the minimum optimal slope of 0.0012414. An ideal proportional and linear validation was obtained, stabilizing the learning in a transition from 150 interactions and being necessary less interactions after each learning, later the parameters are fed back to the PID control to be used in the movement of the mobile robotic table for computer integrated manufacturing (CIM).

The artificial neural network learning system was also evaluated analytically in percentages to determine the accuracy achieved.

With the values that are predicted $\widehat{Z}^t$ and with the current data Z_i the desired error (e_i) was found:

$$e_i = Z_i - \widehat{Z}^t \tag{7}$$

Percentage mean error *PME* using the number of neurons in the output layer n (PME).

$$PME = \left(\frac{1}{n}\sum_{i=1}^{n}\frac{e_i}{z_i}\right)x10_0 \tag{8}$$

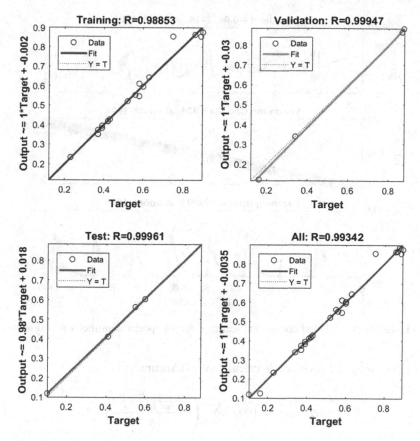

Fig. 10. Real value regression and prediction with neural network model ratio (200:5:5).

Percentage mean absolute error (*PMAE*)

$$PMAE = \left(\frac{1}{n}\sum_{i=1}^{n}\frac{|e_i|}{z_i}\right) x10_0 \qquad (9)$$

Percentage mean squared error (PMSE)

$$PMSE = \frac{1}{n}\left(\sum_{i=1}^{n}\frac{e_i^2}{z_i^2}\right) x10_0 \qquad (10)$$

Percentage root mean squared error (PRMSE)

$$PRMSE = \sqrt{\frac{1}{n}\left(\sum_{i=1}^{n}\frac{e_i^2}{z_i^2}\right)} x10_0 \qquad (11)$$

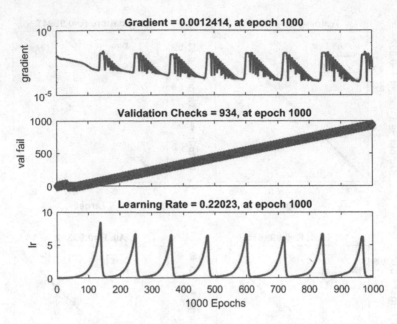

Fig. 11. Gradient-validation checks and learning rate vs epochs in artificial neural networks

The percentage of accuracy is calculated by (Accuracy)

$$Accuracy = 100 - \sum \left(\frac{\frac{|actual\, data\, forecast|}{actual\, data * 10_0}}{n} \right) \qquad (12)$$

Table 1 shows the predictions analyzed by modifying the training, validation and related test set parameters (a:b:c) in percentage, analytically achieving an ideal accuracy for the system.

Table 1. Prediction of errors in ANN models by varying their ratio

Ratio	PME (%)	PMAE (%)	PMSE (%)	PRMSE (%)	Accuracy (%)
100:25:25	0.065378	4.86969E−05	3.55E−02	0.188511	40.8009
150:15:15	0.045432	3.38401E−05	1.75E−02	0.157145	72.2147
200:05:05	0.033451	2.49153E−05	1.11E−03	0.04264	98.7142

The percentage mean squared error (PMSE) is 1.1E−0.3%, since it is not linear, it is necessary to find the value percentage root mean squared (PRMSE) of 0.04264%, being proportional to the system and obtaining a real linear precision of 98.7142%.

The equipment from its inception in the modeling, construction and control (PID-ANN) stages, allows it to be used in computer integrated manufacturing (CIM), meeting requirements to reduce errors, speed, flexibility and integration.

The accuracy of the 98.7142% exceeds that established by the norms and standards (MAP) manufacturing automation protocol and the equipment complies with the OSI (open system interconnection) model.

Since the equipment can be used in computerized numerical control (CNC) or material distribution table, in future jobs it is projected to observe its performance applied to the enterprise resource planning (ERP) modules, with the evaluating of design parameters, factory control, material requirements, classification of orders to customers and the influence on production costs.

References

1. Malte, B., et al.: How virtualization, decentralization and network building change the manufacturing landscape. Int. J. Inf. Commun. Eng. **8**, 37–44 (2014)
2. Moonet, J., et al.: Innovation in knowledge-intensive industries: the double-edged sword of competition. J. Bus. Res. 2060–2070 (2013)
3. Gwiazda, J.A., et al.: Modular industrial robots as the tool of process automation in robotized manufacturing cells. In: IOP Conference Series: Materials Science and Engineering, vol. 95, pp. 012104-1–6 (2015)
4. Cong, S., Liang, Y.: PID-like neural network nonlinear adaptive control for uncertain multivariable motion control systems. IEEE Trans. Ind. Electron. 3872–3879 (2009)
5. Beitao, G., et al.: Adaptive PID controller based on BP neural network. International Joint Conference on Artificial Intelligence, pp. 148–150 (2009)
6. Maleki, E.: Artificial neural networks application for modeling of friction stir welding effects on mechanical properties of 7075-T6 aluminum alloy. In: IOP Conference Series: Materials Science and Engineering, pp. 12–34 (2015)
7. González, F., Guarnizo, J.G., Benavides, G.: Emulation system for a distribution center using mobile robot, controlled by artificial vision and fuzzy logic. IEEE Latin Am. Trans. **12**, 557–563 (2014)
8. Vikram, C., Sunil, K., Singla, D.: Comparative analysis of tuning a PID controller using intelligent methods. Acta Polytechnica Hungarica 235–249 (2014)
9. Boulogne, L.H.: Performance of neural networks for localizing moving objects with an artificial lateral line. In: IOP Conference Series: Materials Science and Engineering, pp. 1–13 (2017)
10. Hassan, A.K.I.: Modeling insurance fraud detection using imbalanced data classification. J. Comput. High. Educ. 1–7 (2015)
11. Kamir, et al.: Modeling insurance fraud detection using imbalanced data classification. Int. J. Innov. Eng. Technol. 1–9 (2015)
12. Oprea, C., et al.: A data mining based model to improve university management. J. Sci. Arts 1–6 (2017)

Hybrid Neural Networks for Time Series Forecasting

Alexey Averkin[1](✉) and Sergey Yarushev[2]

[1] Federal Research Centre of Informatics and Computer Science of RAS,
Moscow, Vavilova, 42, Moscow, Russia
averkin2003@inbox.ru
[2] Plekhanov Russian University of Economics,
Stremiannoy per., 34, Moscow 107113, Russia

Abstract. The paper presents research in the field for hybrid neural networks for time series forecasting. A detailed review of the latest researches in this field is described. The paper includes detailed review of studies what compared the performance of multiple regression methods and neural networks. It is also consider a hybrid method of time series prediction based on ANFIS. In addition, the results of time series forecasting based on ANFIS model and com-pared with results of forecasting based on multiple regression are shown.

Keywords: Time series forecasting · Neural networks · ANFIS
Time series prediction

1 Introduction

The recent intensive research of artificial neural networks (ANN) showed that neural networks have a strong capability in predicting and classification problems. ANN successfully used for various tasks in many areas of business, industry and science [1]. Such high interest to neural networks was caused by the rapid growth in the number of articles published in scientific journals in various disciplines. It suffices to consider several large data-bases to understand the huge number of articles published during the year on the topic of neural networks. It is thousands of articles. A neural network is able to work with input variables in parallel and consequently handle large sets of data quickly. The main advantage of neural networks is the ability to find patterns [2]. ANN is a promising alternative in the toolbox professionals involved in forecasting. In fact, the non-linear structure of the neural networks is partially useful to identify complex relationships in most real world problems.

Neural networks are perhaps the universal method of forecasting in connection with those that they cannot only find the non-linear structures in problems, they can also simulate the processes of linear processes. For example, the possibility of neural net-works in the modeling of linear time series were studied and confirmed by a number of researchers [3–5].

One of the main applications of ANN is forecasting. In recent years, there was an increasing interest in forecasting using neural networks. Forecasting has a long history,

© Springer Nature Switzerland AG 2018
S. O. Kuznetsov et al. (Eds.): RCAI 2018, CCIS 934, pp. 230–239, 2018.
https://doi.org/10.1007/978-3-030-00617-4_21

and its importance reflected in the application in a variety of disciplines from business to engineering.

The ability to predict the future is fundamental to many decision-making processes in planning, developing strategies, building policy, as well as in the management of supply and stock prices. As such, forecasting is an area in which a lot of efforts have invested in the past. In addition, it remains an important and active area of human activity in the present and will continue to evolve in the future. Review of the re-search needs in the prediction presented to the Armstrong [6].

For several decades in forecasting dominated by linear methods. Linear methods are simple in design and use, and they are easy to understand and interpret. However, linear models have significant limitations, owing to which they cannot discern any nonlinear relationships in data. Approximations of linear models with complex one that are not linear relationships do not always give a positive result. Earlier in 1980, there have been large-scale competition for forecasting, in which most widely used linear methods were tested on more than 1000 real-time series [7]. Mixed results showed that none of the linear model did show the best results worldwide, and that can be interpreted as a failure of the linear models in the field of accounting with a certain degree of non-linearity, which is common for the real world.

Predicting financial markets is one of the most important trends in research due to their commercial appeal [8]. Unfortunately, the financial markets are dynamic, non-linear, complex, non-parametric and chaotic by nature [9]. Time series of multi-stationary, noisy, casual, and have frequent structural breaks [10]. In addition, the financial markets also affects a large number of macroeconomic factors [11, 12], such as political developments, global economic developments, bank rating, the policy of large corporations, exchange rates, investment expectations, and events in other stock markets, and even psychological factors.

Artificial neural networks are one of the technologies that have received significant progress in the study of stock markets. In general, the value of the shares is a random sequence with some noise, while artificial neural networks are powerful parallel processors of nonlinear systems, which depend on their own internal relations.

Development of techniques and methods that can approximate any nonlinear continuous function without a priori notions about the nature of the process itself seen in the work of Pino [13] It is obvious that a number of factors demonstrates sufficient efficiency in the forecast prices, and most importantly a weak point in this is that they all contain a few limitations in forecasting stock prices and use linear methods, the relative of this fact, although previous investigation revealed the problem to some extent, none of them provides a comprehensive model for the valuation of shares. If we evaluate the cost and provide a model in order to remove the uncertainty, it is largely can help to increase the investment attractiveness of the stock exchanges. Conduct research to get the best method of forecasting financial time series is currently the most popular and promising task.

2 Latest Researches in Hybrid Time Series Prediction

One approach to solving the difficult-solvable problems of the real world is a hybridization of artificial intelligence and statistical methods to combine the strengths and eliminate weaknesses in the case of each method separately [14].

Bisoi and Dash used a hybrid dynamic neural network (DNN) trained by sliding mode algorithm and differential evolution (DE) [15]. They used this model to predict stock price indices and stock return volatilities of two important Indian stock markets, namely the Reliance Industries Limited (RIL) from one day ahead to one month in advance. The DNN comprises a set of first order IIR filters for processing the past inputs and their functional expansions and its weights are adjusted using a sliding mode strategy known for its fast convergence and robustness with respect to chaotic variations in the inputs. Extensive computer simulations are carried out to predict simultaneously the stock market indices and return volatilities and it is observed that the simple IIR–based DNN–FLANN model hybridized with DE produces better forecasting accuracies in comparison to the more complicated neural architectures.

Kumar proposed Reservoir inflow forecasting using ensemble models based on neural networks, wavelet analysis and bootstrap method [16]. The aim of this study is to develop an ensemble modeling approach based on wavelet analysis, bootstrap resampling and neural networks (BWANN) for reservoir inflow forecasting. In this study, performance of BWANN model is also compared with wavelet based ANN (WANN), wavelet based MLR (WMLR), bootstrap and wavelet analysis based on multiple linear regression models (BWMLR), standard ANN, and standard multiple linear regression (MLR) models for inflow forecasting. This study demonstrated the effectiveness of proper selection of wavelet functions and appropriate methodology for wavelet based model development. Moreover, performance of BWANN models is found better than BWMLR model for uncertainty assessment, and is found that in-stead of point predictions, range of forecast will be more reliable, accurate and can be very helpful for operational inflow forecasting.

Wang proposed a hybrid forecasting model-based data mining and genetic algorithm-adaptive particle swarm optimization [17]. This study proposes a hybrid-forecasting model that can effectively provide preprocessing for the original data and improve forecasting accuracy. The developed model applies a genetic algorithm-adaptive particle swarm optimization algorithm to optimize the parameters of the wavelet neural network (WNN) model. The proposed hybrid method is subsequently examined in regard to the wind farms of eastern China. The forecasting performance demonstrates that the developed model is better than some traditional models (for example, back propagation, WNN, fuzzy neural network, and support vector machine), and its applicability is further verified by the paired-sample T tests.

Singh suggests Big Data Time Series Forecasting Model - a novel big data time series forecasting model which is based on the hybridization of two soft computing (SC) techniques with fuzzy set and artificial neural network [18]. The proposed model is explained viz., the stock index price data set of State Bank of India (SBI). The performance of the model is verified with different factors, viz., two-factors, three-factors, and M-factors. Various statistical analyzes signify that the proposed model can take far better decision with the M-factors data set.

3 Neural Networks Versus Multiple Regression

In their work Cripps and Engin [19] compared the effectiveness of forecasts made by multiple regression and an artificial neural network using back propagation. The comparison is made on the example of predicting the value of residential real estate. Two models were compared using different datasets, functional specifications and comparative criteria. The same specifications of both methods and the real information can explain why other studies, which have compared the ANN and multiple regression, yielded different results.

For an objective comparison of models it is necessary to identify possible problems of each model, which could distort the performance of the method. Research shows that multiple regression and ANN are determined. In addition, some studies on models based on multiple regression are also determined and applied to the data set used in this study.

A comparison used a standard feed-forward neural network with training on the basis of back-propagation. The same experiments were performed with many variations in methods of training neural networks. Different neural network architectures, such as ARTMAP (adaptive resonance theory), GAUSSIAN, neural-regression, were also studied. After hundreds of experiments and changes of architecture, a standard method of reverse spread showed a better performance than other architectures of neural networks. Several other studies on comparison of neural networks and multiple regression obtained different results. Thus, ironically, the main issue and the purpose of this study is to determine the causes why some researchers got the best result using multiple regression, while others have come to the conclusion that using neural network is better.

Some studies show the superiority of the ANN multiple regression in solving the problem of real estate market forecasting [20, 21]. Other studies [22], however, showed that ANN are not always superior to regression. Due to the ability to train ANN to learn and recognize complex images without being programmed by certain rules, they can easily be used with a small set of statistical data. In contrast, with the regression analysis, neural network does not need to form predetermined function al based determinants. This function of ANN is important because several studies [23, 24] found that the property age has a nonlinear dependence on its value (for a set of data used in their studies). Other studies have shown that in addition to age, the living area also has a non-linear relationship with the value of [25]. Based on the results of previous studies and theoretical capacity of ANN, one would expect that ANN have better performance than a multiple regression.

When using multiple regression, it is necessary to solve methodological problems of the functional form because of incorrect specification, linearity, and multicollinearity and heteroscedasticity. The possibility of a nonlinear functional form in most cases can be translated to a linear-nonlinear relationship before we proceed to using a regression analysis [26]. As noted earlier, some studies found that the age and living area have a nonlinear relationship with the value of the property. Multicollinearity does not affect the predictive capabilities of multiple regression, as well as at ANN [27], because the conclusions are drawn together in a certain area of observation. Multicollinearity, however, makes it impossible to affect separation of supposedly independent variables. Heteroscedasticity arises when using a cross of the intersection data. In addition to the model of methodological problems, the lack of relevant explanatory variable is another source of error when using multiple regression and ANN. This is often due to a lack of data.

When using a neural network of direct distribution with training on the basis of back propagation of errors, it is necessary to solve the following methodological problems: the number of hidden layers, the number of neurons in each hidden layer, the sample of training data, the size of the sample, the sample of test data and the corresponding size of the sample as well as overtraining. As a general rule, the level of training and the number of hidden neurons affect the storage and generality of prediction with the model. The more widely training and the bigger the number of hidden neurons used, the better the production of correct predictions on the training set in the model. On the other hand, ANN is less likely to predict a new data (summary), i.e. ANN ability to generalize reduces when there is overtraining, which can occur when the dimension of the hidden layer is too big. To avoid overtraining, it is advisable to use the heuristic method described in Hecht-Nielsen's article [28]. Despite limitations, there is some theoretical basis to facilitate the determination of the number of hidden layers and neurons in use. In most cases there is no way to determine the best number of hidden neurons without training and evaluation of multiple networks generalization errors. If ANN has only a few hidden neurons, then an error and training error of generalization will be high due to its statistical accuracy. If a neural network has too many hidden neurons, there will be a few errors while training, but the error of generalization will be high because of the re-education and high dispersion [29].

If the training sample is not a representative set of data (statistics), there is no basis for ANN training. Typically, a representative training set is generated using a random sample of the dataset. When a training data set is too small, then INS will have a tendency to memorize that training models are too specific and the extreme points (noise) will have an extraordinary impact on the quality model. This can be correct, however, with the help of K-fold cross-validation method of teaching.

The authors compared multiple regression and ANN. An attempt made to conduct valid comparisons of predictive abilities for both models. Multiple comparative experiments were carried out with different sets of training data, the functional specifications and in different periods of the forecast. 108 trials were conducted. For comparison, two criteria: MAPE (mean absolute percentage error) and PE (prediction error). Based on the results of prediction, both of these criteria may differ. Thus, ANN must treated with caution to each indicator for which data are used criteria to assess forecasting accuracy, as well as to the size of the sample used for different samples of a data model. When using the sample size from moderate to large, ANN performs better in both criteria with respect to the multiple regression. For these purposes, the size of data samples from 506 to 1506 cases (a total of 3906 cases) for ANN outperformed multiple regression (using both criteria). In general, a complication of the functional specification ANN training sample size should be increase to ensure that ANN has worked better than a multiple regression. A multiple regression shows better results (using the criterion of the mean absolute relative error) than ANN when using the small size of the data for training. For each of the functional specification model performance multiple regression to some extent remains constant at different sample sizes, while the performance is significantly improves ANN with increasing size of training data.

Fluctuations in the performance of an ANN model are associated with the large number of possible options and the lack of a methodical approach to the choice of the best parameters. For example, experiments must be conduct to determine the best way to represent the data model specification, the number of hidden layers, the number of neurons in each hidden layer, learning rate and the number of training cycles. All of these actions are intended to identify the best model of a neural network. Failure to conduct such experiments can lead to a badly specified ANN model.

If other input variables, such as a fireplace, floors, finishing materials, lot size, connected communication and the type of funding are included, the results may vary. Research findings provide an explanation why previous studies give different results when comparing multiple regression and ANN for prediction. Prognostic efficiency depends on the evaluation criteria (MAPE and FE) used in conjunction with the size and parameters of the model training. Fluctuations of an ANN model performance may be due to the large number of settings selected by experimentation and depend on the size of the training sample.

4 Hybrid Time Series Forecasting with Using ANFIS

ANFIS (adaptive neuro-fuzzy system) [30] is a multilayer feed forward network. This architecture has five layers such as fuzzy layer, product layer, normalized layer, defuzzy layer and total output. The fixed nodes are represented by circle and the nodes represented by square are the adapted nodes. ANFIS has advantages over the mixture of neural network and fuzzy logic. The aim of mixing fuzzy logic and neural networks is to design an architecture, which uses a fuzzy logic to show knowledge in fantastic way, while the learning nature of neural network to maximize its parameters. ANFIS put forward by Jang in 1993 integrate the advantages of both neural network and fuzzy systems, which not only have good learning capability, but can be also easily inter- preted. ANFIS has been used in many applications in many areas, such as function approximation, intelligent control and time series prediction (Fig. 1).

In this work, we create adaptive neuro-fuzzy system for time series prediction. For example, we used data from Russian government company Rosstat - personnel engaged in research and development in the sector of non-profit organizations. We used system, based on multiple regression for forecasting this indicator and comparison results (Fig. 2) with results (Fig. 3) what we made from ANFIS forecasting system.

From presented figures, we can see that ANFIS model has better performance than the multiple regression model.

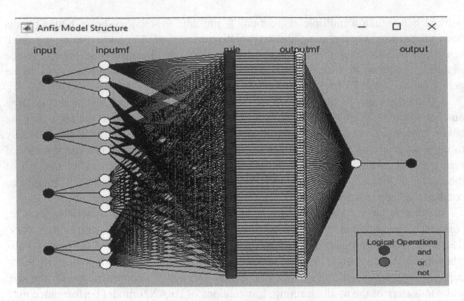

Fig. 1. Structure of ANFIS model for time series forecasting

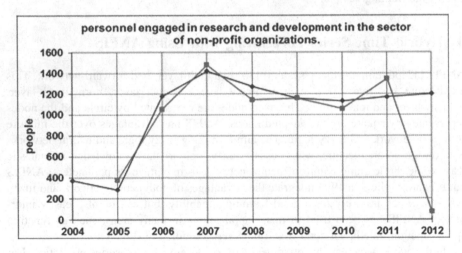

Fig. 2. Forecasting results with multiple regression model

5 Conclusions and Future Work

In this research, we make detailed review of models for time series forecasting such as multiple regression and based on artificial neural networks. A thorough analysis of studies was conducted in this field of research. And the results of studies let to identify the qualitative method to make predictions. Particular attention was given to methods based on comparison of multiple regression and artificial neural networks. An example of a study in which conducted a practical comparison of predictive capability of ANN

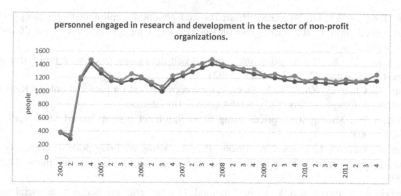

Fig. 3. Forecasting results with ANFIS model

and multiple regression for example predicting the value of residential real estate was described in details. Result of this study demonstrated the superiority of neural networks for prediction quality obtained in comparison with the multiple regression. There were also identified some difficulties encountered in predicting by both methods. Thus, based on the research results, it can be concluded that in order to get the most superiority as predictions using ANN over multiple regression should be administered as soon as possible largest training data [31]. The larger the training set, the qualitative the forecast making with the neural network.

In conclusion, it should be noted that in order to solve the problem of predicting and get the best results one should develop hybrid methods of time series prediction based on neural networks and the statistical techniques. Specifically, based on theory of artificial immune systems, complementary features formation and negative selection can be applied.

References

1. Widrow, B., Rumelhart, D., Lehr, M.A.: Neural networks: applications in industry, business and science. Commun. ACM **37**(3), 93–105 (1994)
2. Chang, P.C., Wang, Y.W., Liu, C.H.: The development of a weighted evolving fuzzy neural network for PCB sales forecasting. Expert Syst. Appl. **32**, 86–96 (2007)
3. Hwang, H.B.: Insights into neural-network forecasting of time series corresponding to ARMA (p, q) structures. Omega **29**, 273–289 (2001)
4. Medeiros, M.C., Pedreira, C.E.: What are the effects of forecasting linear time series with neural networks? Eng. Intell. Syst. 237–424 (2001)
5. Zhang, G.P.: An investigation of neural networks for linear time-series forecasting. Comput. Oper. Res. **28**, 1183–1202 (2001)
6. Armstrong, J.S.: Research needs in forecasting. Int. J. Forecast. **4**, 449–465 (1988)
7. Mariachis, S., et al.: The accuracy of extrapolation (time series) methods: results of a forecasting competition. J. Forecast. **1**(2), 111–153 (1982)
8. Majhi, R., Panda, G.: Stock market prediction of S&P500 and DJIA using bacterial foraging optimization technique. In: 2007 IEEE Congress on Evolutionary Computation, pp. 2569–2579 (2007)

9. Tan, T.Z., Quaky, C., Ng, G.S.: Brain inspired genetic complimentary learning for stock market prediction. In IEEE Congress on Evolutionary Computation, vol. 3, pp. 2653–2660, 2–5 September 2005

10. Oh, K.J., Kim, K.-J.: Analyzing stock market tick data using piece-wise non linear model. Expert Syst. Appl. **22**(3), 249–255 (2002)

11. Miao, K. Chen, F., Zhao, Z.G.: Stock price forecast based on bacterial colony RBF neural network. J. Qingdao Univ. (20), 50–54 (2007). (in Chinese)

12. Wang, Y.W.: Mining stock prices using fuzzy rough set system. Expert Syst. Appl. **24**(1), 13–23 (2003)

13. Pino, R., Parreno, J., Gomez, A., Priore, P.: Forecasting next day price of electricity in the Spanish energy market using artificial neural networks. Eng. Appl. Artif. Intell. **21**, 53–62 (2008)

14. Averkin A.N., Inrushes, S.A.: Hybrid methods of time series prediction in financial markets. In: International Conference on Soft Computing and Measurements. Saint Petersburg State Electro technical University, St. Petersburg. "LETI" name V. I. Ulyanovsk (Lenin), T. 1. Section 1–3, pp. 332–335 (2015)

15. Bios, R., Dash, P.K.: Prediction of financial time series and its volatility using a hybrid dynamic neural network trained by sliding mode algorithm and differential evolution. Int. J. Inf. Decis. Sci. **7**(2), 166–191 (2015)

16. Kumar, S., et al.: Reservoir inflow forecasting using ensemble models based on neural networks, wavelet analysis and bootstrap method. Water Resour. Manage. **29**(13), 4863–4883 (2015)

17. Wang, J., et al.: Hybrid forecasting model-based data mining and genetic algorithm-adaptive particle swarm optimization: a case study of wind speed time series. IET Renew. Power Gener. **10**, 287–298 (2015)

18. Singh, P.: Big data time series forecasting model: a fuzzy-neuro hybridize approach. In: Acharjya, D.P., Dehuri, S., Sanyal, S. (eds.) Computational Intelligence for Big Data Analysis. ALO, vol. 19, pp. 55–72. Springer, Cham (2015). https://doi.org/10.1007/978-3-319-16598-1_2

19. Nguyen, N., Cripps, A.: Predicting housing value: a comparison of multiple regression analysis and artificial neural networks. JRER **22**(3), 314–336 (2001)

20. Tsukuda, J., Baba, S.-I.: Predicting Japanese corporate bankruptcy in terms of financial data using neural networks. Comput. Ind. Eng. **27**(1–4), 445–448 (1994)

21. Do, Q., Grudnitski, G.: A neural network approach to residential property appraisal. Real Estate Apprais. **58**, 38–45 (1992)

22. Allen, W.C., Zumwalt, J.K.: Neural Networks: A Word of Caution, Unpublished Working Paper. Colorado State University (1994)

23. Grether, D., Mieszkowski, P.: Determinants of real values. J. Urban Econ. **1**(2), 127–145 (1974)

24. Jones, W., Ferri, M., Gee, L.M.: A competitive testing approach to models of depreciation in housing. J. Econ. Bus. **33**(3), 202–211 (1981)

25. Goodman, A.C., Thibodeau, T.G.: Age-related heteroskedasticity in hedonic house price equations. J. Hous. Res. **6**, 25–42 (1995)

26. Kmenta, J.: Elements of Econometrics. Macmillan Publishing, New York (1971)

27. Neter, J., Wasserman, W., Kutner, M.H.: Applied Linear Statistical Models, 3rd edn. MacGraw-Hill, New York (1990)

28. Hecht-Nielsen, R.: Kolmogorov's mapping neural network existence theorem. Paper presented at IEEE First International Conference on Neural Networks, San Diego, CA (1987)

29. Geman, S., Bienenstock, E., Doursat, R.: Neural networks and the bias/variance di-lemma. Neural Comput. **4**, 1–58 (1992)
30. Jang, J.S.R.: ANFIS: adaptive-network-based fuzzy inference systems. IEEE Trans. Syst. Man Cybern. **23**, 665–685 (1992)
31. Yarushev, S.A., Ephremova, N.A.: Hybrid methods of time series prediction. In: Hybrid and Synergistic Intelligent Systems: Proceedings of the 2nd International Symposium on Pospelov, pp. 381–388. BFU Kant, Kaliningrad (2014)

Resource Network with Limitations on Vertex Capacities: A Double-Threshold Dynamic Flow Model

Liudmila Yu. Zhilyakova[(✉)]

V. A. Trapeznikov Institute of Control Sciences of Russian Academy
of Sciences, 65, Profsoyuznaya Street, Moscow 117997, Russia
zhilyakova.ludmila@gmail.com

Abstract. The paper presents a double-threshold flow model based on the model called "resource network" [1]. Resource network is a non-classical diffusion model where vertices of directed weighted graph exchange homogeneous resource along incident edges in discrete time. There is a threshold value in this model, specific for every topology: when the total resource amount is large (above this threshold value), the certain vertices start accumulating resource. They are called "attractor vertices". In a novel model described in this paper, attractor vertices have limits on their capacities. Thus, another set of vertices ("secondary attractors") accumulates the remaining surplus of resource. Another threshold value appears in such a network. It is shown that the process of resource redistribution, when its amount is above the second threshold, is described by a non-homogeneous Markov chain. The vertex' ability of being an attractor (primary, secondary, etc.) can be used as a new integer measure of centrality in networks with arbitrary semantics.

Keywords: Graph dynamic models · Threshold models · Flows in networks
Non-homogeneous Markov chains · Resource networks · Centrality index

1 Introduction

We describe a flow network model in which vertices of directed weighted graph exchange homogeneous resource. The investigations of flows in networks were started in 1962 by Ford and Fulkerson in their classical static flow model [2]. Since then many static and dynamic modifications appeared (see e.g. an overview [3]). They investigated directed flows of homogeneous or heterogeneous resource from a source to a sink with different kinds of limitations. Diverse network flow models are used to solve a number of tasks in various subject areas, such as the control problem for resource allocation in virtual networks [4], modeling the distribution of substances in a water environment [5], and many others.

Resource network is a non-classical flow model. It inherits and combines some properties of two classes of models. The first one is *random walks and diffusion on graphs* [6]. The second one is an integer threshold model *chip-firing game*. This model gave rise to many non-trivial mathematical results [7–9]. Moreover, it proved to be an appropriate mathematical formalism for describing processes of self-organized

© Springer Nature Switzerland AG 2018
S. O. Kuznetsov et al. (Eds.): RCAI 2018, CCIS 934, pp. 240–248, 2018.
https://doi.org/10.1007/978-3-030-00617-4_22

criticality [10, 11] and in particular for describing Abelian sandpile model [12]. Description and analysis of some non-classical network models related to resource networks can be found in [13].

The resource network model was proposed in [14]. This pioneering work described a simple particular version of a model, a uniform complete resource network, i.e. a network represented by a complete graph with identical capacities of all edges. Since then many classes of graphs were investigated [15].

2 Model Description

2.1 Resource Network Without Limitations (The Basic Model)

A *resource network* is given by a directed graph $G = (V, E)$ with vertices $v_i \in V$ and edges $e_{ij} = (v_i, v_j) \in E$. All edges have nonnegative weights r_{ij}. The weights denote *throughput capacities* of edges.

Resources $q_i(t)$ are nonnegative numbers assigned to vertices v_i they change in discrete time t. In the basic model, in contrast to the model in question, vertices v_i can store unlimited amount of resource. A *state* $Q(t)$ of the network at time t is a vector $Q(t) = (q_1(t), \ldots, q_n(t))$ that contains values of resources at every vertex at this time.

A state $Q(t)$ is called *stable* or *equilibrium state* iff $Q(t) = Q(t + 1) = \ldots$.

Any equilibrium state for resource network reachable from some state $Q(0)$ will be called *limit state* and denoted by Q^*.

Matrix $R = (r_{ij})_{n \times n}$ is the throughput matrix of a network.

Values $r_i^{in} = \sum\limits_{j=1}^{n} r_{ji}$ and $r_i^{out} = \sum\limits_{j=1}^{n} r_{ij}$ are the total input and output capacities of vertex v_i, respectively. A loop's throughput, if it exists, is included in both sums.

Value $r_{sum} = \sum\limits_{i=1}^{n} \sum\limits_{j=1}^{n} r_{ij}$ is the total throughput capacity of the network.

We denote the total amount of resource in the whole network by W.

The network satisfies the *conservation law*: during its operation, resource does not come from outside and is not spent:

$$\forall t \ \sum_{i=1}^{n} q_i(t) = W.$$

Resource distributes among vertices of a network according to one of two rules with threshold switching. At every time, the vertices use one of the rules depending on the amount of their resource.

Resource Distribution Rules. At time step t, vertex v_i transmits to vertex v_k through edge e_{ik} amount of resource, equal to

$$r_{ik} \text{ if } q_i(t) > r_i^{out} \tag{1}$$

$$\frac{r_{ik}}{r_i^{out}} q_i(t) \text{ if } q_i(t) \le r_i^{out}. \tag{2}$$

The meaning of these rules is as follows. The first rule is applied when a vertex contains more resource than it can transmit to its adjacent vertices through output edges; then it sends "everything it can", i.e. each outgoing edge holds a resource amount equal to its throughput capacity; and totally the vertex gives away resource amount $r_i^{out} = \sum_{j=1}^{n} r_{ij}$.

By rule two, a vertex gives away its entire resource. Notice, that if resource at a vertex equals to its output capacity $q_i(t) = r_i^{out}$, then rules 1 and 2 are identical.

The model is parallel. At every time all vertices with non-zero resource send it into all outgoing edges according to rule 1 or 2; adjacent vertices receive the resource through incoming edges.

The set of vertices with resource $q_i(t)$ not exceeding r_i^{out} is called *zone $Z^-(t)$*. The vertices belonging to $Z^-(t)$ zone operate according to rule 2. *Zone $Z^+(t)$* is the set of vertices with resource exceeding their output capacity; they operate according to rule 1. For the limit state Q^* these zones are denoted by Z^{-*} and Z^{+*}.

Resource Flow. Resource transmitted by vertex v_i along the edge e_{ij} at time t arrives at vertex v_j at time $t + 1$. We assume that in time interval $(t, t + 1)$ the resource is flowing through the edge e_{ij}. This resource is called *flow $f_{ij}(t)$*. The flow in the network in time interval $(t, t + 1)$ is defined by matrix $F(t) = (f_{ij}(t))_{n \times n}$ taken at the left bound of this interval.

The total flow value at time t is the sum: $f_{sum}(t) = \sum_{i=1}^{n} \sum_{j=1}^{n} f_{ij}(t)$.

The edges cannot pass through amount of resource greater than their capacities, which implies that $f_{ij}(t) \le r_{ij}$ and $f_{sum}(t) \le r_{sum}$ for every t.

Let us denote the total outgoing flow for vertex v_i as $f_i^{out}(t) = \sum_{j=1}^{n} f_{ij}(t)$. It is obvious that $f_i^{out}(t) \le r_i^{out}$.

The value $f_j^{in}(t+1) = \sum_{i=1}^{n} f_{ij}(t)$ is the resource ingoing to vertex v_j at the next moment of time; $f_j^{in}(t+1) \le r_j^{in}$. In addition, define that $f_j^{in}(0) = 0$.

If there exists a limit state Q^* in the resource network, then the flow in this state is also called the *limit flow*. We denote the limit flow matrix by $F^* = \left(f_{ij}^*\right)_{n \times n}$.

Threshold Value. *The threshold value T of total resource amount defines the bound between operating rules.* If $W \le T$ then all the vertices beginning at some moment of time t' pass to the zone $Z^-(t)$ and operate according to rule 2; if $W > T$ then the zone $Z^+(t)$ is nonempty and does not change, and starting from some t'', the vertices belonging to $Z^+(t)$ operate according to rule 1.

In [15], the existence and uniqueness of the threshold value T is proved. It is calculated by the formula

$$T = \min_{i \in \{1,\dots,n\}} \frac{r_i^{out}}{q_i^{1*}},$$

where q_i^{1*} are the components of the limit state vector $Q^{1*} = (q_1^{1*}, \dots, q_n^{1*})$ at $W = 1$.

The limit state vector at $W = T$ is denoted by $\tilde{Q}$:

$$\tilde{Q} = (\tilde{q}_1, \dots, \tilde{q}_n)$$

Since the limit state at $W \leq T$ is unique [15], vector $\tilde{Q}$ is proportional to vector Q^{1*} with factor T:

$$\tilde{Q} = T \cdot Q^{1*}.$$

Classification of Vertices. If $W > T$ some vertices begin to accumulate the surplus of resource.

Let us denote:

$$\Delta r_i = r_i^{in} - r_i^{out}.$$

By the sign of Δr_i, vertices are divided into three classes:

- *receiver vertices* for which $\Delta r_i > 0$;
- *source vertices* for which $\Delta r_i < 0$;
- *neutral vertices* for which $\Delta r_i = 0$.

If there is at least a pair of vertices with non-zero value of Δr_i let us call the network non-symmetric. Further, we will consider only regular non-symmetric networks.

The vertices of a network are characterized by a tuple

$$\rho = \left((r_1^{in}; r_1^{out}), (r_2^{in}; r_2^{out}), \dots (r_n^{in}; r_n^{out}) \right)$$

The vertices that can reach the Z^{+*} zone when $W > T$ will be called *attractor vertices* (*attractors*). In non-symmetric regular network, all attractors are receiver vertices. The opposite is not true: receiver vertices are not necessarily attractors.

The ability of vertices to attract resource is called *attractivity*.

Attractivity Criterion. Vertex v_j is an attractor iff it meets the condition

$$j = \arg\min_{i \in \{1,\dots,n\}} \frac{r_i^{out}}{q_i^{1*}}$$

In model with capacity limitations, these vertices will be called *primary attractors*.

2.2 Resource Network with Limitation on Attractor Capacities

Problem Statement. The set of attractor vertices in a random regular network is relatively small. In fact, there is only one attractor vertex in the overwhelming majority of networks. So, at any arbitrarily large value of total resource W, its surplus will be accumulated at one or several vertices in the limit state. All the rest vertices will have constant small amount of recourse at any W. In order to enable other vertices to accumulate resource, the capacity limitation in attractors is introduced.

The capacity limitation is the maximum amount of resource that attractor v_i can store over the value r_i^{out} (for any attractor vertex v_i the equality $\tilde{q}_i = r_i^{out}$ holds, where $\tilde{q}_i$ is the corresponding component of the limit state at $W = T$).

If the total resource in a network is large enough, then after saturation of the attractors with limited capacities it will be distributed among some set of other vertices. New vertices attracting the resource surplus are called *secondary attractors* (Fig. 1).

It is clear that the process of restriction can be continued further, introducing attractors of higher ranks in a network. Because the process of limiting primary, secondary, etc. attractors is the same, consider only the transition from primary to secondary attractors.

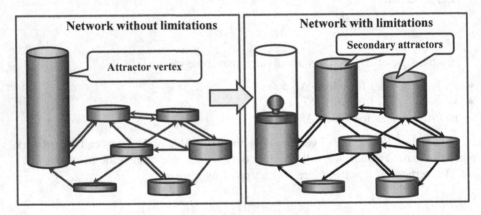

Fig. 1. The two limit states of the same network: without and with capacity limitations on attractor vertex

Properties of Networks with Capacity Limitations. The behavior of networks with limitations is more complicated than the behavior of the basic model. The limitations cause changes in transition matrices, and a homogeneous Markov chain describing the network operation turns into non-homogeneous (changing in time) Markov chain [16].

There exists a second threshold value T^{II} (apart from T) in a regular non-symmetric resource network with capacity limitations on attractor vertices $r_i^{out} + p$, $i = 1, ..., l$, where p is extra-capacity over r_i^{out}, l is a number of attractor vertices in a network. The network operates depending on the total resource value W, in one of four modes with three threshold switching defined by the constants T, lp and T^{II}:

(1) for $W \leq T$, the entire network operates according to rule 2, starting from some finite moment t; the limit state and flow vectors exist and are unique, and the following equalities hold: $F^{in*} = F^{out*T} = Q^* = WQ^{1*}, f^*_{sum} = W$;

(2) for $T < W \leq T + lp$, the total limit flow in a network does not change as the total resource value W increases, the limit flow exists, and $F^{in*} = F^{out*T} = \tilde{Q}, f^*_{sum} = T$;

(3) for $T + lp < W \leq T^{II} + lp$, the total flow increases with increasing W, and $f^*_{sum} = W - lp$;

(4) for $W > T^{II} + lp$, the total limit flow in a network does not change as the total resource value W increases, the total limit flow exists, and $f^*_{sum} = T^{II}$.

The formula for the second threshold value will be given after introducing some additional concepts.

When $W \in (T, T^{II}]$ the network state vector changes according to the formula

$$Q(t) = Q(t-1)R'S'^{-1}(t). \tag{1}$$

Here R' is a stochastic matrix corresponding to matrix R, and $S'^{-1}(t)$ is a matrix describing the reverse flows of resource that does not fit into the attractors. Matrix $R'S'^{-1}(t)$ changes over time, and besides that, it is not necessarily stochastic, in some cases it can be pseudo-stochastic, i.e., some of its elements may turn out to be negative. It is proved that for any network there exist invariant transformations that preserve the values of the non-attractive components of the limit state vector, which transform quasi-stochastic matrices $R'S'^{-1}(t)$ into stochastic ones.

Let $W = T^{II}$. Applying the limit transition at $t \to \infty$ to (1) we obtain

$$\hat{Q} = \hat{Q}R'\hat{S}'^{-1}.$$

The limit of powers of matrix $R'\hat{S}'^{-1}$ exists, and the following equality holds:

$$\left(R'\hat{S}'^{-1}\right)^\infty = \frac{1}{T^{II}}\left(\mathbf{1} \cdot \hat{Q}\right).$$

Now let us build a new network with throughput matrix R_{ext_att}:

$$R_{ext\_att} = \mathrm{diag}(r_1^{out}, \ldots, r_n^{out})R'\hat{S}'^{-1}.$$

We will call it an *extended network* because it has the extended set of attractor vertices in comparison with the initial network. It is proved that its set of attractors coincides with the set of primary and secondary attractors of the initial network, and the threshold value T_{ext_att} equals to the second threshold value of the initial network T^{II}. It follows that we can find T^{II} and formulate the criterion of the second attractivity.

Non-attractive vertex v_j of a regular non-symmetric network defined by throughput matrix R is a secondary attractor of this network with limitations on attractor capacities iff in the extended network with matrix R_{ext_att} it meets the condition

$$j = \arg \min_{i \in \{1,\ldots,n\}} \frac{r_i^{out}}{q_i^{1*\prime\prime}} \tag{2}$$

The second threshold value T^{II} is found by the formula

$$T^{\mathrm{II}} = \min_{i \in \{1,\ldots,n\}} \frac{r_i^{out}}{q_i^{1*\prime\prime}},$$

where $q_i^{1*\prime\prime}$ are elements of the limit probabilities of homogeneous Markov chain with the limit stochastic matrix $\left(R'\hat{S}'^{-1}\right)^{\infty}$.

The known results for Markov chains imply the equality

$$Q_{ext\_att}^{1*} = \frac{1}{T^{\mathrm{II}}}\hat{Q}.$$

Moreover, because of the uniqueness of the non-negative eigenvector of a stochastic matrix, we obtain that vector $Q_{ext\_att}^{1*}$ coincides with vector $Q^{1*\prime\prime}$.

3 Attractivity Rank as a Centrality Index

There are many structural indicators in graph theory characterizing every vertex or edge (local parameters) and graph structure as a whole (global parameters). The term *centrality* refers to a class of local parameters. *Centrality indices* identify the most important (in many different senses) vertices of a graph. The notion of centrality was introduced in social network analysis. In general, centrality is an integer- or real-valued function defined on the vertices that provides their ranking. The more centrality value of vertex is the more significant it is in some aspect.

There exist many different centrality indices, because this vertex feature can have a wide number of meanings. The most commonly used centrality indices are the *degree centrality* (the number of edges incident to a vertex), *closeness centrality* (the average length of the shortest paths between a vertex and all other vertices in the network), *betweenness centrality* (the number of the shortest path containing the given vertex between all pairs of vertices in the network), etc. These indices share one common property: they are all explicitly derived from the graph structure.

The property of attractivity also can be considered as a centrality index in a social network. After the secondary attractivity, we can consider the attractivity of higher ranks. Under k-rank attractivity, we mean the property of a vertex to attract surpluses of resource under the limitation of attractors $k - 1$ times. Thus, attractivity centrality index is an integer-valued function characterizing the vertices by their ability to attract resource. This index, in contrast to the one mentioned above, is an implicit indicator of the structure. In this sense, it more closely resembles the eigenvector centrality (see formula (2)). Nevertheless, it was proved [15] that with the same eigenvector, any vertex can become an attractor under transformation the throughput matrix R, which is invariant with respect to the eigenvector.

In a social network, consider the weight matrix R that defines the influence of agents on each other. For this network, the attractivity centrality index will help to identify obvious and hidden leaders of small and large groups, organizing classes of attractors of different orders. Continuing the iterative process of limiting attractors of higher orders, we can rank the entire network by the influence of its agents.

4   Conclusion

In this article, we presented results of research on our new model resource network with capacity limitation on attractor vertices. It is shown that there are four intervals for the total resource amount in a network characterizing by network's different behavior. These intervals are separated by three values: T, lp and T^{II}. Here, T is a threshold value of the "basic" resource network, i.e. of the corresponding network without limitations, T^{II} is a second threshold value appearing due to the limitations, l is a number of primary attractor vertices in a network, and p is the limit extra-capacity over r_i^{out} in every attractor.

The transient processes are studied and their convergence is proved on each of the intervals. The formula for the second threshold value is obtained and the criterion of the secondary attraction is found. Thus, all the boundary constants defining the network behavior are obtained. The vectors of limit states and flows are found for every interval of total resource value and for every topology and weight matrix R of a regular non-symmetric network.

The rank of attractivity as a new centrality index is proposed, its study will be the subject of our further research.

Another direction for further research is the investigation of the double-threshold resource networks based on graphs with more complex structure, like multigraphs, graphs with nonstandard reachability, etc. [17–22]. With such an extension of the model, the graph vertices can exchange several different resources simultaneously (but perhaps not independently) which is inherent to social networks (exchange of various services, participation of users simultaneously in several interest groups, etc.).

Acknowledgments. This work was partially supported by the Russian Foundation for Basic Research, project no. 17-07-00541.

References

1. Kuznetsov, O.P., Zhilyakova, L.Y.: Bidirectional resource networks: a new flow model. Dokl. Math. **82**(1), 643–646 (2010)
2. Ford Jr., L.R., Fulkerson, D.R.: Flows in Networks. Princeton University Press, Princeton (1962)
3. Ahuja, R.K., Magnanti, T.L., Orlin, J.B.: Network Flows: Theory, Algorithms and Applications, p. 846. Prentice Hall, New Jersey (1993)
4. Szeto, W., Iraqi, Y., Boutaba, R.: A multi-commodity flow based approach to virtual network resource allocation. In: Proceedings of Global Telecommunications Conference. GLOBECOM 2003, vol. 6. pp. 3004–3008. IEEE (2003)

5. Zhilyakova, L.Yu.: Using resource networks to model substance distribution in aqueous medium. Autom. Remote Control **73**(9), 1581–1589 (2012). https://doi.org/10.1134/S0005117912090111
6. Blanchard, Ph., Volchenkov, D.: Random Walks and Diffusions on Graphs and Databases: An Introduction. Springer Series in Synergetics. Springer-Verlag, Berlin (2011). https://doi.org/10.1007/978-3-642-19592-1
7. Björner, A., Lovasz, L., Shor, P.: Chip-firing games on graphs. Eur. J. Comb. **12**, 283–291 (1991)
8. Björner, A., Lovasz, L.: Chip-firing games on directed graphs. J. Algebraic Comb. **1**, 305–328 (1992)
9. Biggs, N.L.: Chip-firing and the critical group of a graph. J. Algebraic Comb. **9**, 25–45 (1999). Kluwer Academic Publishers, Netherlands
10. Bak, P., Tang, C., Wiesenfeld, K.: Self-organized criticality. Phys. Rev. A **38**, 364–374 (1988)
11. Bak, P.: How Nature Works: The Science of Self-Organized Criticality. Copernicus, New York (1996)
12. Dhar, D.: The Abelian sandpile and related models. Phys. A: Stat. Mech. Appl. **263**(1–4), 4–25 (1999)
13. Zhilyakova, LYu.: Dynamic graph models and their properties. Autom. Remote Control **76**(8), 1417–1435 (2015). https://doi.org/10.1134/S000511791508007X
14. Kuznetsov, O.P.: Uniform resource networks. I. Complete graphs. Autom. Remote Control **70**(11), 1767–1775 (2009). Moscow
15. Zhilyakova, L.Yu., Kuznetsov, O.P.: Resource Network Theory: Monograph. RIOR: INFRA-M, Moscow (2017). (Science). 283 p. https://doi.org/10.12737/21451
16. Zhilyakova, L.Yu.: Resource Networks with Limitations on Vertex Capacities: Monograph. RIOR, Moscow (2018). 160 p. https://doi.org/10.12737/1745-6
17. Erusalimskii, Ya.M.: Flows in Networks with Nonstandard Reachability, Izv. Vyssh. Uchebn. Zaved., Severo-Kavkaz. Region, Estestv. Nauki, no. 1, pp. 5–7 (2012)
18. Jungnickel, D.: Graphs, Networks and Algorithms, 3rd edn. Springer, Berlin (2007). 650 p.
19. Basu, A., Blanning, R.W.: Metagraphs and Their Applications. Springer, New York (2007). 172 p. https://doi.org/10.1007/978-0-387-37234-1
20. Chartrand, G., Lesniak, L., Zhang, P.: Graphs & Digraphs, 6th edn. CRC Press, Boca Raton (2016). 625 p.
21. Gross, J.L., Yellen, J., Zhang, P. (eds.): Handbook of Graph Theory. Discrete Mathematics and Applications. CRC Press, Boca Raton (2014). 1630 p.
22. Knuth, D.E.: Art of Computer Programming, Volume 4, Fascicle 4: Generating All Trees-History of Combinatorial Generation. Addison-Wesley Professional, Boston (2006). 128 p.

Advanced Planning of Home Appliances with Consumer's Preference Learning

Nikolay Bazenkov[✉] and Mikhail Goubko

Trapeznikov Institute of Control Sciences,
65 Profsoyuznaya st., Moscow 117997, Russia
n.bazenkov@yandex.ru
http://www.ipu.ru/staff/bazenkov

Abstract. For modern energy markets it is typical to use dynamic real-time pricing schemes even for residential customers. Such schemes are expected to stimulate rational consumption by the end customers, provide peak shaving and overall energy efficiency. But under dynamic pricing planning a household's energy consumption becomes complicated. So automated planning of household appliances is a promising feature for developing smart home environments. Such a planning should adapt to individual user's habits and preferences over comfort to cost balance. We propose a novel approach based on learning customer preferences expressed by a utility function. In the paper an algorithm based on inverse reinforcement learning (IRL) framework is used to infer the user's hidden utility. We compare IRL-based approach to multiple state-of-the art machine learning techniques and the proposed earlier parametric Bayesian learning algorithm. The training and test datasets are generated by the simulated user's behavior with different price volatility settings. The goal of the algorithms is to predict a user's behavior based on the existing history. The IRL and Bayesian approaches showed similar performance and both of them outperforms modern machine learning algorithms such as XGBoost, random forest etc. In particular, the preference learning algorithms significantly better generalize to data generated with parameters different from the training sample. The experiments showed that preference learning approach can be especially useful for smart home automation problems where future situations can be different from those available for training.

Keywords: Preference learning · Smart grid · Consumer simulation
Bayesian learning · Inverse reinforcement learning

1 Introduction

Price liberalization is considered as a powerful tool for increasing efficiency of energy systems. At present there are energy markets where electricity is sold at the real-time dynamic prices even to residential customers [6]. These measures are intended to stimulate rational energy consumption and provide load

© Springer Nature Switzerland AG 2018
S. O. Kuznetsov et al. (Eds.): RCAI 2018, CCIS 934, pp. 249–259, 2018.
https://doi.org/10.1007/978-3-030-00617-4_23

peak shaving and investments to energy saving technologies. But optimization of energy consumption at the household level becomes a complicated task under real-time pricing. Some automation of the cost saving process may be an attractive feature for emerging smart home technologies like famous NEST thermostat [15].

We propose a model of a smart device adapting to a user's behavior. Specifically we focus at the automated bread maker which should provide the user with fresh hot bread every morning. The bread making process is automated and the user needs only to load the ingredients and set the desired finish time. The user's satisfaction or comfort reaches maximum if the bread is cooked just before the breakfast while the cost is low at night. The problem is that each user has individual hidden preferences over comfort/cost balance. Since we suppose the electricity prices varying from day to day the optimal choice may also be different for different days. Our goal is to devise an algorithm which can recommend the finish time balancing the user's comfort and the cost.

We model user preferences as a utility function and propose an algorithm of learning this function from the history of past user's choices. After the utility function is learned it can be used to recommend the best option to the user. Different approaches can be used to infer the utility function. Here we study an algorithm based on Inverse Reinforcement Learning [10]. To evaluate the quality of the algorithm we use the simulation model of smart home environment developed in [7]. We compare the IRL-based algorithm with Bayesian learning proposed in [7] and state of the art machine learning techniques like XGBoost.

The proposed approach may be applied to other problems arising in smart energy management. For instance, in a household with a local energy storage the control system can optimally choose when to store energy, sell it back to the market and use for the household's activities. Augmenting such systems with a model of user behavior may significantly increase the perceived comfort.

The article has the following structure. Section 2 gives a brief overview of related research in smart energy systems and preference learning. The model of smart home environment is described in Sect. 3. The preference learning problem and algorithms are described in Sect. 4. The experimental evaluation is provided in Sect. 5. The results are summarized in Sect. 6.

2   Related Work

Automated scheduling of residential energy consumption has received much attention during recent years. Hot topics include smart grids organization [4], peak shaving [3], distributed generation [9]. Typically a sort of optimization problem is formulated, either linear [5], mixed integer linear [3] or combinatorial [13]. User preferences are formalized in two ways: either as an utility function over possible schedules [5], or as a set of constraints imposed on feasible schedules while total electricity bill is minimized [2].

Energy-unaware user behavior may increase energy waste up to thirty percents [11]. So, different techniques are needed to take into account the user's behavioral patterns and comfort requirements in design of smart home and smart building environments.

3   Model of Smart Home Environment

We consider a home appliance performing a regular task which must be finished until certain hour (Type 2 according to [8]). The general ideas are illustrated with a simplified example of an automated bread maker able to recommend the optimal finish time accordingly to the user's preferences over comfort/cost balance. Here we briefly describe the model of the device operation developed in [7].

The model simulates the behavior of a hypothetical rational consumer who has some eating schedule: breakfast, lunch (not every day) and dinner. A certain amount of bread is required for each of them. The consumer's comfort depends on the bread freshness which decays with a constant rate. So, the comfort is at maximum when the bread cooking has been finished just before the breakfast. The electricity price changes from day to day, but we consider the reliable day-ahead price forecast is available to the device. This assumption is not highly restrictive because there are lots of academic work devoted to short-term price forecasting [14] and commercial services available for established energy markets (see, for example [1]).

Formally, the consumer's decision making problem is to choose the finish time Δ that maximize the following utility function:

$$u(\Delta, y) = d(\Delta, y) - c(\Delta, y). \tag{1}$$

Here y is the vector representing the current situation which includes: current bread stock σ, day of week θ, the day-ahead electricity price forecast. The comfort $d(\Delta, y)$ represents the user's total satisfaction during the next day, and $c(\Delta, y)$ is the cost incurred by electricity consumption. We assume the user is aware about the cost of each possible finish time. For example, the cost may be shown on the interface of the device. Figure 1 gives an idea of what such an interface may look like.

The comfort is defined as

$$d(\Delta, y) = \sum_{t=0}^{T} d_0 \cdot \phi(t) \min\{\sigma(t), e(t)\}, \tag{2}$$

where d_0 is the "default satisfaction", $\phi(t)$ is the bread freshness, $\sigma(t)$ is the bread stock available at time t and $e(t)$ is the bread consumption. Planning horizon T equals to the end of the next day.

The cost is defined as

$$c(\Delta, y) = \sum_{t=0}^{\Delta} c_0(\Delta - t)p(t), \tag{3}$$

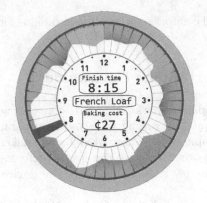

Fig. 1. Mock of the device's interface

where $c_0(\tau)$ is the device's power consumption at the given cooking stage and $p(t)$ is the electricity price for time t.

Figure 2 illustrates typical comfort, cost and utility functions. The three peaks of comfort function correspond to the finish time set exactly to breakfast, lunch, dinner. The comfort is lower as long as finish time is set earlier because the bread freshness decays. The model which was used to generate this comfort function is described in [7].

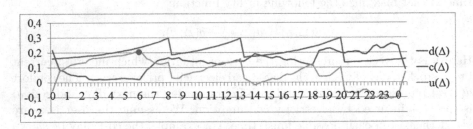

Fig. 2. Simulated user preferences: comfort (blue), cost (red) and utility (green). Horizontal axes shows the possible finish times. The red dot marks the optimal finish time. (Color figure online)

4   Preference Learning

4.1   Problem Formulation

The final goal is to build the device smart enough to predict the choice maximizing the user's utility (1) and recommend it to the user. So the problem is formulated as follows. Predict the optimal choice for the user, given the history of the previous choices $\{(y^i, \Delta^i)\}$, $i = \overline{1, n}$ (n – the number of observations).

We discretize the day into $N = 96$ 15-min intervals. So, $\Delta^i \in \{1, \ldots, N\}$ is the number of the chosen interval. The description of the situation y^i includes:

1. Current bread stock σ_0^i. This parameter can be measured directly or calculated with reasonable accuracy from the time have passed since the last breadmaker run.
2. Day of week w^i following the day when the user made the decision.
3. Costs $c_1^i, \ldots, c_N^i$ for every alternative finish time.

One may apply some general-purpose model like decision trees or multivariate regression and predict the choice directly without modeling such internal details as the utility function. However, the goal of the paper is to examine how much we can gain by building the internal representation of the user's decision making process. Specifically, we use the history of the user's actions to infer the approximation $d^*(\Delta, y)$ of the hidden comfort function and then predict further choices as

$$\Delta^*(y) = \arg\max_{\Delta}[d^*(\Delta, y) - c(\Delta, y)]. \tag{4}$$

In [7] the Bayesian model was proposed which made use of a relatively simple parametric approximation of the comfort function (2). Here we employ another, non-parametric approach called Inverse Reinforcement learning and compare it to previously used Bayesian approach and some conventional ML techniques.

4.2   Inverse Reinforcement Learning

Inverse reinforcement learning (IRL) problem was first formulated in [10] in the following way. If we observe the behavior of a rational agent seeking to maximize its payoff which is unknown to us, how can we reconstruct this hidden payoff function?

The approach proposed in [10] is that the payoff function should be the best rationalization of the observed behavior. Formally, the function should maximize some criterion of the agent's rationality. In our work the payoff function is equivalent to the utility function. One of the criteria proposed in [10] is so-called contrast criteria formulated like that:

$$\max \sum_{i=1}^{n} [u(\Delta^i, y^i) - \max_{\Delta \neq \Delta^i} u(\Delta, y^i)] \tag{5}$$

$$u(\Delta^i, y^i) \geq u(\Delta, y^i) \; \forall \Delta \neq \Delta^i \tag{6}$$

The problem (5) provides the following explanation for the observed history of the user's actions. For every historical entry (Δ^i, y^i) the utility of the choice Δ^i significantly outperforms the next choice. The condition (6) means that every choice must be rational i.e. it's utility must be greater than for every alternative choice available. In this paper the utility function is the difference between the hidden comfort function and the observable costs. So, it is the comfort function that should be learned.

The criterion requires a reasonable representation of the comfort function that depends on the current situation. We employed the following simplified discretization of the state and action spaces. The current bread stock σ_0 falls into one of the 4 possible intervals: $[0, 0.15), [0.15, 0.3), [0.3, 0.45), [0.45, 0.6]$. We also simplified the day of week taking only binary values: weekday (0) or weekend (1). For industrial applications the situation model should be more complex, including differences in the weekdays, holidays, seasons etc.

This discretization provides us with 8 distinct basic situations. For every basic situation $y' = (\sigma^0, w)$ we learn its own comfort function represented as $N = 96$ separate values $d_1, \dots, d_N$. Each value d_k is the level of the user's comfort if the baking program is finished at the k-th 15-min interval in the day.

The straightforward application of Eqs. (5) and (6) implies that every basic situation is distinct from others and should be grounded in its own historical data. However, this approach requires enough data for every situation which may not be the case in the real world. So, we assume that the comfort functions for different situations are similar and the similarity is defined by the kernel function:

$$K(y, y') = \exp(-\beta \rho(y, y')), \tag{7}$$

where $\rho(y, y') = \gamma |w - w'| + |\sigma_0 - \sigma_0'|$ is the weighted L_1 distance. Coefficient β defines the similarity decay with the growth of the distance and γ specifies the relative weight of the difference in the day of week.

For each basis situation y' we reconstruct comfort function $d = (d_1, \dots, d_N)$ as the solution to the following problem.

$$\max_{d_1, \dots, d_N} \sum_{i=1}^{n} [d(\Delta^i) - \max_{\Delta \neq \Delta^i} \{d(\Delta) - c^i(\Delta)\}] K(y^i, y') - \alpha ||Ld||^2 \tag{8}$$

$$d(\Delta) - c^i(\Delta) \leq d(\Delta^i) - c^i(\Delta^i) - \varepsilon \ \forall \ \Delta \neq \Delta^i \tag{9}$$

$$d_k \geq 0. \tag{10}$$

Equation (8) is the modified contrast criterion (5). Kernel smoothing $K(y^i, y')$ allows us to use every available observation, not only those that fall into the basic situation y'. Here the regularization term $\alpha ||Ld||^2$ can be chosen as zero order regularization with $L = E$ or first-order regularization. The coefficient $\alpha \in [0, 1]$ controls the influence of the regularization.

Constraints (9) weakens the rationality assumption (6). We admit the user may be indifferent to gains in utility less than ε.

The problem (8)–(10) can be reduced to quadratic programming and solved by any specialized software.

5   Experimental Evaluation

5.1   Data Description

The data were generated by the simulation model developed in [7]. The eating schedule is $e(t) = a$ from 0 to 3 times per day with breakfast at $8^{00} + rnd[0, 1h]$, lunch at $13^{00} + rnd[0, 1h]$, dinner at $20^{00} + rnd[0, 1h]$; $rnd[0, 1h]$ is a random number between 0 and 1 h. At other time the bread is not consumed. The probability of having breakfast and dinner is 0.9, probability of lunch is 0.1 in weekday, and 0.8 in weekend.

The electricity price is modeled as

$$p(t) = p_0(t) \exp(\xi RW(t)) \tag{11}$$
$$RW(t+1) = 0.9RW(t) \pm 1, \tag{12}$$

where $p_0(t)$ is the baseline price, ξ is the price volatility and $RW(t)$ is the random walk process. The baseline price is calculated according to E-1 three-zonal tariff by Pacific Gas and Electric Company for 2015 [12], the evening price equals $p_e = \$0.45$ per kW h, the daytime price equals $0.5p_e$ and nighttime price equals $0.1p_e$.

The model of consumer behavior was used to generate three datasets with different price volatility. The datasets description is provided in Table 1. The training set was 80% of the data with moderate volatility. The rest 20% was used as a test set as well as low- and high-volatility data.

Table 1. Data generated by simulation

Dataset, price volatility	# of observations	Min. finish time	Max. finish time	Finish time std, hours
Low	163	6^{00}	13^{00}	1.15
Moderate	231	6^{00}	13^{00}	1.54
High	127	5^{30}	13^{15}	1.47

5.2 Performance Comparison

The IRL algorithm was implemented in MATLAB with IBM CPLEX Optimizer used to solve the problem (8).

The IRL algorithm was trained on the 80% of data with moderate volatility with 5-fold cross-validation to choose optimal combination of hyperparameters α, β and γ. The best performance was shown by the range of the parameters: $\alpha \in \{0.001, 0.1, 0.25\}$, $\beta = 20$, $\gamma \in [1, 2]$. Different regularization terms were examined, zero- and first-order showed the same quality.

Figure 3 shows the distributions of finish time in the training data. Each plot represents a histogram of the chosen finish times for each situation. Each vertical bar corresponds to the number of finish times that fall at the given time interval.

In the Fig. 4 the learned comfort functions for all basis situations are shown. Note that the function for left bottom situation ($\sigma_0 \in [0.45, 0.6]$, $w =$ weekday) was inferred only from the data that fall into other situations with kernel smoothing because no data were present for this particular situation.

The IRL algorithm was compared with previously proposed parametric Bayesian method, denoted as ABC, approximate Bayesian Computation.

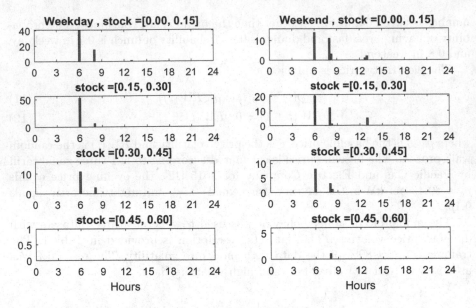

Fig. 3. Finish time distribution for moderate price volatility. Each plot shows the histogram of the finish times for the basis situation described by the current bread stock (four rows) and weekday/weekend variable (left/right column).

The algorithm [7] employs a simple parametric model of comfort function. The function is represented by several triangles and the parameters are the peak locations and the angles of the slopes. Both IRL and ABC are preference learning methods because they use the internal model of the decision making. We also evaluate the quality of modern machine learning methods available in scikit-learn toolbox. The quality measure was mean absolute error (MAE) in prediction of the finish time chosen by the consumer.

$$Err = \sum_{i=1}^{n} |\Delta^i - \Delta^*(y^i, c^i(\cdot))| \tag{13}$$

Here Δ^i is the actual choice of the consumer in the situation y^i and $\Delta^*(y^i, c^i(\cdot))$ is the predicted choice.

In the Table 2 the preference learning algorithms are compared with conventional ML methods. Note that all the algorithms were trained on the data with moderate price volatility. The proposed IRL algorithm outperforms all conventional methods and showed the results comparable to parametric Bayesian learning. The IRL also showed the best result on the high volatility data. This emphasizes its high generalization ability that may be especially valuable in smart home environments.

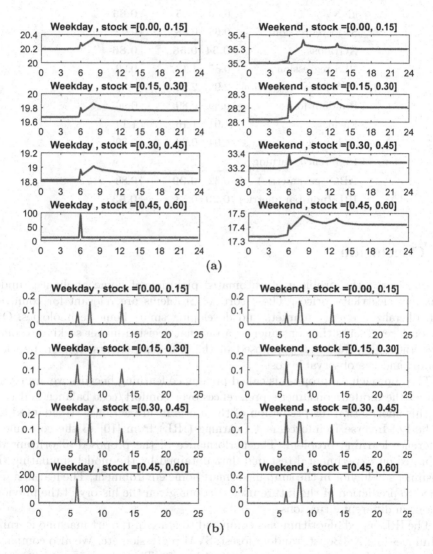

Fig. 4. Comfort functions for each basis situation described by the current bread stock (four rows) and weekday/weekend (left/right column): (a) First-order regularization; (b) Zero-order regularization

Table 2. Error in finish time prediction

Price volatility	Low	Moderate	High
Algorithm	Mean absolute error, hours		
kNN	0.80	0.95	**0.85**
SVM regression	0.65	0.88	0.92
XGBoost	**0.34**	**0.58**	**0.86**
PLS regression	0.87	1.13	0.99
OLS linear regression	2.93	2.93	5.72
Random forest	0.59	0.89	0.96
Mean	1.01	1.16	1.14
Ridge regression	0.94	0.93	0.89
Preference learning			
ABC (parametric)	0.32	**0.39**	0.83
IRL (non-parametric)	**0.29**	0.53	**0.63**

6 Conclusion

We considered the problem of automated planning of home appliances under dynamics electricity prices. These sorts of problems are relevant for countries with liberalized energy markets and developing smart home technologies. Our approach represents the consumer as a rational decision-maker seeking to maximize the utility function composed of the two components: hidden personal comfort and the observable cost.

The approach we propose is called preference learning because we are trying to infer the hidden consumer's preferences over comfort/cost balance and then use this information for prediction of the user's future actions. We adapted the method of Inverse Reinforcement Learning (IRL) from [10] to the consumer's preference learning problem. The performance of the proposed algorithm was evaluated on experimental test bed data obtained by the model simulating the consumer's behavior in the simplified smart home environment. The learning was aimed to prediction of the future user's actions given the history of the previous actions in different situations.

The IRL-based algorithm was compared to state of the art machine learning techniques like XGBoost, random forest, SVM regression etc. We also compared it to parametric Bayesian learning proposed in [7]. The experiments showed the IRL algorithm outperforms modern ML algorithms under the studied parameter values and is approximately as effective as Bayesian learning. Given the IRL doesn't require a pre-defined parametric model the results are promising for smart home applications.

Acknowledgements. The work was partially supported by Russian Foundation of Basic Research (project no. 17-07-01550).

References

1. AleaSoft: Energy Forecasting (2017). https://aleasoft.com/energy-price-forecasting/. Accessed 13 Oct 2017
2. Bradac, Z., Kaczmarczyk, V., Fiedler, P.: Optimal scheduling of domestic appliances via MILP. Energies 8(1), 217–232 (2015). https://doi.org/10.3390/en8010217. http://www.mdpi.com/1996-1073/8/1/217
3. Caprino, D., Vedova, M.L.D., Facchinetti, T.: Peak shaving through real-time scheduling of household appliances. Energy Build. 75, 133–148 (2014). https://doi.org/10.1016/j.enbuild.2014.02.013. http://www.sciencedirect.com/science/article/pii/S0378778814001248
4. Chan, S.C., Tsui, K.M., Wu, H.C., Hou, Y., Wu, Y.C., Wu, F.F.: Load/price forecasting and managing demand response for smart grids: methodologies and challenges. IEEE Sig. Process. Mag. 29(5), 68–85 (2012). https://doi.org/10.1109/MSP.2012.2186531
5. Conejo, A.J., Morales, J.M., Baringo, L.: Real-time demand response model. IEEE Trans. Smart Grid 1(3), 236–242 (2010). https://doi.org/10.1109/TSG.2010.2078843
6. Eid, C., Hakvoort, R., de Jong, M.: Global trends in the political economy of smart grids. WIDER WP 22 (2016)
7. Goubko, M.V., Kuznetsov, S.O., Neznanov, A.A., Ignatov, D.I.: Bayesian learning of consumer preferences for residential demand response. IFAC-PapersOnLine 49(32), 24–29 (2016). https://doi.org/10.1016/j.ifacol.2016.12.184. Cyber-Physical Human-Systems CPHS 2016. http://www.sciencedirect.com/science/article/pii/S2405896316328567
8. Li, N., Chen, L., Low, S.H.: Optimal demand response based on utility maximization in power networks. In: 2011 IEEE Power and Energy Society General Meeting, pp. 1–8 (2011). https://doi.org/10.1109/PES.2011.6039082
9. Lujano-Rojas, J.M., Monteiro, C., Dufo-Lpez, R., Bernal-Agustn, J.L.: Optimum residential load management strategy for real time pricing (RTP) demand response programs. Energy Policy 45, 671–679 (2012)
10. Ng, A.Y., Russell, S.J.: Algorithms for inverse reinforcement learning. In: Proceedings of the Seventeenth International Conference on Machine Learning, ICML 2000, pp. 663–670. Morgan Kaufmann Publishers Inc., San Francisco (2000). http://dl.acm.org/citation.cfm?id=645529.657801
11. Nguyen, T.A., Aiello, M.: Energy intelligent buildings based on user activity: a survey. Energy Build. 56(Suppl. C), 244–257 (2013). https://doi.org/10.1016/j.enbuild.2012.09.005. http://www.sciencedirect.com/science/article/pii/S0378778812004537
12. Pacific Gas and Electric Company: Electric schedule EV (2015). http://www.pge.com/tariffs/tm2/pdf/ELEC_SCHEDS_EV.pdf. Accessed 31 Mar 2016
13. Volkova, I.O., Gubko, M.V., Salnikova, E.A.: Active consumer: optimization problems of power consumption and self-generation. Autom. Remote Control 75(3), 551–562 (2014). https://doi.org/10.1134/S0005117914030114
14. Weron, R.: Electricity price forecasting: a review of the state-of-the-art with a look into the future. HSC Research Reports HSC/14/07, Hugo Steinhaus Center, Wroclaw University of Technology (2014). https://EconPapers.repec.org/RePEc:wuu:wpaper:hsc1407
15. Yang, R., Newman, M.W.: Learning from a learning thermostat: lessons for intelligent systems for the home. In: Proceedings of ACM UNICOMP, pp. 93–102 (2013)

Implementation of Content Patterns in the Methodology of the Development of Ontologies for Scientific Subject Domains

Yury Zagorulko[✉], Olesya Borovikova, and Galina Zagorulko

A.P. Ershov Institute of Informatics Systems, Siberian Branch of the Russian Academy of Sciences, Acad. Lavrentiev avenue 6, 630090 Novosibirsk, Russia
{zagor,olesya,gal}@iis.nsk.su

Abstract. The paper considers an approach to the development and implementation of ontology design patterns used in constructing the ontologies of scientific subject domains. Using such patterns allows us to provide a uniform and consistent representation of all the entities of the ontology under development and to save resources and avoid errors typical for ontological modeling. The paper pays a special attention to the development and implementation of this kind of ontology design patterns as content patterns playing an important role in the methodology proposed by the authors for the development of scientific subject domain ontologies.

Keywords: Ontology · Ontology design patterns · Content patterns
Methodology for the development of ontology of scientific subject domain

1 Introduction

Ontologies are widely used to formalize the knowledge contained in scientific subject domains (SSD). The ontology not only helps to produce a convenient representation of all the necessary concepts of a modeled domain, but also helps to ensure their uniform and consistent description.

As ontology development is rather complex and time-consuming process, various methodologies and approaches are proposed to facilitate it. Recently, the approach based on the application of ontology design patterns (ODPs) obtained new attention [1, 2]. According to this approach, each ODP is a documented description of proven solution for a typical ontological modeling problem. ODPs are created in order to simplify and facilitate the process of building ontology and to help developers avoid the typical errors uring ontological modeling.

Although the use of ontology design patterns makes it possible to save human resources and improve the quality of the ontology being developed, there exists presently only one methodology for constructing ontology, namely the eXtreme Design methodology (XD methodology) [2] proposed within the framework of the project NeOn [3], which openly declares the application of the ODPs.

© Springer Nature Switzerland AG 2018
S. O. Kuznetsov et al. (Eds.): RCAI 2018, CCIS 934, pp. 260–272, 2018.
https://doi.org/10.1007/978-3-030-00617-4_24

Among other tools using ontology design patterns we may point, for example, a plug-in for the ontology development in the NeOn project, and another plug-in for the WebProtégé ontology editor [4].

However, these tools cover only some of the possible tasks dealing with patterns. For instance, there are absolutely no tools to support the construction, search and retrieval of the patterns from an ontology, and a very few tools supporting the collection, discussion and dissemination of the patterns. The catalogs of ontology design patterns [5–7], which have been developed recently, can be attributed to the latter group of tools.

The present paper considers an approach to the implementation of this kind of the ODPs as content patterns that are playing an important role in the methodology proposed by the authors for the development of the ontologies for scientific subject domains [8, 9]. Used in this methodology patterns appear as a result of solution of the ontological modeling problems that the authors of the paper encountered in the process of developing ontology for a wide range of scientific subject domains [10–14].

The rest of the paper is structured as follows. The second chapter contains a brief introduction to the ontology design patterns. The third chapter is devoted to the issues of content patterns development and implementation. It presents a fragment of the OWL implementation of a content pattern for describing scientific activity. The main advantages of the application of ontology design patterns for developing the ontologies of scientific subject domains are discussed in the Conclusion.

2   The Ontology Design Patterns

As mentioned above, ontology design patterns are created to describe solutions to typical problems arising in ontology development.

Depending on the type of the problems for which the patterns are created, they distinguish among structural ODPs, correspondence ODPs, content ODPs, reasoning ODPs, presentation ODPs and lexico-syntactic ODPs [1].

The most popular patterns for the development of ontologies directly by knowledge engineers are structural ODPs, content ODPs and presentation ODPs. Let us consider them in more detail.

Structural patterns either fix ways to solve problems caused by the limitations of the expressive capabilities of ontology description languages or specify the general structure and the kind of ontology. Patterns of the first type are called logical patterns, and patterns of the second type are referred to as architectural patterns. The latter type contains proposals for ontology organization in general, including, for example, structures such as taxonomy and modular architecture.

The content patterns define ways of representing typical ontology fragments that can serve as a basis for building the ontologies for a class of subject domains. Based on these patterns, you can describe more complex (composite) content patterns, supplementing them with missing components and introducing additional relationships. However, it should be remembered that the use of composite patterns may generate assembly and integration problems.

The presentation patterns define recommendations for the naming, annotating, and graphical representation of ontology elements. The application of these patterns should make ontology to be more "readable", convenient and easy to use.

The naming patterns include conventions on the naming rules for namespaces, files, and ontological elements. For example, when declaring an ontology, it is welcome to use capital letters and the singulars in the name of a class, name of the parent class in the names of its subclasses (as a suffix), and small letters in the names of properties.

The annotation patterns specify annotation rules for ontology elements, such as providing classes and their properties with comments and labels in several languages.

At the moment, there is no common standard for describing patterns, but they are most often described in the format proposed by the Association for Ontology Design & Patterns (ODPA) [6]. In accordance with this format, a pattern description includes its graphical representation, text description, a set of scenarios and usage examples, links to other patterns in which it is used, as well as information about the name of the pattern, its author and scope.

The methodology of extreme ontology design [2] also suggests providing each content pattern with a set of Competency Questions. These can be used both at the stage of developing patterns and for finding patterns suitable for the development of a specific ontology.

The development and use of content patterns relies on the five basic operations [1]: import, specialization, generalization, composition and expansion.

Import is including a content pattern in the ontology under development. This is the basic mechanism for reusing such patterns. The elements of imported patterns cannot be modified.

Specialization is creating sub-classes of a content pattern's class and/or sub-properties of some content pattern's properties.

Generalization means that a content pattern *CP1* is a generalization of another content pattern *CP2* if *CP1* imports *CP2*, and at least one ontology element from *CP1* generalizes an element from *CP2*.

Composition is associating classes (properties) of a content pattern with the classes (properties) of other content patterns by means of axioms.

Expansion is adding new classes, properties and axioms to ontology with the aim of meeting the requirements not addressed by the reused content patterns.

3   Development and Application of Content Patterns in Building the Ontologies of SSDs

When creating ontology of SSDs, it is important to ensure the possibility of a uniform and consistent presentation of appropriate scientific concepts and their properties. One way to solve this problem is to use content patterns describing the concepts typical for most scientific subject domains and scientific activities performed in them. A set of such patterns is provided by a methodology for constructing ontology developed within the technology for creating thematic intelligent scientific Internet resources [8].

It should be noted that this methodology, in addition to the content patterns, also includes the presentation patterns and structural logical patterns used for constructing content patterns [9].

The need to use structural logical patterns in this methodology was attributed to the absence in the OWL language [15] some expressive means for representing complex entities and constructions required to build the SSD ontologies. For example, logical structural patterns help to represent attributed relationships (binary relations with attributes) and domains (sets of elementary values).

The patterns of latter type were introduced to specify the ranges of values called domains in the relational data model and characterized by their name and a set of atomic values. Domains are convenient to use in the descriptions of possible values of the properties of a class, when the entire set of such values is known in advance. Using domains not only allows users to control the input of information; it can also make this operation more convenient by providing users with the opportunity to select property values from a given list of values (see Subsect. 3.3).

Because of a limited length of the paper, in this section we will only dwell on the development and implementation of content patterns.

3.1 Development of Content Patterns

Within the framework of the methodology for constructing SSD ontologies, we have developed content patterns for presenting the following concepts: the Division of Science, Object of Research, Subject of Research, Research Method, Scientific Result, Person, Organization, Activity (Scientific activity), Project, Publication, etc.

When developing content patterns for each of them, we defined a set of Competency Questions describing their. These questions helped us to identify the composition of the obligatory and optional ontological elements of the pattern and describe the requirements for them. The requirements were presented in the form of axioms and limitations.

Let us consider some content patterns in more detail. First let us note that in the pictures of the patterns presented below classes are shown in the form of ellipses, while their individuals and attributes are in the form of rectangles. An *ObjectProperty* connection is shown by a solid straight line and a *DataProperty* connection, by a dash line. At the same time, mandatory classes and attributes (individuals) are represented by figures surrounded by a thick line.

As a rule, scientific activity is realized through projects, programs, expeditions and examinations.

The graphical representation of a pattern intended for the description of scientific activity is shown in Fig. 1. This pattern sets the following requirement: when any kind of activity is described, it is necessary to give a reference to the object of research to which this activity is devoted, and to the division of science in which it is being carried out.

Figure 2 shows a pattern designed to describe a scientific result. This pattern sets the following requirement: when we describe a scientific result, we should give a reference to the activity under which it was obtained.

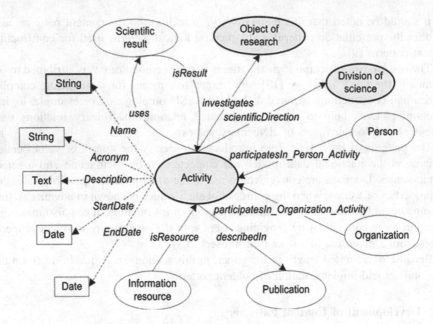

Fig. 1. Pattern for the description of a scientific activity.

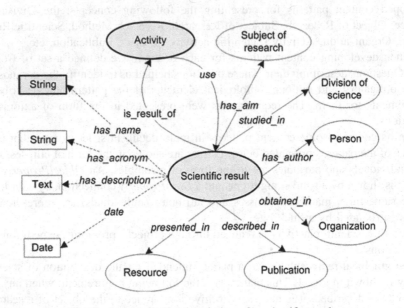

Fig. 2. Pattern for the description of a scientific result.

Figure 3 shows a pattern for describing the object of research, which includes the following classes as obligatory: Subject of Research, Activity, and Division of Science. The instances of these classes should be connected with the object of research by such relations as *has_aspect, investigated_in* and *studied_in,* respectively. The object of research can be structural (i.e. include other objects of research).

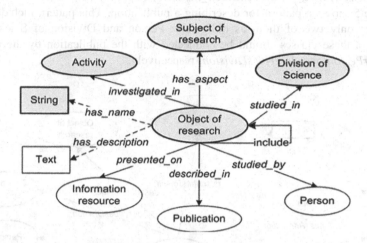

Fig. 3. Pattern for the description of an object of research.

The pattern presented in Fig. 4 is intended to describe the research methods used in a scientific activity.

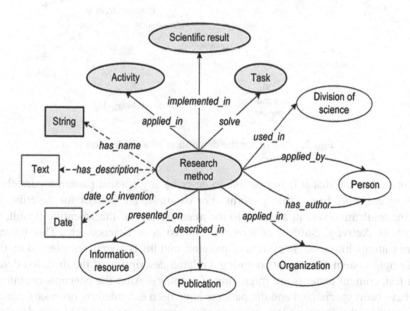

Fig. 4. Pattern for the description of a research method.

Elements of the description of the pattern of a research method are represented by such obligatory ontology classes as Activity, Scientific Result, Task, and others, and such relations as *used_in, implemented_in, solve,* etc.

Figure 5 shows a pattern for describing a publication. This pattern includes many classes but only two of them as obligatory: Person and Division of Science. The instances of these classes should be connected with the publication by the relations *hasAuthorPerson* and *describesDivision,* respectively.

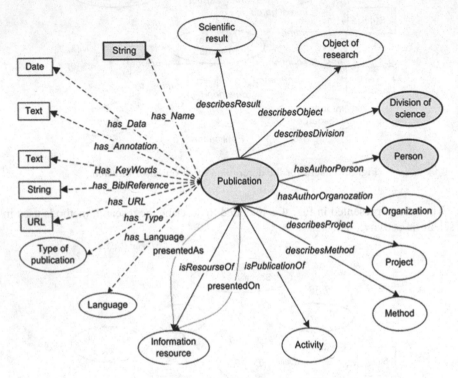

Fig. 5. Pattern for the description of a publication.

It is easy to see that it is impossible to describe any content pattern separately, i.e. not using concepts from other patterns. For example, the pattern for describing the scientific result involves, in addition to the main concept of the Scientific Result, such concepts as Activity, Subject of Research, Division of Science, etc. This unites all content patterns into a single system of patterns, and the concepts represented by them, into a single system of concepts defining a related description of the modeled domain.

In fact, content patterns are fragments of ontology. After the concepts contained in them have been specialized and the patterns have been extended by necessary concepts and properties, they become integral parts of the ontology of the SSD being developed.

3.2   Implementation of Content Patterns

All content patterns of the methodology under consideration are realized by means of the OWL language.

Below is a fragment of the OWL implementation of the content pattern for describing scientific activity in the Turtle format. (Note that the names of the onto-logical elements of the patterns are taken from the namespace of the ontology inir: http://www.semanticweb.org/IIS/ontologies/INIR_Ontology#.)

Let us describe the content components of this pattern.

First of all, this is the set of classes used in it and the axiom of disjointness defined for these classes (AllDisjointClasses):

```
:InformationResource rdf:type owl:Class ;
:ObjectOfResearch rdf:type owl:Class ;
:Organization rdf:type owl:Class ;
:Person rdf:type owl:Class ;
:Publication rdf:type owl:Class ;
:BranchOfScience rdf:type owl:Class ;
:ResearchResult
[ rdf:type owl:AllDisjointClasses ;
  owl:members
( :Activity
:InformationResource
:ObjectOfResearch
:Organization
:Person
:Publication
:BranchOfScience
:ResearchResult)] .
```

The set of properties of the *Activity* class is as follows:

```
:scientificDirection _Activity_BranchOfScience rdf:type owl:ObjectProperty ;
rdfs:domain :Activity ;
rdfs:range :BranchOfScience ;
:investigates_Activity_ObjectOfResearch rdf:type owl:ObjectProperty ;
rdfs:domain :Activity;
rdfs:range :ObjectOfResearch;
:describedIn_Activity _Publication rdf:type owl:ObjectProperty ;
rdfs:domain :Activity ;
rdfs:range :Publication ;
:participatesIn_Person_Activity rdf:type owl:ObjectProperty ;
rdfs:domain :Person;
rdfs:range :Activity;
:isResult_ResearchResult_Activity rdf:type owl:ObjectProperty ;
rdfs:domain :ResearchResult;
rdfs:range :Activity;
:isResource _InformationResource_Activity rdf:type owl:ObjectProperty ;
rdfs:domain :InformationResource;
rdfs:range :Activity;
:participatesIn _Organization _Activity rdf:type owl:ObjectProperty ;
rdfs:domain :Organization;
rdfs:range :Activity;
:Activity_Name rdf:type owl:DatatypeProperty ;
rdfs:domain :Activity;
rdfs:range rdfs:Literal ;
:Activity_Acronym rdf:type owl:DatatypeProperty ;
rdfs:domain :Activity;
rdfs:range rdfs:Literal ;
```

Next, for each instance (individual) of the Activity class, the *FunctionalProperty* property is imposed on requiring only one start date and one end date:

```
:Activity _EndDate rdf:type owl:DatatypeProperty, owl:FunctionalProperty ;
rdfs:domain :Activity;
rdfs:range rdfs:Literal ;
:Activity _StartDate rdf:type owl:DatatypeProperty, owl:FunctionalProperty ;
rdfs:domain :Activity;
rdfs:range rdfs:Literal ;
```

Finally, for each instance of the *Activity* class, by imposing constraints on properties, the following requirements for the number and type associated with this class of concepts are described:

- at least one direction of scientific activity must be specified;
- at least one object of research should be specified;
- activities must involve at least one person;
- an activity must have at least one name.

```
:Activity rdf:type owl:Class ;
owl:equivalentClass [ rdf:type owl:Restriction ;
owl:onProperty :scientificDirection _Activity_BranchOfScience;
owl:someValuesFrom :BranchOfScience] ,
[ rdf:type owl:Restriction ;
owl:onProperty :investigates_Activity_ObjectOfResearch;
owl:someValuesFrom :ObjectOfResearch] ,
[ rdf:type owl:Restriction ;
owl:onProperty [ owl:inverseOf :participatesIn_Person_Activity] ;
owl:minQualifiedCardinality "1"^^xsd:nonNegativeInteger ;
 owl:onClass :Person] ,
[ rdf:type owl:Restriction ;
owl:onProperty :Activity_Name;
owl:minCardinality "1"^^xsd:nonNegativeInteger] ;
```

3.3   Use of Content Patterns

The use of content patterns in the methodology described above is supported by a special editor allowing the users (knowledge engineers) to populate the ontology with the objects (individuals) of classes for which content patterns are implemented.

When populating an ontology with the help of the editor, the user selects from a tree of ontology classes a class (concept) an object of which he desires to create to supplement the ontology, and the editor exploits the name of this class to find the corresponding pattern. After that, the editor, using the information from the pattern, builds a form containing the fields for filling in all the properties of the object created.

It should be noted that when describing such an object, the user should take into account and the editor should control the restrictions on the values and cardinality of its properties. The restrictions are used not only to control the input information, but also as a potential source of the property values of the objects. Thus, the user can select values for each such property, either from the domain description specified by the corresponding structural pattern as the value range of this property or from the list of individuals of the class indicated as the range of this property. Such a list is dynamically formed from the individuals of the class presented in the ontology.

The content patterns considered above were applied to create an ontology of the subject domain "Decision support in weakly formalized areas". The use of these patterns was very effective, since this subject domain includes a rich set of methods and tasks, and also relies on the results obtained in many other scientific disciplines (Theory of Decision Making, Operations Research, Artificial Intelligence, Cognitive Psychology, System Analysis, etc.).

The ontology of this SSD contains detailed classifications and descriptions of tasks and methods for solving decision support problems. It identifies five main groups of methods, namely, modeling methods, data analysis methods, expert methods, reasoning methods, and auxiliary methods.

Figure 6 shows an example of using the pattern for the description of research method to describe the hierarchy analysis method (the analytic hierarchy process) [16] proposed by Saaty in the 1980s.

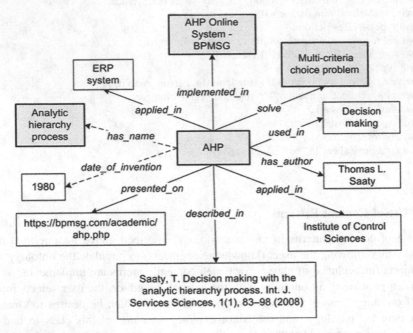

Fig. 6. Example of using the pattern for describing the research method.

4   Conclusion

The paper addresses the issues of applying ontology design patterns to the development of the ontologies of scientific subject domains. It is shown that the use of these patterns provides a uniform and consistent representation of all the entities of an ontology being developed, saving human resources and assuring a lower number of errors committed in ontology development.

The authors have devised an approach to the implementation of content patterns playing a leading role in the methodology proposed to develop SSD ontologies. This methodology and content patterns described in the paper have shown their practical usefulness in the development of the ontologies of various subject areas.

The work was financially supported by the Russian Foundation for Basic Research (Grant No. 16-07-00569) and the Siberian Branch of the Russian Academy of Sciences (Block 36.1 of the Complex Program of the SB RAS Basic Scientific Research II.1).

References

1. Gangemi, A., Presutti, V.: Ontology design patterns. In: Staab, S., Studer, R. (eds.) Handbook on Ontologies. IHIS, pp. 221–243. Springer, Heidelberg (2009). https://doi.org/10.1007/978-3-540-92673-3_10
2. Blomqvist, E., Hammar, K., Presutti, V. Engineering ontologies with patterns: the eXtreme design methodology. In: Hitzler, P., Gangemi, A., Janowicz, K., Krisnadhi, A., Presutti, V. (eds.) Ontology Engineering with Ontology Design Patterns. Studies on the Semantic Web. IOS Press (2016)
3. NeOn Project (2016). http://www.neon-project.org/nw/Welcome_to_the_NeOn_Project. Accessed 30 June 2018
4. Hammar, K.: Ontology design patterns in WebProtégé. In: Proceedings of 14th International Semantic Web Conference (ISWC-2015). CEUR Workshop Proceedings (2015)
5. Dodds, L., Davis, I.: Linked data patterns (2012). http://patterns.dataincubator.org/book. Accessed 30 June 2018
6. Association for Ontology Design & Patterns (2018). http://ontologydesignpatterns.org. Accessed 30 June 2018
7. Ontology Design Patterns (ODPs) Public Catalog (2009). http://odps.sourceforge.net. Accessed 30 June 2018
8. Zagorulko, Y., Zagorulko, G.: Ontology-based technology for development of intelligent scientific internet resources. In: Fujita, H., Guizzi, G. (eds.) SoMeT 2015. CCIS, vol. 532, pp. 227–241. Springer, Cham (2015). https://doi.org/10.1007/978-3-319-22689-7_17
9. Zagorulko, Y., Borovikova, O., Zagorulko, G.: Application of ontology design patterns in the development of the ontologies of scientific subject domains. In: Kalinichenko, L., Manolopoulos, Y., Skvortsov, N., Sukhomlin, V. (eds.) Selected Papers of the XIX International Conference on Data Analytics and Management in Data Intensive Domains (DAMDID/RCDL 2017), vol. 2022, pp. 258–265. CEUR Workshop Proceedings, (CEUR-WS.org) (2017). http://ceur-ws.org/Vol-2022/paper42.pdf. Accessed 30 June 2018
10. Zagorulko, Y., Zagorulko, G.: Ontology-based approach to development of the decision support system for oil-and-gas production enterprise. In: Proceedings of the 9th SoMeT_10 New Trends in Software Methodologies, Tools and Techniques, pp. 457–466. IOS Press, Amsterdam (2010)
11. Zagorulko, Y., Borovikova, O., Zagorulko, G.: Knowledge portal on computational linguistics: content-based multilingual access to linguistic information resources. In: Selected topics in Applied Computer Science, Proceedings of the 10th WSEAS International Conference on Applied Computer Science (ACS 2010). Iwate Prefectural University, Japan, 4–6 October 2010, pp. 255–262. WSEAS Press (2010)
12. Zagorulko, Y., Borovikova, O.: Technology of ontology building for knowledge portals on humanities. In: Wolff, K.E., Palchunov, Dmitry E., Zagoruiko, Nikolay G., Andelfinger, U. (eds.) KONT/KPP -2007. LNCS (LNAI), vol. 6581, pp. 203–216. Springer, Heidelberg (2011). https://doi.org/10.1007/978-3-642-22140-8_13
13. Zagorulko, Y., Zagorulko, G.: Architecture of extensible tools for development of intelligent decision support systems. In: Proceedings of the Tenth SoMeT_11 New Trends in Software Methodologies, Tools and Techniques, pp. 253–263. IOS Press, Amsterdam (2011)
14. Braginskaya, L., Kovalevsky, V., Grigoryuk, A., Zagorulko, G.: Ontological approach to information support of investigations in active seismology. In: The 2nd Russian-Pacific Conference on Computer Technology and Applications (RPC-2017), pp. 27–29. IEEE Xplore Digital Library (2017). http://ieeexplore.ieee.org/document/8168060. Accessed 30 June 2018

15. Antoniou, G., Harmelen, F.: Web ontology language: OWL. In: Staab, S., Studer, R. (eds.) Handbook on Ontologies, pp. 67–92. Springer, Berlin (2004). https://doi.org/10.1007/978-3-540-24750-0_4
16. Saaty, T.: Decision making with the analytic hierarchy process. Int. J. Serv. Sci. 1(1), 83–98 (2008)

Protection of Information in Networks Based on Methods of Machine Learning

Sergey G. Antipov, Vadim N. Vagin[✉], Oleg L. Morosin, and Marina V. Fomina

National Research University "MPEI", Moscow, Russia
vagin@appmat.ru

Abstract. The paper considers the possibility of using artificial intelligence methods in information security tasks: methods for generating inductive concepts for analyzing network traffic, as well as methods of argumentation for automated security decision support systems. The approach proposed in the work allows giving quantitative assessments of the quality of the recommendations developed by the system, thereby helping to solve an important task - the task of choosing the way of responding to suspicious activity in the system. Examples of handling dangerous situations occurring in the system are also presented.

Keywords: Information security · Argumentation · Defeasible reasoning
Degree of justification · Resolution of contradictions · Time series
Inductive concept formation

1 Introduction

The modern world business and industry is inextricably linked with the development of the Internet and new network technologies. In the network infrastructure a large amount of personal, military, commercial and government information is circulating. The growth of networks, their complexity and the increasing number of attacks on the network are the reason for serious attitude to the problems of network security. Solving network security problems is a complex task, involving a large number of factors and requiring finding reasonable compromises between maintaining security, stable operation, operating costs and the limitations of the functionality of complex information systems.

2 Network Security Problems

Modern computer networks are large distributed systems of software and devices that interact with each other for the exchange, storage and processing of information. Networks connect different types of devices through communication channels.

It is very important to develop systems that can identify suspicious network activity due to urgent need to ensure the security of computers interacting on the network. Many network security systems use only constant set of security rules, programmed manually on the basis of expert experience and do not have the possibilities of self-learning.

S. O. Kuznetsov et al. (Eds.): RCAI 2018, CCIS 934, pp. 273–279, 2018.
https://doi.org/10.1007/978-3-030-00617-4_25

Methods of intelligent data analysis are currently used to increase the security of the systems through rapid detection of attacks. They can effectively operate in the face of unexpected situations, not described by experts. Let's list the main tasks that can be solved with the help of intelligent data analysis methods and machine learning in the field of information security:

1. Rapid recognition of threats based on the analysis of information stored in the knowledge base of an intelligent system.
2. Optimization of the search of malicious software (malware), which can also be self-learning.
3. Use of information that may seem insignificant for a conventional security system, which reduces the risk of unauthorized access.
4. Protect from malicious software, which can also be self-learning.
5. Generalize information in the learning process, and build new rules for detecting malware on the basis of gained experience for a more powerful protection system.

3   The Application of Argumentation in Network Security Systems

Modern security systems must handle large amounts of information, that is often noisy and contradictory. Often security systems use mechanisms of classical logic (see, for example, [1]) to describe the rules of invasion detection and other suspicious actions of users in information systems. At the same time, false triggering of protection mechanisms is extremely undesirable, since it can lead to significant financial losses. The simplest example is firewalls, containing thousands [1] of rules for detecting suspicious activity, under triggering of which the blocking of data exchange with the user is brought about. Reducing the percentage of false actions of protection systems is an important problem [2], the solution of which will significantly improve the quality of computer security systems.

Here, it is proposed to describe the mechanism for decision making not only in terms of assessing the plausibility of malicious actions, but also in terms of assessing the potential risks from the application of protection mechanisms. For this purpose, it is suggested to use the mechanism of argumentation theory to determine the need for protective measures taking into account possible risks from false triggering protection systems.

Argumentation is a formal approach to decision making that has proved its efficiency in a number of areas. Argumentation is usually understood as the process of constructing assumptions about a certain analyzable problem. Typically, this process involves conflicts detection and ways of problem solutions. Unlike classical logic, the argumentation suggests that there can be arguments for both, and against, any assumption [3]. The fundamentals of the theory of argumentation are considered in [4, 5].

To apply the theory of argumentation in problems requiring the calculation of quantitative assessments of the argument reliability (such as, for example, network security tasks) the mechanism of justification degrees for arguments is applied. Argumentation systems with the justification degrees were considered in [6]. Let's present a simplified example (without justification degrees) of how the developed argumentation system can be used to improve network security systems.

Let us consider an example of the network security problem given in [7]. For clarity, the statement of the problem will be simplified. Let there be a complex information system protected by some security system. The security system in the case of detection of suspicious activity informs about a threat that has arisen, its type and the assessment of threat reality probability. Suppose that there are several open ports in the system and let the web service be running on the port 80 to handle client requests. By default, when a threat is detected on one of the ports, this port is blocked. As a result of port blocking, all services using this port must be stopped. If the web service is stopped, a critical error will occur and the company will incur serious loss. If a threat is not serious - you cannot admit the occurrence of critical errors as a result of the application of protective measures. Network worms belong to the class of not very dangerous network invasions. Suppose that the security system has detected suspicious activity on port 80, similar to the attack of a network worm.

In the formal language, this problem will take the following form, where $A1$–$A5$ are original arguments, $R1$ is the undercutting rule, and $R2$ is the defeasible rule:

A1: *attack(port_80, warm)* – the suspicion of a network worm attack on port 80 is detected;

A2: *use(web_service, port_80)* –port 80 is used by the web *service*;

A3: $\forall x \, \forall y \, (block(x) \, \& \, use(y, x) \to stop(y))$ – when the port is blocked, all services using this port must be stopped;

A4: *stop(web_service)* $\to$ *critical_error* – interrupt the web service is a critical error;

A5: *~serious_attack(warm)* – network worms belong to the class of not very dangerous network invasions;

R1: $\forall y \, \sim serious\_attack(y) \, \& \, critical\_error => \forall x \, attack(x, y)$ @ *block(x)* – the undercutting rule stating that if the threat is not very serious it is impossible to admit the emergence of critical errors as a result of the application of protective measures. The scheme $A => B @ C$ means that argument A is the reason not to consider the defeasible inference link between B and C as plausible.

R2: $\forall x \forall y \, attack(x, y)) \, |=> block(x)$ – the defeasible rule that states that by default when a threat is detected on one of the ports, this port is blocked.

Figure 1 shows the inference graph for this task. As a result of solving this problem by argumentation based on defeasible reasoning, the argument *A6: block (port_80)* became defeated, i.e. a possible attack of the network worm on port 80 is not so great to decide to block all traffic on this port because this will lead to significant costs.

The same argumentation inference could be made with not Boolean statuses (defeated or not) of arguments, but with the justification degrees. The main problem here is how the initial degrees should be chosen. It can be either an expert knowledge or, which is more promising, calculated on the previous experience using machine-learning techniques. This is the field of further research.

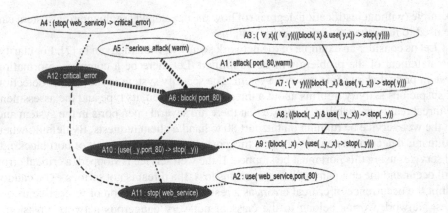

Fig. 1. The example of the argumentation application for decision making about port blocking

4 The Problem of Determining Unauthorized Access in Networks

Information leakages during network exchange are possible due to the use of hidden functionality that negatively affects the safety of the information being processed. The basis for the detection of malicious software functions in the information exchange system will be traffic analysis: if the parameters of typical information exchange over a certain protocol are known, then the abnormal pattern of behavior in a computer network detected during the analysis of traffic exchange under this protocol can indicate that in the analyzed system there is a backdoor, and the anomalies in traffic are caused by the actions of such malware programs.

The process of physical data transmission at a certain interval can be considered as a time series, since the data transmission is the process of data exchange, characterized by transmitted bits per unit of time.

The task is to detect anomalies in collections of time series, for the case when each time series corresponds to a normal information exchange process. The presence of samples in the data sets that do not satisfy some presumed typical behavior may indicate the presence of malicious software. The anomaly detection problem [8] is set up as the task of searching for templates in data sets that do not satisfy some typical behaviors. The anomaly, or "outlier" is defined as an element that stands out from the data set which it belongs to and differs significantly from other elements of the sample.

It is required to build on the basis of available data (considering that the time series correspond to certain situations that in turn relate to several admissible classes) a model that is a generalized description of normal processes and allows us to distinguish normal and anomalous processes. The building of such a model is considered as a problem of generalization of time series. The general statement of generalization and classification problem could be found, for example, in [9]. Classification of time series consists in building a model (based on historical data), that is capable to determine to which class the newly presented time series belongs.

To determine anomalies in sets of time series the two algorithms were developed. The algorithm TS-ADEEP for the case when a learning set contains only one class is given in details in [10]. Its modification TS-ADEEP-Multi for the cases where time series belong to several different admissible classes was presented in [11]. Let's present how these algorithms could be used for the security problem stated above.

5   Experimental Results

The machine experiment should have allowed evaluating the ability of the developed algorithms to find anomalies in the case of monitoring the course of information exchange. The algorithms TS-ADEEP and TS-ADEEP-Multi were preliminarily tested on the data sets presented in [12, 13], and showed good results.

All the data used in experiments were preliminarily processed, consisting of two stages: normalization and subsequent discretization of the normalized time series with the transition to the symbolic data representation, where the alphabet size (the number of used symbols) depending on the task could vary. The process of preliminary data processing is based on the ideas of the algorithm Symbolic Aggregate ApproXimation (SAX) [14] and is described in detail in [10, 15].

As an illustration of the method, FTP file transfer protocol was chosen. The method can be extended to other standard protocols of information interaction, having specifications in the form of standards or widely distributed de facto and having a description in open sources.

Based on the analysis of network traffic when transferring files via FTP in various conditions (including the simultaneous transmission of several files), a data set was obtained, that is a learning sample for creating a model of "reference" data transmission. As test data, among others, there were used specially generated time series simulating the transfer of unauthorized data.

Below are the results obtained for the "Traffic" data set.

Tables 1 and 2 show the results of recognition of anomalies in the transmission of data on the above protocols. The best results are shown in bold. As can be seen from the

Table 1. The accuracy of the anomaly detection in data sets "Traffic" with one class for the TS-ADEEP algorithm

		Dimensionality of time series						
		210	150	100	50	30	20	10
The number of alphabet symbols	5	71, 43	82, 14	60, 71	64, 29	60, 71	46, 43	67, 86
	10	92, 86	96, 43	100	96, 43	82, 14	85, 71	64, 29
	15	92, 86	**100**	**100**	96, 43	92, 86	82, 14	85, 71
	20	92, 86	**100**	**100**	96, 43	92, 86	82, 14	82, 14
	25	92, 86	**100**	**100**	96, 43	92, 86	82, 14	92, 86
	30	92, 86	**100**	**100**	96, 43	92, 86	92, 86	82, 14
	40	92, 86	**100**	**100**	96, 43	92, 86	92, 86	92, 86
	50	92, 86	**100**	**100**	96, 43	92, 86	92, 86	92, 86

presented data, for the problem under consideration it was possible to achieve an accuracy of 100% anomaly classification when selecting the number of alphabet symbols and the dimension of the time series.

Table 2. The accuracy of anomaly detection in data sets "Traffic" with several classes for the TS-ADEEP-multi algorithm

		Dimensionality of time series						
		210	150	100	50	30	20	10
The number of alphabet symbols	5	85, 71	89, 29	57, 14	64, 29	60, 71	46, 43	67, 86
	10	96, 43	96, 43	**100**	96, 43	96, 43	85, 71	67, 86
	15	92, 86	**100**	**100**	96, 43	92, 85	82, 14	67, 86
	20	96, 43	**100**	**100**	96, 43	96, 43	96, 43	75, 00
	25	96, 43	**100**	**100**	96, 43	96, 43	82, 14	82, 14
	30	96, 43	**100**	**100**	96, 43	96, 43	96, 43	92, 86
	40	96, 43	**100**	**100**	96, 43	96, 43	96, 43	92, 86
	50	96, 43	**100**	**100**	96, 43	96, 43	96, 43	96, 43

6 Conclusion

The paper proposes the idea of using argumentation in network security systems, which will make such systems more flexible and will allow us to assess the appropriateness of using certain security mechanisms. The use of arguments with degrees of justification allows one to give numerical estimates to recommendations, which allows solving one more important task - the task of choosing the way of responding to suspicious activity in the system.

Also, methods for detecting anomalies when solving the task of analyzing network traffic for malware detection were considered. Algorithms for finding anomalies have been programmed. The results of the experiment showed high accuracy of detection of anomalies, which indicates good prospects for using the proposed methods and software. It allows us to classify the situation data with high accuracy.

Acknowledgement. This work was supported by grants from the Russian Foundation for Basic Research № 18-01-00201, 17-07-00442a, grant of the President of the Russian Federation MK2897.2017.9.

References

1. Ou, X., Govindavajhala, S., Appel, A.W.: MulVAL: a logic-based network security analyzer. In: USENIX Security (2005)
2. Khosravifar, B., Bentahar, J.: An experience improving intrusion detection systems false alarm ratio by using honeypot. In: Advanced Information Networking and Applications, pp. 997–1004. IEEE (2008)
3. Pollock, J.L.: How to reason defeasibly. Artif. Intell. **57**(1), 1–42 (1992)

4. Vagin, V.N., Morosin, O.L., Fomina, M.V.: Inductive inference and argumentation methods in modern intelligent decision support systems. J. Comput. Syst. Sci. Int. **55**(1), 79–95 (2016)
5. Pollock, J.L.: Defeasible reasoning with variable degrees of justification. Artif. Intell. **133**(1), 233–282 (2001)
6. Vagin, V.N., Morosin, O.L.: Implementation of defeasible reasoning system with justification degrees. Softw. Syst. **109**(1) 2015. (in Russian)
7. Bandara, A.K., Kakas, A., Lupu, Emil C., Russo, A.: Using argumentation logic for firewall policy specification and analysis. In: State, R., van der Meer, S., O'Sullivan, D., Pfeifer, T. (eds.) DSOM 2006. LNCS, vol. 4269, pp. 185–196. Springer, Heidelberg (2006). https://doi.org/10.1007/11907466_16
8. Chandola, V., Banerjee, A., Kumar, V.: Anomaly detection - a survey. ACM Comput. Surv. **41**(3) 2009
9. Vagin, V., Golovina, E., Zagoryanskaya, A., Fomina, M.: Exact and plausible inference in intelligent systems. In: Vagin, V., Pospelov, D. (eds.), FizMatLit, Moscow (2008). (in Russian)
10. Antipov, S., Fomina, M.: Problem of anomalies detection in time series sets. Softw. Syst. **2**, 78–82 (2012). (in Russian)
11. Antipov, S.G., Vagin, V.N., Fomina, M.V.: Detection of data anomalies at network traffic analysis. In: Open Semantic Technologies for Intelligent Systems, Minsk, Belarus, pp. 195–198 (2017)
12. Lichman, M.: UCI Machine Learning Repository. http://archive.ics.uci.edu/ml. Accessed 01 Apr 2018
13. Chen, Y., Keogh, E., Hu, B., et al.: The UCR Time Series Classification Archive, July 2015. www.cs.ucr.edu/~eamonn/time_series_data
14. Lin, J., Keogh, E., Lonardi, S., Chiu, B.: A symbolic representation of time series, with implications for streaming algorithms. In: Proceedings of the 8th ACM SIGMOD Workshop on Research Issues in Data Mining and Knowledge Discovery, pp. 2–11 (2003)
15. Fomina, M., Antipov, S., Vagin, V.: Methods and algorithms of anomaly searching in collections of time series. In: Abraham, A., Kovalev, S., Tarassov, V., Snášel, V. (eds.) Proceedings of the First International Scientific Conference "Intelligent Information Technologies for Industry" (IITI'16). AISC, vol. 450, pp. 63–73. Springer, Cham (2016). https://doi.org/10.1007/978-3-319-33609-1_6

Risk Health Evaluation and Selection of Preventive Measures Plan with the Help of Argumental Algorithm

Oleg G. Grigoriev and Alexey I. Molodchenkov[✉]

Federal Research Center "Computer Science and Control" of RAS,
Vavilova str. 44, kor. 2, Moscow 119333, Russian Federation
oleggpolikvart@yandex.ru, aim@tesyan.ru

Abstract. The paper describes knowledge base principles and method for disease risk evaluation that were used for intelligent healthcare management system creation. The present version of the knowledge base is implemented using a heterogeneous semantic network approach and utilizes expert opinions about risk factors and events influencing an individual's health. Data includes genetic predisposition, lifestyle, and external environment. Data is compiled with the aid of questionnaires, mobile devices, case histories and information from social media. Information from social media is analyzed using data and text mining methods with the goal of evaluating the user's condition. All of the data obtained is accumulated in a single database. The method for risk evaluation and preventive measures plan hypotheses generation is based on an argumentation reasoning algorithm that is modified to the task at hand. All prevention recommendations are based on the principles of P4 medicine. The current version of the system is based on expert knowledge obtained by automated monitoring and analysis of a large number of publications and recommendations on this topic.

Keywords: Knowledge base · Artificial intelligence · Prevention
P4 medicine · Argumentation reasoning · Diseases risk evaluation
Heterogeneous semantic network

1 Introduction

Research and services in the field of prevention and maintenance of human health are currently undergoing active development. This trend is clearly visible in large corporations. It was named as P4 medicine [1, 2]. The main goal of P4 medicine is to determine the state of health of a particular person and propose preventive measures before the appearance of the first symptoms of possible diseases. The high number and multidimensionality of risk factors and adverse situations make it impossible for a doctor to take more than a fraction of them into account. An early prognosis can reduce the risk of pathology. A plan of medical and social activities should reduce the impact of risk factors and situations that contribute to the disease emergence. Due to the active development of machine learning methods and big data analysis, the main emphasis in this area is on genomics and other omics technologies [3].

S. O. Kuznetsov et al. (Eds.): RCAI 2018, CCIS 934, pp. 280–290, 2018.
https://doi.org/10.1007/978-3-030-00617-4_26

However, one should take into account that in addition to genome information, a number of factors related to lifestyle, habits, nutrition, relatives' diseases, etc. could affect the health. All available information should be used to assess the possible diseases risks and to reduce their impact. The paper describes the methodology and the knowledge base of the system for assessing the disease risks and preventive measures plan construction to reduce these risks.

2 System Inference Principles

Differentiating between various conditions in a user's health depends on both direct and indirect relationships between risk factors and diseases [4]. The argument "for" ensures generation of hypotheses that indicate the risks of diseases. This makes it possible to take changes in multiple risk factors and various situation-based characteristics into account. At the same time, protective factors, both biological and external, are counter-arguments that reduce the risk of diseases.

3 System Functioning

Users of the system on their first visit to the site fill out a preliminary questionnaire (in the future, it is proposed to manually fill in only those fields that show changes in dynamics and that cannot be filled out automatically). Then the system switches to other information sources (electronic medical cards, social networks, etc.), which may contain information about a particular individual. In addition, mobile devices collect information about the user's physical condition (pulse rate, blood pressure, etc.). All the information received is fed into the knowledge base and analyzed for identification and evaluation of risk factors and health abnormalities. The presence of counterarguments, characterized by negative links to the risk of pathology, lowers the risk level. A plan of preventive measures is formed in accordance with the hypothetical disease risk, taking into account the user's individual characteristics. The plan also includes lifestyle changes. The described procedure is repeated when necessary.

4 Knowledge Base Structure

The main component of the system for disease risks assessing and preventive measures selection is the knowledge base. The knowledge base is implemented in the form of a heterogeneous semantic network [5, 6], further HSN, and consists of sections, HSN nodes, node properties and connections between nodes.

HSN nodes form the main part of the knowledge base. Nodes can represent specific statements, situations, results of observations, characteristics of user's condition, recommendations for prevention, disease risk factors, lifestyle facts, personal factors. Examples of nodes are "Overweight", "Stress frequency once a day", "Young age", "Alcohol abuse", "Male", "Female", "Smoking", "Duration of smoking up to 5 years, "Strokes in old age among relatives of the 1-st degree kinship", "Insomnia", "High sense of loneliness", etc. Experts group the nodes into sections according to their meaning in the system. There are following sections in the system knowledge base: "Risk factors", "Hidden characteristics", "Disease risks", "Preventive measures", "Situations in user's life", etc. Below we provide explanations for some node groups.

Hidden characteristics are hidden nodes that are responsible for the implementation of "AND", "OR" operations. For the values of disease risk factors, we use the following linguistic scale: very high, high, medium, low, very low. Thus, for each disease, five nodes determine its risk value. For example, the following nodes of the knowledge base characterize risk of a stroke: "Very high stroke risk", "High stroke risk", "Average stroke risk," "Low stroke risk", "Very low stroke risk". Risks for Myocardial Infarction and Depression are represented in the same way.

A node can have properties whose collection is a characteristic of this node. Other nodes associated with this one, and its values can represent the properties of the node. For example, the properties of the node "Alcohol abuse for more than 10 years" are: his name (Alcohol abuse for more than 10 years); value from the range of "Yes", "No", "Unknown"; related nodes - "Abuse of alcohol now" (with their values) and "average risk of depression".

There are two node categories in the knowledge base. The first category includes nodes that reflect health and lifestyle knowledge, which determine the course of reasoning: node-data; the second includes nodes that participate in the formation of solutions: nodes-hypotheses. In the knowledge base of the system, the risk factors, situations in the user's life, user characteristics act as data nodes, and the disease risks and health-improving measures - as hypotheses nodes.

Links between nodes in the knowledge base show the impact of the node on the generation or exclusion of others. Links are possible between any two nodes, although the most important ones are between the data nodes and the hypothesis nodes. The knowledge base includes the following types of links:

(1) RS or ART. This type of connection corresponds to the following statement: "When node 1 is observed, node 2 can be observed (or is usually observed)". This connection is called positive, it can generate node 2, or increase the confidence in the presence of node 2.

(2) TRA. This type of connection corresponds to the following statement: "When node 1 is observed, node 2 is always observed". This connection is also positive, but strongly positive. Establishing such a connection is equivalent to declaring node 2 to be a property of node 1.

(3) SN. This type of connection corresponds to the following statement: "Node 2 is usually absent (or may be absent) when node 1 is observed". This connection is called negative; it reduces the confidence in the presence of node 2.

(4) S. This type of connection corresponds to the following statement: "When node 1 is observed node 2 is always absent". This connection is called excluding, node 2 is excluded from the list of possible hypotheses.

Implemented in the solver argumentation algorithm is used for a set of hypotheses construction and for the final solution. Fragments of knowledge base for stroke and depression are illustrated on Figs. 1 and 2. Black nodes without tests are hidden nodes.

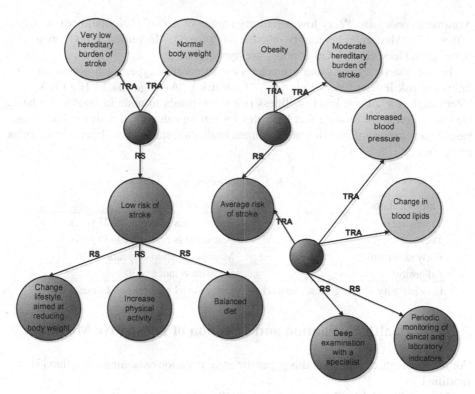

Fig. 1. Fragment of the knowledge base on the example of stroke risk. (Color figure online)

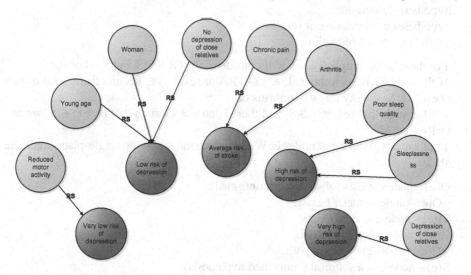

Fig. 2. Fragment of the knowledge base on the example of depression risk.

Arguments nodes are "Very low hereditary burden of stroke", "Normal body weight", "Obesity", "Moderate hereditary burden of stroke", "Increased blood pressure", "Change in blood lipids". Other nodes are hypotheses.

In our system all risks and arguments are categorical type. Each illness has five linguistic risk levels: "Very low risk", "Low risk", "Average risk", "High risk" and "Very high risk". Each level of illness risk corresponds to node in knowledge base. Developed program module that has rules for setting categorical nodes in knowledge base depending on information about human healthcare state. Table 1 gives some rules examples.

Table 1. Rules examples.

Categorical note	Rule conditions
Young age	Age value is between 18–23 years
Teen age	Age value is between 24–39 years
Early menopause	Menopause up to 40 years
Adiposity	Body mass index > 30
Low intensity of using social networks	Using social networks 1–2 times a day

5   Risk Health Evaluation and Selection of Preventive Measures

For the problems described in this paper the argumentation reasoning algorithm [7] was modified.

This modified algorithm consists of the following steps:

- hypothesis generation;
- hypotheses confirmation or rejection;
- set of hypotheses reduction.

For the sake of simplicity, we will consider only TRA, RS and S relations:

If the node e1 is such that (e1, e2) ∈ TRA or (e1, e2) ∈ RS and there is no e such that (e, e1) ∈ S, we say that e1 confirms e2.

If e1 is such that (e1, e2) ∈ S, and there is no e ≠ e2 such that (e, e1) ∈ S, we say that e1 rejects e2.

Let there is a given set of nodes E. We introduce the following single-place predicate symbols:

- O(e) – node e exists (observed or confirmed).
- ¬O(e) – node e doesn't exist.
- M(e) – node e maybe exists.
- H(e) – node e is a hypothesis.
- ¬H(e) – node e is not a hypothesis.
- S(e) – node e is a solution (confirmed hypothesis).
- ¬S(e) – node e is not a solution (confirmed hypothesis).

Let us introduce the following inference rules:

P1. $(O(e_1), (e_1, e_2) \in TRA) \rightarrow S(e_2)$
P2. $(O(e_1), (e_1, e_2) \in RS) \rightarrow H(e_2)$
P3. $(H(e_2), O(e_1), (e_1, e_2) \in S) \rightarrow \neg H(e_2)$
P4. $(H(e_1), (e_1, e_2) \in RS_{)} \rightarrow M(e_2)$
P5. $(H(e_1), \neg O(e_2), (e_1, e_2) \in TRA_{)} \rightarrow \neg H(e_1)$

For convenience reasons, hereinafter, we will use, besides logical, set-theoretic notation too.

For example, we will use the fact that $O = \{e|O(e)\}$. The same way, $H = \{e|H(e)\}$, $M = \{e|M(e)\}$ и $S = \{e|S(e)\}$.

The argumentation algorithm consists of the following steps.

Step 1. Generating a set of hypotheses.

If e is such that exists $O(e)$, then we apply one of the inference rules P1 or P2 to e (depending on whether the pair (e, e2) belongs to TRA or RS relation).
Repeat for every $e \in O$.
$H := H \cup S$

Step 2. Extending the set of arguments.

To all e, such that $e \in H$, we apply the rule P4 and construct the set of all e such that $M(e)$.

Step 3. Testing the arguments. For each e such that $M(e)$, we apply the procedure for confirming the presence of nodes; if $Q(e) = O(e)$, then $O := O \cup \{e\}$. Transition to step 1.

Steps 1–3 are performed until the sets O and H are stabilized, in other words, until the solution of the fixed point equation $ARG(X) = X$, where $X = H \cup O$ is found.

Step 4. Decrease the set of hypotheses using rejecting arguments.

For all e such that $H(e)$, $O(e1)$, and $(e_1, e) \in S$
$H := H \backslash \{e\}$.

Step 5. Decrease of the set of hypotheses using conditioned (absent confirming) arguments.

For all e such that $H(e)$, $(e, e1) \in TRA$, and $\neg O(e1)$ holds, in accordance with P5 rule, we conclude that $\neg H(e)$. We set $H := H \backslash \{e\}$.

Step 6. If the power of the hypotheses set $|N|$ turned out to be less than or equal to one, we set $S = H$ and the algorithm terminates.
Step 7. Differentiation of the hypotheses set.

If $|H| > 1$ and there are two hypotheses in H, whose argument sets are strictly embedded one into another, the hypothesis with a smaller number of arguments e, such that $O(e)$ takes place, is removed from the set of hypotheses. This procedure is applied in pairs to all such hypotheses. The resulting set of hypotheses H is called an explanatory set.

Step 8. Minimizing the explanatory set.

We denote the set of arguments e of the hypothesis h such that O(e) by Arg(h).

If $|H| > 2$ and H contains hypotheses h_1, h_2, h_3, ..., h_n, then any hypothesis h_1 for which $Arg(h_1) \subset Arg(h_2) \cup Arg(h_3) \cup ... \cup Arg(h_n)$ is removed from H. The process is repeated until exhaustion of the set of such hypotheses.

Step 9. S = H. Shutdown.

The algorithm and some of its steps are illustrated on Figs. 3, 4, 5, 6 and 7.

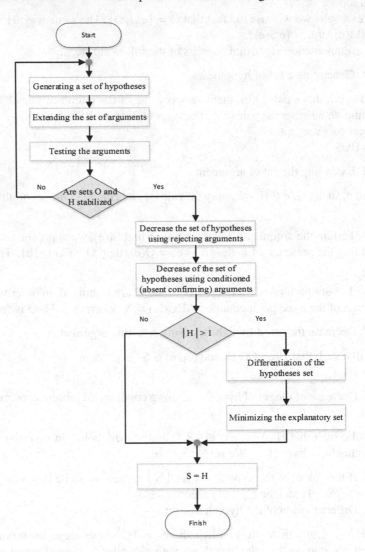

Fig. 3. Argumentation algorithm.

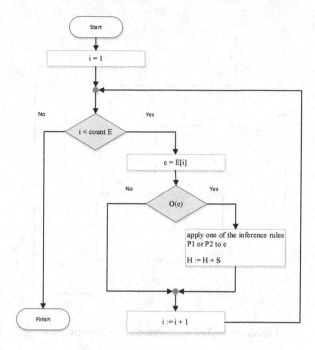

Fig. 4. Step 1 of the argumentation algorithm.

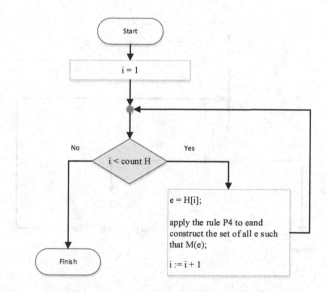

Fig. 5. Step 2 of the argumentation algorithm.

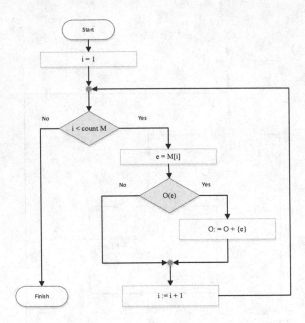

Fig. 6. Step 3 of the argumentation algorithm.

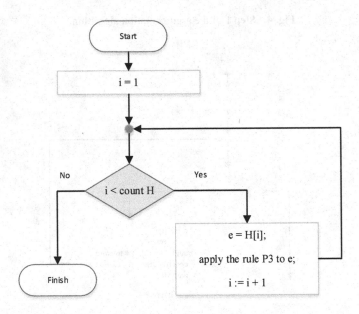

Fig. 7. Step 4 of the argumentation algorithm.

This modification of the argumentation algorithm makes it possible to use all the hypotheses generated with the help of the knowledge base to derive a solution.

One should take into account that, in order to carry out the tasks, it is necessary to form two consistent solutions based on two disjoint subsets of the hypotheses set. Namely, it is necessary:

(1) evaluate user's disease risks,
(2) propose a set of preventive measures aimed at reducing these risks.

We denote by Input the procedure for entering information about the user health status. We denote by Proc the procedure that divides the set of all hypotheses of the knowledge base into two subsets: the risks of diseases (we denote it by RiskH) and preventive measures (we denote it by ProfH).

We denote by ARG(IH) the function for the argumentation algorithm execution, where IH is the set of hypotheses based on which solutions are constructed. In other words, the set H defined above consists of the following elements: H = {e|IH(e)}. The function ARG(IH) returns the solution IS, which is a subset of IH.

The procedure for necessary solutions constructing consists of the following steps:

(1) Running the Input procedure to enter user health information.
(2) Running the Proc procedure for splitting the set of hypotheses of the knowledge base into two subsets of RiskH and ProfH.
(3) RiskS = ARG(RiskH) – formation of user disease risks set.
(4) ProfS = ARG(ProfH) – formation of preventive measures set.

Thus, the System Solver issues two solutions using a single knowledge base.

Let's illustrate the work of the algorithm by example.

Knowledge base contains preventive recommendations, that are added to solution depending on risk level and arguments that indicate this risk. A fragment of the knowledge base for assessing stroke risk, where circles indicate nodes and arrows – links, is depicted in Fig. 1. Data nodes are Blue-colored, hypotheses groups for stroke risk evaluation (RiskH set) - deep blue and preventive recommendations hypotheses groups (ProfH set) are purple. Suppose that the user has "Very low hereditary burden of stroke" and "Normal body weight". Then, by algorithm work completion, RiskS set will contain "Low risk of stroke", and ProfS set - "Increase physical activity", "Rational nutrition".

6 Conclusion

The paper describes the main components of the knowledge base and the methodology of automatic construction of a preventive measures plan for preventing or reducing disease risks.

At the present stage, the knowledge base focuses on detecting and assessing those personal risks of diseases such as stroke, myocardial infarction and depression, which characterize pathological conditions of various body systems. The knowledge base is a heterogeneous semantic network. This makes it possible not only to describe weakly structured subject areas, but also to explain the formulated solutions.

The modified algorithm of argumentation reasoning for disease risk assessment and issuing recommendations on disease prevention is given. The algorithm work is illustrated in a small fragment of the knowledge base, using the example of stroke risk. The methods and algorithms described in the article formed the basis for a pilot version of the system for constructing a plan of preventive measures to reduce the disease risks.

The application of methods of artificial intelligence makes it possible to solve the problem of health maintenance using the personalized approach of P4 medicine, thus ensuring the user's involvement in the care of their own health.

Acknowledgments. The Ministry of Education and Science of the Russian Federation financially supported this work. Grant No. 14.604.21.0194 (Unique Project Identifier RFMEFI60417X0194).

References

1. Flores, M., Glusman, G., Brogaard, K., Price, N.D., Hood, L.: P4 medicine: how systems medicine will transform the healthcare sector and society. Per. Med. **10**(6), 565–576 (2013)
2. Hood, L.: P4 medicine and scientific wellness: catalyzing a revolution in 21st century medicine. Mol. Front. J. **1**(02), 132–137 (2017)
3. Morley, J.E., Vellas, B.: Patient-centered (P4) medicine and the older person. J. Am. Med. Dir. Assoc. **18**(6), 455–459 (2017)
4. Kobrinskii, B.A.: Logic of the argument in decision-making in medicine. Autom. Doc. Math. Linguist. **35**(5), 1–8 (2001)
5. Osipov, G.S.: Formulation of subject domain models: part I. Heterogeneous semantic nets. J. Comput. Syst. Sci. Int. **30**(5), 1–12 (1992)
6. Makarov, D.A., Molodchenkov, A.I.: Creation of systems of acquisition of knowledge for construction of medical expert systems on the basis of a kernel of software MedExp. J. Inf.-Meas. Control. Syst. **7**(12), 86–89 (2009)
7. Osipov, G.S.: Methods for Artificial Intelligence, 2nd edn., p. 296. Fizmatlit, Moscow (2016)

Author Index

Printed in the United States
By Bookmasters